Evolutionary Genetics

Evolutionary Genetics

Concepts, Analysis, and Practice

Glenn-Peter Sætre and Mark Ravinet

OXFORD
UNIVERSITY PRESS

Great Clarendon Street, Oxford, OX2 6DP,
United Kingdom

Oxford University Press is a department of the University of Oxford.
It furthers the University's objective of excellence in research, scholarship,
and education by publishing worldwide. Oxford is a registered trade mark of
Oxford University Press in the UK and in certain other countries

© Glenn-Peter Sætre & Mark Ravinet 2019
Illustrations by Kaj Clausen

The moral rights of the authors have been asserted

First Edition published in 2019

Published in the United States of America by Oxford University Press
198 Madison Avenue, New York, NY 10016, United States of America

British Library Cataloguing in Publication Data
Data available

Library of Congress Control Number: 2018963935

ISBN 978–0–19–883091–7 (hbk.)
ISBN 978–0–19–883092–4 (pbk.)

DOI: 10.1093/oso/9780198830917.001.0001

Printed and bound by CPI Group (UK) Ltd, Croydon, CR0 4YY

Links to third party websites are provided by Oxford in good faith and
for information only. Oxford disclaims any responsibility for the materials
contained in any third party website referenced in this work.

To Maria, Helene, and Camilla—my girls—Glenn-Peter Sætre

To Silje, Maya, and Milo—for making it worthwhile—Mark Ravinet

Preface

Genetic research is at a historical crossroads. Large quantities of genetic data are being generated at an ever-increasing pace and are being used and interpreted by students and professionals from a range of disciplines, including biology, medicine, archaeology, informatics, and forensic science. Working in genetics today requires a broad skillset, from the lab bench to the computer terminal. Most importantly, though, a thorough understanding of evolutionary principles is essential for making sense of genetic data. Accordingly, evolutionary genetics is becoming a key subject for the life sciences and beyond.

DNA is an essential molecule, occurring in every living organism on this planet. It orchestrates a multitude of complex biochemical processes and pathways that ultimately result in life, allowing organisms to interact with each other in similarly complex ways. DNA is a self-replicating molecule. Every time a cell divides, its DNA is copied, base for base, with great precision, but not absolute perfection. Any copy of DNA in an organism living today is the descendent of an unbroken chain of copies ranging back to the dawn of life on Earth. A range of deterministic and stochastic processes has molded this remarkable molecule, and thus the organisms it encodes, for nearly four billion years. The anomalies, the replication errors or *mutations*, represent the very raw material for organic evolution. Processes, such as natural selection and genetic drift determine the fate of the novel varieties that arise from mutations. Over time these processes cause organisms to change and diverge from each other, giving rise to all the endless varieties of species that have occupied our planet, including the present-day biota and ourselves.

Thinking in terms of evolutionary genetics involves abstractions that challenge our intuition.

First, evolutionary biology is a scientific discipline with a historical dimension. One must be able to conceptualize dynamic processes that operate at very different time scales: From instantaneous germ-line mutations and recombination events, through processes of selection and drift that alter the genetic make-up of a population over hundreds and thousands of generations, via processes of lineage splitting and speciation where the time scale may be in the orders of hundreds of thousands, or millions of years, all the way to large scale patterns of evolutionary change that span the history of life on Earth. Second, evolutionary processes involve several different levels of organization. From changes at the molecular level, to the effects these have on biochemical pathways, cells, organs, individual organisms, populations, and species.

Our ambition with this book is to help you make these abstractions and to ultimately make sense of the diverse and rich field of evolutionary genetics. We want you to understand fundamentally important processes such as mutation, natural selection, genetic drift, and speciation, and their consequences. Therefore, in the first half of the book we have focused on explaining a limited set of evolutionary models, but in some depth. Our aim is to motivate you to move beyond just understanding concepts, and to introduce you to how such concepts are applied. So, in the second half of the book, we focus more and more on how high-throughput sequencing data is changing the way evolutionary biology research is carried out. We are well aware that understanding evolutionary concepts and analysis is best done with a practical approach. Practical experience is second-to-none when it comes to developing an understanding of how to use genetic data to analyze and address interesting questions in the life sciences and how to interpret results in meaningful ways. So, along with the main text we provide study questions to motivate you to think and reflect on the concepts in each chapter. Furthermore, we provide a series of online, computer-based practical sessions where you will be able to reinforce your understanding and gain hands-on experience with analyzing and interpreting genetic data with an evolutionary perspective. Crucially, the practical computing assignments are centered on the R programming language. This might seem a little daunting at first but it is now an essential skillset for any evolutionary geneticist. It is a tool we use every day in our own research and one we cannot stress the importance of enough.

The book and associated teaching material is based on an undergraduate course at the University of Oslo, Norway. We assume that the readers have some general knowledge in biology and genetics. Moreover, we assume some knowledge and experience in algebra, probability theory, and statistics. Inexperienced readers should not worry; we have aimed to build arguments from basic principles throughout the book.

Acknowledgments

There are many people without whom this book would not have been possible. It is a real privilege to have reached the point where we can thank them publicly in the pages of this text. First of all, we thank Kaj Clausen for excellent illustrations—you absolutely made a difference to the quality of the book and to how we took the time to explain difficult concepts and ideas. It has been an absolute pleasure working with you! We must also give thanks to Richard I. Bailey, Angelica Cuevas, Tore O. Elgvin, Fabrice Eroukhmanoff, Giada Ferrari, Caroline Ø. Guldvog, Jo S. Hermansen, Thomas F. Hansen, Øistein H. Holen, Jun Kitano, Melissah Rowe, Anna Runemark, Camilla Lo Cascio Sætre, Bastiaan Star, Thomas Svennungsen, Cassandra N. Trier, and Kjetil L. Voje for fruitful discussions and for reading and providing invaluable comments and suggestions on earlier drafts of the chapters. Your suggestions and insight helped us guide our writing and we are indebted to you for improving the book in this way. Of course, it goes without saying that any errors or mistakes are our own! Thanks too to our many colleagues that have provided photographs, data, information, and figures so that we could include them in the book: Simon Griffith, Darren Irwin, Chris Jiggins, Stephen Leslie, John Novembre, Sarah Pryke, Ilik Saccheri, Ole Seehausen, Alex Twyford, Anja Marie Westram, and Graham Young. We are also grateful to our employers at the University of Oslo for their support during this project. Thanks, in particular to Rein Aasland, Tom Andersen, Anne Krag Brysting, Finn-Eirik Johansen, and Nils Christian Stenseth for the arrangements, encouragements, and help that made it possible to devote the necessary motivation and time to write the book. A special thanks to Nils Christian Stenseth for inspiration, unstoppable enthusiasm, and for having made the Centre for Ecological and Evolutionary Synthesis (CEES) what it is—a most stimulating work environment where nothing seems impossible! Thanks to all past and present members of the "sparrow group" who we were not able to mention, as well as our former supervisors, collaborators, and co-workers over the years, for all we have learned from joining forces in science. A special thanks to our former PhD and postdoc supervisors Tore Slagsvold and Hans Ellegren (GPS), and Chris Harrod, Roger K. Butlin, Jun Kitano, and Kerstin Johannesson (MR). We are also grateful to all those that helped us in developing the online computer tutorials, in particular Bernt Christian Helén and Alexandra Treimo. We should also thank Anne Krag Brysting, Svenja Christiansen, Mari Engelstad, Siv Nam Khang Hoff, Nina Knudtzon, Dabao Sun Lu, William Brynildsen Reinar, Carina Rose, and Camilla Lo Cascio Sætre, who helped us teach and perfect the R tutorials for the first time. Your feedback and that of our students have been invaluable in shaping how we designed the practical component of the book. Special thanks must also go to Bethany Kershaw, Ian Sherman, and the staff at Oxford University Press for believing in the book project and helping us in realizing it. Your assistance has been really invaluable. Warm thanks to all our friends for beautiful friendship, including necessary distractions and encouragements, and our loved ones and families for being the most important persons in our lives: Maria, Helene, Camilla, Sigurd, Henrik, and Leonie—I love you (GPS); Silje, Maya, Milo, and all the Ravinets—thank you for reminding me what matters most (MR).

Contents

CHAPTER 1

The foundations of evolutionary genetics

The fields of evolutionary biology and genetics were founded as separate disciplines in the mid-1800s, largely due to the works of Charles Darwin and Gregor Mendel respectively. It would, however, take decades before biologists finally understood the deep connection between the principles of heredity and the processes of evolution and that neither could be fully understood without the other. In this chapter we shall take a brief look at the history of these disciplines. We will learn how the key concepts, ideas, and technologies, fundamental to all of biology, were discovered and how these have affected our ways of thinking. We foreshadow many important concepts here that will be handled at greater depth in later chapters. Hence, we recommend reading this chapter a second time, after finishing the other chapters, to fully appreciate the history of this multi-faceted scientific field. The chapter ends with a brief review of important methods in contemporary evolutionary genetics and a hint towards future developments in the field. Many of the methods introduced here will become familiar to you through the assignments associated with the textbook and as you gain hands-on experience with evolutionary genetic analysis.

1.1 Concepts and ideas in a historical context

1.1.1 Evolutionary thoughts before Darwin

Charles Darwin is rightfully credited as the founder of modern evolutionary biology. Yet, he was not the first to suggest that organisms may change over time. The idea that one organism can transform into another can be traced back to the writings of early Greek philosophers such as Anaximander of Miletus and Empedocles hundreds of years BCE (Kirk et al. 1983). These ideas were not universally acclaimed, however, and the more influential philosophers, Plato and Aristotle, held that all living things were fixed by divine design (Taylor et al. 1999). Christianity reinforced this latter philosophy of essentialism and for more than 2000 years there was very little progress in evolutionary thinking (Mayr 1982).

In modern times, Jean Baptiste Lamarck formulated the most influential pre-Darwinian theory of evolution in his *Philosophie Zoologique* (1809). Lamarck held that simple organisms could arise spontaneously from non-living matter and evolve into more complex ones through a mysterious innate life force. Simple organisms would have arisen relatively recently, whereas complex ones, such as higher vertebrates, would have had a longer history of striving for perfection and complexity (Figure 1.1). An influential religious idea at the time was that God's creation could be ordered hierarchically into the great chain of being (*scala naturae*), from lifeless matters such as minerals, via simple organisms such as weeds, to trees, animals of increasing complexity, humans, angels, and eventually God himself. Lamarck's theory thus appeared to provide a semi-scientific explanation for this popular belief.

Evolutionary Genetics: Concepts, Analysis, and Practice. Glenn-Peter Sætre & Mark Ravinet, Oxford University Press (2019).
© Glenn-Peter Sætre & Mark Ravinet 2019. DOI: 10.1093/oso/9780198830917.001.0001

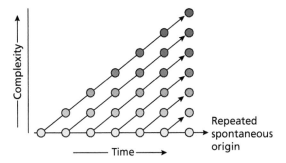

Figure 1.1 Lamarck's theory of evolution. Simple lifeforms (yellow) originate spontaneously out of lifeless matter from time to time and evolve towards greater complexity (increasingly red) through an innate life force. Complex lifeforms of today have therefore had a longer evolutionary history than simple ones. The theory appears to yield a logical explanation for the once popular idea of the great chain of being, suggesting that God's creation can be ordered hierarchically from lifeless matter, via simple life forms, to advanced beings.

Lamarck is probably most famous for his erroneous idea of **inheritance of acquired characteristics**. He stated that organisms adapt to their environment by the use and disuse of organs and structures, like when skin thickens and muscles grow following hard physical labor. He believed that such changes could be passed on to the offspring, and so could explain the striking observation that organisms appear adapted, as if designed, for the lives they live. Although Lamarck failed to produce a viable theory for the mechanisms of evolution and adaptation, his theory set the tone for subsequent thinking and most certainly inspired Charles Darwin (Gould 2002).

1.1.2 Evolution by natural selection—Darwin

Charles Darwin (1809–1882) published his monumental book, *On the Origin of Species* in 1859. Here a convincing case for evolution was set. Darwin made the argument that all organisms that have ever roamed our planet have descended from a single (or a few) primitive ancestor(s)[1] at the dawn of time. Darwin's view on the history of life on Earth thus differed markedly from that of Lamarck. Apparently simple organisms, such as bacteria and yeast, would have had an equally long and prosperous

[1] Chemical similarities between all types of living cells and a basically universal genetic code strongly suggest that all living organisms on Earth descend from a single common ancestor.

evolutionary history as more complex ones. Hence, there was no mysterious innate drive for complexity in nature. This also shattered the illusion of the great chain of being. Instead, Darwin suggested an entirely materialistic, scientific explanation to account for evolutionary change, namely **natural selection**. According to Darwin, natural selection has been the main mechanism causing gradual evolutionary change, adaptation, speciation, and in some lineages, increasing complexity. The theory was backed by a massive amount of empirical evidence, observations from comparative anatomy, paleontology, biogeography, embryology, and animal breeding. Importantly, Darwin also discussed possible weaknesses and problems with his theory.

Darwin had only a vague idea about heredity. He noticed that offspring tend to resemble their parents more than they resemble the average individual in the population. In short, he understood that "something" was inherited, but he had no mechanistic explanation for the process. Despite this, his observations were sufficient for him to develop the theory of evolution by natural selection, one of the most important scientific ideas ever published. Darwin's main arguments can be summarized as follows:

1) In all organisms, the inherent potential for ever-increasing population growth is checked by the limitation of essential resources, such as food. As a result, there will be competition for these resources and a struggle for survival and reproduction.
2) Individuals in a population vary from one another in phenotypic traits and much of this variation is heritable. Some of this (heritable) variation will affect the likelihood that an individual survives and reproduces under prevailing conditions—they affect **fitness**.
3) Consequently, natural selection will take place in which individuals more suited to the environment will win in the struggle for life at the expense of those less suited. These survivors will pass on their favorable traits to the next generation.
4) Over time, this process of natural selection will cause the population to evolve gradually and become better adapted to the environment. Ultimately, evolutionary change will accumulate in different evolving lineages forming new species.

Darwin's book founded the field of evolutionary biology and has had a huge impact on society and contemporary thinking. Evolution soon became widely accepted as a natural phenomenon. Darwin's contribution marked a farewell to the philosophy of essentialism that had dominated Western thinking for centuries. However, the significance of natural selection in evolution continued to be a matter of debate for decades to come and many scientists did not share Darwin's belief in the importance of its role.

To be fair, Darwin shares the credit for the theory of natural selection with another great scientist, namely Alfred Russel Wallace (1823–1913). He independently conceived the idea of natural selection and in 1858 he wrote a letter to Darwin where he laid out his theory. At the time, Darwin had been working on what he called his "great book" for many years already and so the letter from Wallace was a great shock to him. Friends and colleagues of Darwin arranged for a fair solution to the dilemma Darwin found himself in. In a meeting in the Linnean Society the same year, Darwin's and Wallace's theories were presented together and later that year, a transcript of the presentation was published as a joint paper in which the theory of natural selection was laid out (Darwin & Wallace 1858).

1.1.3 Mendel and the origin of genetics

The Augustinian monk and scientist Gregor Mendel (1822–1884) discovered the basic principles of genetic heredity in the 1850s and 1860s (Mendel 1865). In a series of crossing experiments on the garden pea (*Pisum sativum*), conducted in the gardens of St. Thomas Monastery in Brno in present day Czech Republic, he obtained and interpreted results that laid the foundation for the field of genetics (Edelson 1999).

Mendel demonstrated that heritable factors, which later became known as **genes**, behave in predictable ways. For instance, when he crossed plants with yellow peas with plants with green peas he noticed that all the offspring in the following (F1) generation carried yellow peas. Then, when these F1 individuals self-pollinated to produce the next (F2) generation, green pea-bearing plants reappeared at a 1:3 ratio. Mendel coined the term **recessive** for

the factor causing green peas and that for yellow peas **dominant**. The 1:3 ratio in the F2 generation can be explained if each individual has two discrete, heritable units that segregate independently during gamete production (meiosis) and are randomly combined as new pairs at fertilization. Using modern vocabulary, we call these heritable units **alleles**, and we call the set of two alleles that each individual carries a **genotype**. Using convention from the field, we can denote the allele for yellow color A_1 and that for green color A_2. Hence, the yellow and green parent generation possesses the **homozygous** genotypes A_1A_1 (yellow) and A_2A_2 (green), respectively. The individuals in the F1 generation would therefore have the **heterozygous** genotype A_1A_2 with one allele inherited from each parent. These offspring would all carry yellow peas because the allele for yellow color is dominant and thus overrides the effect of the recessive allele for green color. Crossing of F1 heterozygotes when alleles segregate independently would then (ideally) result in 25% A_1A_1 homozygotes, 50% A_1A_2 heterozygotes, and 25% A_2A_2 homozygotes. Only the A_2A_2 genotype would carry green peas (Figure 1.2), which is what Mendel found.[2]

Independent segregation of the alleles of a gene during gametogenesis is referred to as **Mendel's first law**, or the law of segregation. Additional crossing experiments suggested to Mendel that the alleles of different genes also segregate independently of each other during gametogenesis. For instance, that the alleles for pea color segregate independently from those for flower color. This **Mendel's second law**, or law of independent assortment, is not universally true, however. For instance, if a pair of genes occur in close physical proximity on the same chromosome (i.e. they are tightly linked) the alleles at the two genes will tend to be inherited as a unit rather than independently. We shall explore deviations from Mendel's second law further in later chapters.

Mendel's elegant experiments and insights into the laws of heredity were unfortunately not appreciated by his contemporaries. Few if any understood the significance of his discoveries. His paper was forgotten and Darwin never got to read it.

---

[2] Ronald Fisher and others have argued that Mendel's results were suspiciously close to the theoretical predictions.

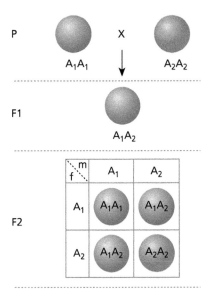

Figure 1.2 One of the classical crossing experiments conducted by Gregor Mendel. When parent plants (P) that are respectively homozygous for yellow (A_1A_1) and green peas (A_2A_2) are crossed, they produce heterozygous offspring that all carry yellow peas (A_1A_2). The allele for yellow color (A_1) is dominant and thus overrides the effect of the recessive allele for green color (A_2). When heterozygous plants self-fertilize to produce the F2 generation, gametes carrying A_1 and A_2 are randomly combined. Hence, 25% of the offspring is expected to get the green homozygous genotype that produces green peas (A_2A_2), 25% will be homozygous for the yellow allele (A_1A_1), and the rest will be heterozygous.

Hence, the link between heredity and evolution was for later generations of biologists to discover.

1.1.4 The mutation confusion

Toward the end of the nineteenth century, crossing experiments similar to those Mendel had conducted many years earlier were independently replicated by Hugo de Vries, Carl Correns, and Erich von Tschermak. All three authors published their results in 1900 and credited Mendel.[3] Thus, the Mendelian laws of heredity finally received international attention (Henig 2009).

However, it took some time for the scientific community to understand the relationship between the

principles of heredity and Darwin's theory of evolution by natural selection. Instead, the rediscovery of Mendel's laws reinforced opposing schools of thought that had begun to develop in the late nineteenth century: the biometric school of Francis Galton and others who used statistical methods to study continuous variation in phenotypes, and the saltation school of William Bateson and Hugo de Vries, who focused on discrete phenotypic varieties. It was de Vries who developed the **mutation theory of evolution** (de Vries 1900, 1903). Working experimentally with the evening primrose (*Oenothera lamarckiana*), de Vries discovered that novel, significantly deviating phenotypic forms could arise instantaneously, a phenomenon he termed **mutation**.[4] This inspired the theory that new species could arise instantaneously through large-scale mutations that alter the phenotype. The particulate nature of inheritance of discrete phenotypes as discovered by Mendel seemed to fit the mutation theory of evolution perfectly. Accordingly, de Vries and the early Mendelians and experimental geneticists rejected Darwin's theory of gradual change through natural selection.

1.1.5 Theoretical population genetics

Comprehensive integration of Mendelian genetics into evolutionary theory occurred with the work of the theoretical population geneticists Ronald Fisher (1890–1962), Sewall Wright (1889–1988), and J. B. S. Haldane (1892–1964) between 1918 and 1932.

In 1918, Fisher published a paper in which he demonstrated that continuous variation in a phenotypic trait is compatible with Mendelian inheritance (Fisher 1918). Continuous variation can result if genetic variation at several genes contributes to the phenotype. This finally resolved the long dispute between biometricians and Mendelians, at least in theory. Moreover, Fisher's paper set the stage for the field of **quantitative genetics**, a subdiscipline of evolutionary genetics that relies on statistical analysis of continuously varying phenotypic traits, assuming that several Mendelian genes affect the traits.

---

[3] de Vries, who published first, only mentioned Mendel in a footnote, whereas Correns pointed out Mendel's priority after having read de Vries. It is uncertain whether von Tschermak truly understood Mendel's results (Henig 2009).

[4] Later work has shown that the phenotypic novelties de Vries observed in the evening primrose were actually caused by genome duplication events and not genetic mutations.

Fisher made several significant contributions to evolutionary genetic theory, statistics, and experimental design. Commonly regarded as the foundation of population genetics, *The Genetical Theory of Natural Selection* (Fisher 1930) outlines a comprehensive mathematical theory of natural selection acting on genes in (large) populations. Fisher argued that mutations are an integral part of the evolutionary process. For instance, he envisioned that natural selection would ensure that the average individual in a population would be close to a phenotypic optimum. Therefore, a mutation with a large phenotypic effect would most likely impede fitness and be removed by selection because it would either overshoot the optimum or move the phenotype away from it, depending on the direction of change (Figure 1.3). By contrast, mutations with sufficiently small phenotypic effects would have a 50% probability of enhancing fitness and be favored by positive natural selection. Accordingly, evolution would proceed in small gradual steps, much like Darwin had stated. Fisher's book also contains important contributions to sex ratio theory,

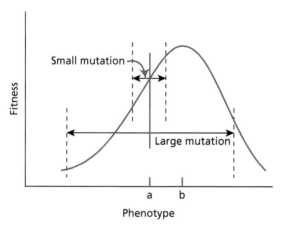

Figure 1.3 Graphical model that explains Fisher's argument for gradual evolution through small mutation steps. The figure depicts the relationship between Darwinian fitness and the size of a phenotypic trait. Individuals in the population currently have trait size *a*, which is somewhat smaller than the optimum *b*. A large mutation would always lead to lower fitness and be selected against, whereas a sufficiently small mutation would have a 50% chance of enhancing fitness. Carriers of the mutant allele would in the latter case be favored by natural selection and the population would evolve closer to the optimum.

sexual selection, the evolution of mimicry, and the maintenance of polymorphisms through balancing selection (heterozygote advantage), to name but a few of the topics he addressed.

Fisher's contribution to statistics is equally impressive. He pioneered the analysis of variance (ANOVA – see chapter 6), the method of maximum likelihood (we will learn about maximum likelihood in chapter 9), and the derivation of several important statistical sampling distributions, and also coined the term **null hypothesis**. He is thereby one of the most important contributors to the modern science of statistics.

Sewall Wright also made several lasting contributions to evolutionary genetics. Notably, he introduced the inbreeding coefficient F_{IS} and other F-statistics that we will explore in chapter 3. He also introduced the metaphor of **adaptive landscapes** (e.g. Wright 1931, 1932). According to this metaphor, a population climbs to the nearest fitness maximum in phenotype/genotype space by natural selection. Evolutionary stasis will occur when a population has reached a local fitness maximum because evolving to another peak would involve crossing a valley of low fitness, which would be opposed by selection. Wright was also the main developer of the theory of **genetic drift**—evolution caused by random events rather than by differential fitness. He envisaged scenarios in which genetic drift and natural selection could both affect the course of evolution (Figure 1.4). For instance, although selection alone would not allow the population to evolve from one adaptive peak to another, genetic drift may cause the population to stray from the optimum and across a low-fitness valley. Once the population has crossed the valley, selection would act to raise the population to the new adaptive peak.

J. S. B. Haldane is regarded as the third founder of population genetics. His book, *The Causes of Evolution* (1932), included important mathematical investigations on how natural selection affects gene frequencies and how selection, mutation, and gene flow interact. He is also noteworthy for pioneering the theory of kin selection (see chapter 5) and for his thinking on the genetics of speciation (see chapter 8).

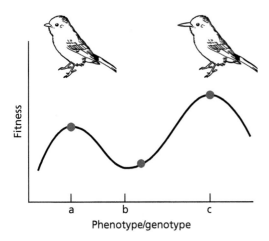

Figure 1.4 Wright's adaptive landscape metaphor showing the relationship between fitness and beak length in a hypothetical bird species, with local fitness maxima at *a* and *c* and a minimum at *b*. According to the metaphor, natural selection is a climbing process so a population will evolve to a local fitness maximum (left dot). Although other peaks may be higher, natural selection alone cannot move the population there because of the fitness valley in between. However, genetic drift (evolution caused by random events) might lead the population astray from the local optimum, eventually beyond the fitness minimum (central dot). At this point, natural selection would reinforce the evolutionary trend and the population would evolve to the higher fitness peak (right dot).

The work of these three scientists set the stage for the modern synthesis of evolutionary theory with genetics and they are rightfully regarded as its main architects.

1.1.6 The modern synthesis of evolution

Although Fisher, Wright, and Haldane laid the foundations for modern evolutionary genetics in 1918–1932, old controversies between biological sub-disciplines lingered over the first third of the twentieth century. Moreover, the mathematical language of these authors was hard to access and understand for empirically oriented biologists. But in 1936–1950, a general consensus in evolutionary biology developed, referred to as the **modern synthesis of evolution** (Huxley 1942).

Four major conclusions became generally accepted during this period (Dobzhansky 1937; Huxley 1942; Mayr 1982):

1) Evolution is a gradual process as Darwin had suggested, consistent both with known genetic mechanisms and observational evidence of naturalists.
2) Evolutionary understanding requires population-level thinking. Genetic variation exists within a population, generated by mutations, recombination, and gene flow. Evolution is due to changes in allele frequencies caused by natural selection and genetic drift acting within populations.
3) Natural selection is the dominant force in evolution. Selection acts on the phenotype of individuals in populations according to prevailing conditions and causes changes in the frequency of alleles in the genes that affect those phenotypes.
4) Macroevolution, the large-scale evolutionary trends and patterns occurring at the species level and above, is fully compatible with micro-evolution (i.e. the changes in allele frequencies that take place within populations).

The role of genetic drift in evolution remained a matter of controversy, but the main authors of the modern synthesis tended to follow Fisher, who thought the process to be of little importance relative to selection.

Theodosius Dobzhansky (1900–1975) was a major contributor to the integration of genetics and evolutionary theory during the modern synthesis. An empirically oriented geneticist and evolutionary biologist, Dobzhansky was instrumental in explaining the population genetic insights of Fisher, Wright, and Haldane to a broader audience (Dobzhansky 1937). His work on natural populations convinced him that Fisher, Wright, and Haldane had underestimated the amount of genetic variation that is normally present in a population, and that the evolutionary potential in nature therefore is greater than previously appreciated. Furthermore, Dobzhansky conducted pioneering work on the genetics of speciation (see chapter 8).

Other influential scientists associated with the modern synthesis include Julian Huxley, who synthesized a range of biological sub-disciplines in the light of evolutionary theory (Huxley 1942); Ernst Mayr, on systematics and speciation (Mayr 1942); George Ledyard Stebbins, on plant evolution (Stebbins 1950); and George Gaylord Simpson on paleontology (Simpson 1944).

1.1.7 Toward molecular genetics

Darwin's theory of evolution by natural selection was founded without any workable knowledge of heredity. Similarly, the field of genetics emerged without any knowledge of the cellular or molecular nature of genes.

The first breakthrough in understanding the cellular basis of heredity occurred early in the twentieth century, when Walter Sutton (1902) and Theodor Boveri (1904) identified **chromosomes** as the likely carriers of genetic material. Sutton discovered that grasshopper chromosomes occur in matched pairs inherited from either parent, and that these chromosomes separate during meiosis. In other words, just like the color of peas, they behave according to the Mendelian laws of heredity, an observation not lost on Sutton. Studying sea urchins, Boveri determined that all chromosomes must be present for proper embryonic development to take place, which further supported the chromosome theory. Subsequently, Thomas Hunt Morgan (1866–1945) confirmed and refined the chromosome theory through a series of elegant experiments on *Drosophila* fruit flies. Morgan was the first to map a gene for a particular phenotypic trait to a specific chromosome when he demonstrated that the mutation *white*, associated with white eye-color, showed sex-limited inheritance and therefore had to be located on the X chromosome (Morgan 1911) (Figure 1.5). He further developed the ideas of **genetic linkage** and **crossing over** (see chapter 2) and suggested that the frequency of crossing over between a pair of genes would indicate the physical distance between them, and thus laid the foundation for gene mapping. Because of the work by Morgan and co-workers, the physical mechanism of Mendelian inheritance was well understood and accepted by 1915 (Morgan et al. 1915). However, chromosomes are composed of both DNA and protein, so which one of them was the molecule of heredity?

Because of their great complexity and variation, proteins were the favored candidates for the heredity molecules for a considerable period of time. Deoxyribonucleic acid (DNA) had already been isolated in 1869 but scientists failed to identify its importance for many decades. DNA was thought to be a relatively simple molecule (e.g. Levene 1917) and so an unlikely candidate for carrying genetic information.

One famous experiment by Avery and co-workers was instrumental in establishing the role of DNA in heredity. Specifically, Avery et al. (1944) demonstrated that **transformation**, the process by which bacteria incorporate hereditary material from the environment into the cell, is due to an uptake of DNA and not protein. Avery and co-workers worked with a virulent and non-virulent strain of *Streptococcus pneumoniae*, a bacterium that, depending on the strain, can cause pneumonia in mammals. It was already known that a non-virulent strain could transform into a virulent one if the former was kept in contact with dead (heat-killed), virulent bacteria, suggesting that the non-virulent bacteria could somehow incorporate hereditary material from the dead bacteria. Mice injected with a mixture of the non-virulent and dead, virulent strain would develop pneumonia and die. Avery and co-workers treated the heat-killed, virulent bacteria with either a protease, an enzyme that degrades protein, or a DNAase, an enzyme that degrades DNA, before mixing them with a live strain of non-virulent bacteria. They found that protease did not affect the ability of the dead strain to transform the non-virulent one, whereas DNAase did. They concluded that the genetic material involved in transformation was most likely DNA.

The ultimate breakthrough in DNA research came with Watson and Crick (1953) and their double-helix model of DNA (Clark & Pazdernik 2012) (Figure 1.5). At the time of the Watson–Crick model, certain characteristics of DNA had already been deduced. For example, Levene (1917) had proposed that DNA was composed of a series of **nucleotides**, and that each nucleotide was composed of one of four nitrogen-containing bases, a sugar molecule, and a phosphate group. However, he erroneously deduced that DNA is a short molecule and that the nucleotides were linked in the same order and were thus invariable. Additionally, Chargaff (1950) had discovered that the nucleotide composition varies between species and that the amount of the nitrogen-containing base adenine (A) closely mirrors the amount of thymine (T), and that the amount of guanine (G) mirrors the amount of cytosine (C). However, he failed to deduce that this could imply chemical bonds between A and T and between G and C. Of crucial importance to Watson and

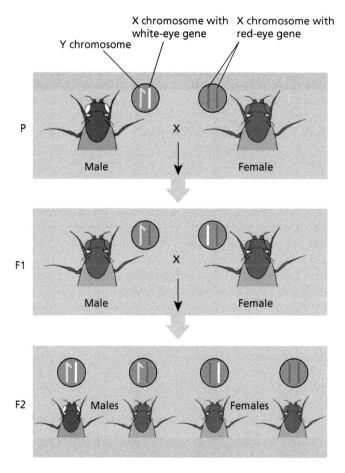

Figure 1.5 Sex-limited inheritance of eye color in *Drosophila* fruit flies, as demonstrated by Morgan (1911), is consistent with Mendelian inheritance of a gene with two alleles (red and white) located on the X chromosome. Note that in this example, the white allele is recessive.

Crick's model was the X-ray crystallography work by Rosalind Franklin[5] and Maurice Wilkins, which was compatible with a three-dimensional helical structure of the molecule. Watson and Crick's model, which is essentially the same as our current understanding of the molecule, can be summarized as follows (Figures 1.6 and 1.7):

1) DNA is composed of sequences of four different nucleotides. Each nucleotide consists of a sugar, deoxyribose, bound to a phosphate group and a nitrogen-containing base, which is either A, T, G, or C.

2) The nucleotides are arranged in a double-stranded helix in which the sugar components (deoxyribose) and the phosphate groups constitute the backbone. The nitrogen-containing bases (A, T, G, or C) point inwards and each base is connected to a complementary base (A with T, G with C) on the other strand by hydrogen bonds.

3) The two strands are anti-parallel, meaning that they run in opposite directions so that the 5' (five prime) end of one strand is paired with the 3' end of the complementary strand.

4) The outer edges of each base are exposed and available for hydrogen bonding to other molecules, including proteins involved in **DNA replication** and **transcription**.

---

[5] It has been argued that Watson and Crick should have included Franklin as co-author on the famous paper where the DNA model was presented, as it relied so heavily on her work.

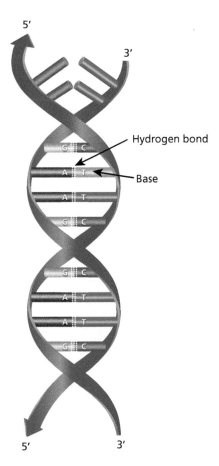

Figure 1.6 Schematic representation of Watson and Crick's double-helix model of the DNA molecule. See main text for further explanation.

The DNA double-helix model of Watson and Crick paved the way for the **central dogma of molecular biology** that seeks to explain the flow of genetic information in living organisms (Crick 1958, 1988).[6] The key components of the central dogma include:

1) **DNA replication**. During cell division (mitosis and meiosis), genetic information from the mother cell is transferred to the daughter cells by means of DNA replication (Figure 1.8a). During this process, the double helix is unwound and each strand acts as a template to

⁶ Francis Crick used the word "dogma" to signify that the theory is larger and of greater significance than a simple hypothesis. At the time, he was not aware of the negative associations many people would have to that word (Crick 1988).

dictate the complementary nucleotide sequence of two new strands. Several proteins are involved in initiating and regulating the process, including DNA polymerase that assembles free deoxy nucleotides (dATP, dTTP, dGTP, and dCTP) to build the new strand. Enzymatic DNA-repair mechanisms ensure that the copying process is (nearly) perfect, so that the daughter cells inherit identical copies of the DNA from the mother cell. However, errors in DNA replication during meiosis do occur and are one of the main sources of genetic mutation.

2) **Transcription**. This is the process in which the genetic information of DNA is transcribed in the form of ribonucleic acid (RNA). RNA resembles DNA except that it is a single-stranded molecule; the sugar of the backbone is ribose rather than deoxyribose; and it includes the base uracil (U) rather than T (that is A, C, G, and U). As in DNA replication, several proteins are involved in initiating and regulating the process. In eukaryotes, there are several classes of transcribed RNA molecules that have different functions in the cell. They are involved in gene regulation, RNA processing, catalysis of chemical reactions, and, notably, protein synthesis. There are three main classes of RNA involved in protein synthesis: messenger RNA (mRNA) holds the genetic code for synthesizing the protein; ribosomal RNA (rRNA) makes up the RNA component of the ribosomes (the other component being protein) where protein synthesis takes place; and transfer RNA (tRNA), which carries amino acids to the ribosome. The ribosomes are found in the cytoplasm of the cell whereas DNA is located in chromosomes in the nucleus. Hence, the genetic information of protein coding genes is transcribed into mRNA and transported out of the nucleus (Figure 1.8b).

3) **Translation**. This is the process in which the mRNA is translated into a protein (Figure 1.8c). Triplets of mRNA nucleotides, **codons**, are read by the ribosome. Each codon specifies a specific amino acid (Table 1.1). tRNAs have anticodons that are complementary to the mRNA codons, and each tRNA carries a single amino acid into the ribosome–mRNA complex. The complementary codons and anticodons of mRNA and tRNA

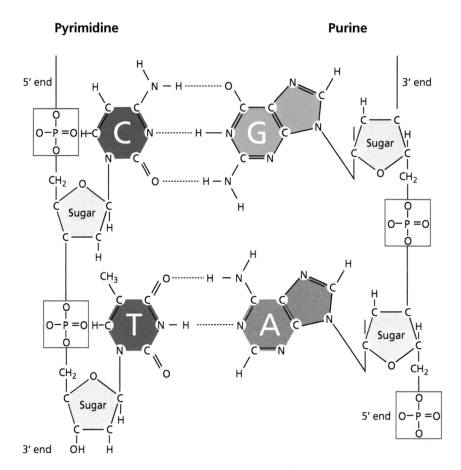

Figure 1.7 Molecular structure of DNA. Each nucleotide consists of a sugar (deoxyribose) bound to a phosphate group and a nitrogen-containing base (A, T, C, or G). The bases are of two main types, pyrimidines (C and T) that are single-ringed molecules, and purines (A and G) that are double-ringed. Pairs of bases on each strand are connected through hydrogen bonds.

cause the two molecules to temporarily bind. As the mRNA is processed, amino acids get linked into a growing polypeptide chain (a protein), which eventually folds into its correct configuration. Translation is initiated with the start codon of the mRNA (AUG) and it ends with a stop codon (UAA, UGA, or UAG).

With the structure of DNA and the central dogma of molecular biology worked out, the field of molecular genetics was founded.

1.1.8 Diverging fields

Following Watson and Crick's DNA model and the formulation of the central dogma of molecular biol-

ogy, the field of molecular genetics underwent a period of rapid advancement, which continues to accelerate to this day (see section 1.2). However, during the 1960s, 1970s, and 1980s, molecular geneticists were mostly concerned with studying the structure and function of genes at the molecular level, using laboratory model organisms such as yeast and mice. With the exception of the search for mutations causing genetic disease, and hence low fitness, molecular geneticists rarely placed their research in an evolutionary framework.

The field of evolutionary biology also saw several important new advances during this period, although molecular genetic data was difficult to attain and used by only a few. Instead research focused on phenotypic studies of plants and animals

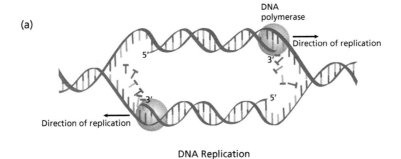

(a)

DNA
polymerase

Direction of replication

5'

3'

3'

5'

Direction of replication

DNA Replication

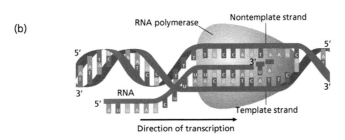

(b)

RNA polymerase

Nontemplate strand

5'

5'

3'

RNA

3'

3'

5'

Template strand

Direction of transcription

Transcripton

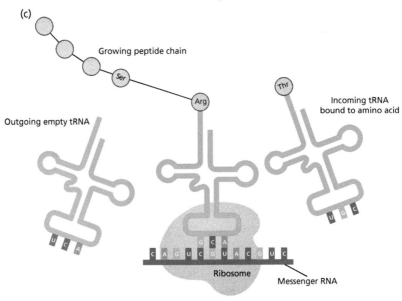

(c)

Growing peptide chain

Ser

Arg

Thr

Incoming tRNA
bound to amino acid

Outgoing empty tRNA

Ribosome

Messenger RNA

Translation

Figure 1.8 The three main pathways of genetic information transfer according to the central dogma of molecular biology. **(a)** DNA replication. Transfer of genetic information from a mother cell to its daughter cells through DNA copying. The double helix is unwound by enzymatic breakage of the hydrogen bonds between the two strands. Each of the freed strands (blue) now act as templates for building new complementary strands (red) as dictated by the nucleotide sequence. DNA polymerase assembles free nucleotides to build the new strands. **(b)** Transcription. Transfer of genetic information from DNA to messenger RNA (mRNA). As in replication, DNA is unwound. RNA polymerase assembles free ribonucleotides to build an RNA strand, which is complementary to the template strand of the DNA. The resulting mRNA is transported out of the nucleus to a ribosome in the cytoplasm where protein synthesis takes place. **(c)** Translation. Genetic information is translated from mRNA into protein in a ribosome. Triplets of nucleotides (codons) dictate the amino acid sequence of the protein. Transfer RNAs (tRNA) carry specific amino acids. The anticodon of the tRNA binds it temporarily to the complementary codon of the mRNA and the amino acid is bound to a growing polypeptide chain. The polypeptide chain eventually folds into a mature protein.

Table 1.1 The genetic code table (standard code).

Amino acid	Codon
Isoleucine	AUU, AUC, AUA
Leucine	CUU, CUC, CUA, CUG, UUA, UUG
Valine	GUU, GUC, GUA, GUG
Phenylalanine	UUU, UUC
Methionine & START	AUG
Cysteine	UGU, UGC
Alanine	GCU, GCC, GCA, GCG
Glycine	GGU, GGC, GGA, GGG
Proline	CCU, CCC, CCA, CCG
Threonine	ACU, ACC, ACA, ACG
Serine	UCU, UCC, UCA, UCG, AGU, AGC
Tyrosine	UAU, UAC
Tryptophan	UGG
Glutamine	CAA, CAG
Asparagine	AAU, AAC
Histidine	CAU, CAC
Glutamic acid	GAA, GAG
Aspartic acid	GAU, GAC
Lysine	AAA, AAG
Arginine	CGU, CGC, CGA, CGG, AGA, AGG
STOP	UAA, UAG, UGA

in natural populations, or on mathematical population genetics and quantitative genetic modeling following the tradition of Fisher and Haldane. It was a golden age for research into the evolutionary significance (in terms of genetic fitness) of animal behavior, including the evolution of social behavior (e.g. Hamilton 1964a,b), animal conflict (e.g. Maynard Smith & Price 1973), and sexual selection (e.g. Lande 1981; Andersson 1982), topics that are handled in chapters 4 and 5. Although largely detached from the parallel advances in molecular genetics, this period reinforced and refined a gene-centric view of evolution; a development that received widespread public attention because of the popular books by Richard Dawkins, such as *The Selfish Gene* (1976) and *The Blind Watchmaker* (1986).

Some evolutionary geneticists did, however, rely on molecular genetic data. By the 1960s, improved methods of gel electrophoresis allowed variation in the amino acid sequence of proteins to be visualized and this suggested much greater variation was present than expected. The implications of these empirical findings had a profound influence on those working on theoretical population genetics. Mooto Kimura (1924–1994), a mathematically oriented evolutionary geneticist, was startled by the magnitude of protein variation in natural populations. Since the modern synthesis, the notion that Darwinian natural selection is the dominant force in evolution was widely accepted by evolutionary biologists: directional selection explained changes in allele frequencies; and balancing selection, via mechanisms such as heterozygote advantage, explained the maintenance of genetic polymorphisms. Kimura (1968, 1983) presented an alternative theory, the **neutral theory of molecular evolution**, which has had a profound impact on evolutionary genetics. The theory states that:

1) The majority of evolutionary changes at the molecular level are caused by random genetic drift acting on alleles that are selectively neutral, and not by directional (positive) selection.
2) Most of the molecular variation present in a species at any given time is selectively neutral and is not maintained by balancing selection.
3) Most mutations are deleterious and are rapidly removed by purifying (negative) selection. Such mutations would therefore be of little significance in evolution. They would rarely be encountered in a population sample and would be transient.

We explore the neutral theory and its implications in depth in chapter 7, but two points are worth mentioning here. First, the neutral theory concerns evolution at the *molecular* level. Kimura and other neutralists acknowledge that phenotypic evolution cannot be understood without invoking divergent (positive) natural selection. Accordingly, the molecular basis for these phenotypic changes also has to involve positive selection. Second, selection—in terms of negative selection—is an integral component of the neutral theory.

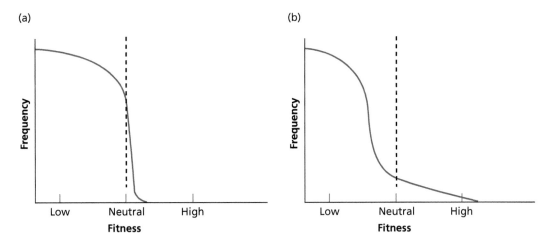

Figure 1.9 Frequency distribution of non-synonymous mutations according to their effects on fitness as expected from **(a)** the neutral theory of molecular evolution and **(b)** the traditional selection theory. The two theories differ with respect to the expected relative frequency of (near) neutral versus positive mutations.

Kimura's work initiated a long lasting and, at times, heated debate between his followers (neutralists) and those who maintained the traditional view that selection is paramount also at the molecular level (selectionists). It is important to remember that the debate was concerned with variation at the level of amino acids. Advances in molecular genetics and genomics have revealed that only a small fraction of, for example, the human genome consists of DNA that codes for protein (see chapter 2). There is hardly any debate around the notion that much of the noncoding DNA predominantly evolves neutrally.[7] The more interesting empirical question is to what extent **non-synonymous mutations**, those that alter the amino acid sequence of a protein, evolve by selection or drift. Neutralists and selectionists would agree that the majority of such mutations would have a negative impact on fitness because there are more ways of impairing than improving the function of an already well-functioning protein. The controversy is instead concerned with the relative frequency of (nearly) neutral non-synonymous mutations and those that enhance the function of a protein (Figure 1.9).

---

[7] The ENCODE Consortium (www.encodeproject.org) has challenged the view that non-genic DNA is void of function. This view has, however, been severely criticized (Doolittle 2013).

1.2 Contemporary evolutionary genetics: toward a new synthesis

1.2.1 Molecular biology meets evolutionary genetics

Technological developments in molecular biology eventually revolutionized empirical research in evolutionary genetics. A series of inventions has made it possible for standard research laboratories to generate large amounts of high-quality genetic data from virtually any species at an ever-increasing speed and decreasing cost. The genetic data of primary relevance in evolutionary genetics are:

1) **DNA sequence** data—the precise order of nucleotides of a DNA segment ranging from gene fragments of a few hundred nucleotides to entire genomes.
2) Sets of **genetic markers**—short segments of DNA in which allelic variants can be identified to yield individual genotypes. Important genetic markers include:

   a. *Microsatellites*—short (2–5 nucleotides), tandem repeated sequence motifs that vary in number of repeats, and hence in length (e.g. ...CACACACA...). The number of alleles (different repeat numbers) segregating in a population typically ranges from 2 to 40 per locus (sometimes more). Most microsatellites are noncoding, although some three-nucleotide repeats are located in protein coding genes. Most organisms have hundreds or thousands of polymorphic microsatellites scattered through their genomes (e.g. Payseur et al. 2011).

b. *Single nucleotide polymorphism (SNPs)*—nucleotide positions where two (or more) nucleotides are segregating in a population (e.g. ... ACCTG<u>G</u>TTGG...versus...ACCTG<u>A</u>TTGG...). SNPs are mostly biallelic and occur at high densities throughout the genome of most species, in both coding and noncoding regions.

Access to large amounts of genetic data has transformed evolutionary genetics from a field rich in theory but poor in data to a more exploratory field; from a hypothesis-driven science where the theories were difficult to empirically test to one in which large amounts of data are often gathered first and interpreted *post hoc*. The population genetic models of Fisher, Wright, and Haldane were of a predictive nature. Given initial conditions, what will the evolutionary outcome be? Comparatively recent theoretical developments, specifically **coalescent theory** (e.g. Kingman 1982a,b), do the opposite. The pattern of genetic variation observed in a current population sample can be traced back to an ancestral gene copy. We can use statistical analysis to infer the evolutionary processes that have shaped the pattern of genetic variation we observe today. We will learn more about this in chapter 7.

The aims of contemporary evolutionary genetics are manifold. The study of how evolutionary processes such as genetic drift and natural selection shape the genetic makeup of natural populations remains central. Further, we aim to reconstruct the evolutionary past of populations and species and establish the genetic relationships between them (chapters 8 and 9). Molecular genetic data are used in studies of both organismal and molecular evolution (chapters 7 and 8).

1.2.2 The genotype–phenotype map: opportunities and challenges

The ongoing synthesis of molecular and evolutionary genetics creates new opportunities and avenues for research. One important goal is to unravel the molecular basis for heritable phenotypic variation. We want to identify the building blocks of adaptive evolution and understand how the different components interact.

It is worth noting, however, that this synthesis of molecular and evolutionary genetics is currently causing confusion. One problem is that the meaning of key concepts has changed over time, and the changes have not always been parallel in adjacent fields. For instance, early evolutionary geneticists would define a gene as *a unit of heredity that affects a phenotypic trait*. In phenotypically oriented research, one still commonly uses this definition. In molecular genetics, a gene is normally defined as *a segment of DNA that specifies a functional protein or RNA product*. Heritable phenotypic variation can be caused by variation in the sequence of a protein coding gene, but may also have different causes. For instance, mutations in the vicinity of a protein coding gene can affect **gene expression**, the amount of gene product a gene produces or when during development the product is being produced, and hence phenotype. The traditional and molecular definitions therefore only partly overlap. In evolutionary genetics, the heritable units that are affected by evolutionary forces remain the center of focus, whether they are coding genes or not. We will try to avoid referring to such an evolving unit as a gene if it does not correspond to a protein- or RNA-specifying DNA. We will instead use the neutral term **locus** to signify genomic regions of interest to evolutionary biologists. A locus simply refers to a specific genomic location; it may signify a protein coding gene, a non-genic mutation site that does or does not affect a phenotype or fitness, a genetic marker, or any other site of interest.

Another and perhaps more important source of confusion is the distinction between factors that affect phenotypic variation and the subset of these factors that are heritable in an evolutionary sense. Molecular biologists are making great progress in unraveling the molecular basis for phenotypic variation. Environmental factors can induce changes in gene expression, for instance through **DNA methylation**, a process in which a methyl group (CH_3) is added to the cytosine (or adenine) nucleotides, causing the gene to be down-regulated or inactivated. Such epigenetic changes in gene expression do not involve changes in the DNA sequence. However, a cell with a methylated gene will produce daughter cells in which the same gene is also methylated, so these changes are heritable in the sense that an epigenetic signal is transferred to daughter cells. Heritable epigenetic changes are important in development because different sets of genes need to be turned on or off in different tissues for appropriate cell function. It is heritable epigenetic signals that make sure that a liver cell gives rise to daughter liver cells and not, say, muscle cells. Only undifferentiated **stem cells**,[8] such as those that are produced during the first rounds of cell division following fertilization, have the potential to transform into any kind of cell. This is because they have not yet been affected by epigenetic signals. Some epigenetic signals can even be inherited across one or two generations. For instance, an environmentally induced change in gene expression in a pregnant woman can be transferred to her female fetus as well as the future offspring of that fetus, because the fetus already carries the egg(s) that will eventually become fertilized. Despite this, to the best of our current knowledge epigenetic inheritance is not directly relevant for evolution because it is a transient phenomenon. Epigenetic change is not evolutionary change. Moreover, most epigenetic signals are lost at gametogenesis; the zygote is a stem cell. Phenotypic plasticity and epigenetics are evolutionarily important in a more indirect sense, however, and we will explore this further in chapters 5 and 10.

1.3 Generating molecular genetic data

In order to study evolutionary genetics empirically we need to generate molecular genetic data from our study organisms and use statistical and analytical tools to interpret the data in meaningful ways. Along with phenotypic and gene expression data, variation in DNA sequences and genetic markers constitute the main units of data in contemporary, empirical, evolutionary genetics. DNA can be collected from organic material, such as tissue samples from your study organism, and can

easily be isolated and purified in the lab. Below we briefly explain the logic of some important techniques that have enabled us to generate such data.

1.3.1 Gel electrophoresis

The technique of gel electrophoresis is used to separate biomolecules such as proteins and DNA by size, physical resistance, or charge. Gel electrophoresis has been the main method for genotyping genetic markers for decades and is an integral component in a range of different DNA- and protein-sequencing techniques. The principle is very simple: charged biomolecules are transported through a resistant medium (e.g. an agarose or polyacrylamide gel) by an electric current. Different molecules will move through the medium at different speeds because of their differences in resistance. For instance, shorter fragments will move faster and hence further per unit of time than longer fragments (Figure 1.10). Gel electrophoresis is still widely used for separating (genotyping) genetic markers that exhibit length polymorphism, such as microsatellites. Different methods for genotyping are continually being developed and improved, such as mass spectrometry, which separates biomolecules by mass. However, gel electrophoresis remains a standard multipurpose tool in most molecular genetic laboratories.

1.3.2 Polymerase chain reaction

Genotyping and DNA sequencing became much easier after Kary Mullis developed the polymerase chain reaction (PCR) method for amplifying targeted DNA segments in 1983 (Bartlett & Stirling 2003). By generating a large number of copies of a targeted DNA segment, such as a microsatellite marker, PCR makes it straightforward to visualize such segments using gel electrophoresis. Similarly, many sequencing technologies require large quantities of template DNA, which can easily be generated by PCR.

The method is based on repeated cycles of DNA denaturation, annealing, and extension. A mixture is created that includes the DNA template of interest; pairs of **oligonucleotides**—short, synthetic DNA sequences, called primers; a thermostable DNA

---

[8] There are two classes of stem cells in mammals: embryonic stem cells that can differentiate into any kind of cell; and adult stem cells found in certain tissues, such as bone marrow, that can differentiate into a limited set of adult cells.

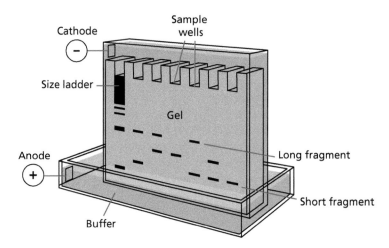

Figure 1.10 Fragment size separation of biomolecules using gel electrophoresis. Samples are loaded in the wells and transported through the gel by an electric current. Shorter fragments travel further per unit of time than longer ones and are therefore separated out. Actual fragment length can be determined by size standards of known length as shown in the left well.

polymerase; and free nucleotides (dATP, dTTP, dGTP, and dCTP). In the first stage, the mixture is heated until the hydrogen bonds of the double-stranded template DNA are broken (denatured) to yield single-stranded DNA, typically at 95°C. Next, the mix is cooled to the point where complementary DNA strands re-anneal. The included primer pair is complementary to the 3'-ends of either strand of the targeted DNA segment and anneals to the DNA template. It is essential to choose the temperature that yields optimal annealing between the template and the primer pair during this phase, normally between 55°C and 65°C. In the final enzymatic DNA extension stage, the free nucleotides are assembled to build a new complementary strand as dictated by the template strand (Figure 1.11). One commonly used enzyme for this extension stage is *Taq* polymerase, isolated from a thermophilic bacterium, *Thermus aquaticus*, which lives in hot springs. Accordingly, the enzyme's optimum activity occurs at a high temperature, around 72°C, and it is not destroyed during the denaturation stage. As the PCR progresses, the DNA fragments generated are themselves used as templates for replication. Thus, a chain reaction occurs in which the DNA template is exponentially amplified.

1.3.3 Restriction enzyme digestion

Certain bacteria produce enzymes that recognize short specific sequences and produce a double-

stranded cut in the DNA in or near the recognition site. Such restriction enzymes can be isolated for use in molecular genetic laboratories. For instance, the restriction enzyme *Sma1* recognizes the sequence GGGCCC/CCCGGG and makes a blunt cut between the three Cs and Gs. So, by digesting genomic DNA with *Sma1*, the DNA will be cut at all sites possessing this sequence and thus become fragmented. Individual sequence variation at recognition sites will cause differences in fragment lengths. Such restriction fragment length variation has many applications in evolutionary genetics, including genotyping procedures based on patterns of variation in fragment lengths and SNP detection. In recent years, restriction site–associated markers (RAD tags) have become a popular method for cheaply and easily generating and genotyping large sets of genome-wide genetic markers. Using high-throughput sequencing technology (see section 1.3.5), the DNA sequence flanking each restriction site can be determined. By mapping this sequence to a reference genome, one can obtain not only the genotype of a large number of SNP markers, but also their genomic location.

1.3.4 Sanger sequencing

An early method of rapidly sequencing DNA was developed by Sanger et al. (1977) and involves *in vitro* DNA replication. It requires: a DNA polymerase;

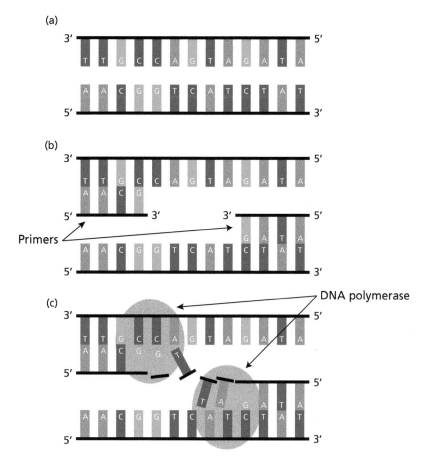

Figure 1.11 Polymerase chain reaction (PCR) is a method for amplification of specific DNA segments by means of repeated cycles of *in vitro* DNA replication. A PCR cycle includes: **(a)** a denaturation phase (usually) at 95°C, in which the hydrogen bonds between the DNA strands are broken to yield single-stranded DNA; **(b)** an annealing phase at 55–65°C, in which short, synthetic DNA primers of ±20 nucleotides anneal to a complementary DNA template (including templates generated in previous cycles); and **(c)** an extension phase at, for example, 72°C in which a thermostable DNA polymerase, such as *Taq*, assembles free nucleotides (dNTPs) into a new DNA strand as dictated by the nucleotide sequence of the DNA template (replication). In each cycle, the number of copies of the targeted DNA is (at least in theory) doubled. A typical PCR consists of an initial denaturation phase; 25–35 cycles of denaturation, annealing, and extension; and a final prolonged extension phase.

a single-stranded DNA template (for instance, a PCR-product); a primer complementary to the 3′-end of the target to be sequenced; free normal nucleotides (deoxyribonucleotide triphosphates (dNTPs); A, C, T, and G); and, importantly, modified dideoxynucleotides (ddNTPs) that cause the DNA polymerase to cease DNA extension when the ddNTP is incorporated. The sequencing reaction therefore results in a large number of extended DNA products of different lengths determined by when during the DNA extension process that the ddNTP is incorporated. In automated sequencers, the ddNTPs are labeled with different fluorescent

colors for each of the four nucleotides. The DNA fragments are size-separated by capillary gel electrophoresis,[9] and a laser detector registers the 3′ fluorescent-labeled ddNTP ends (Figure 1.12). By sequencing the target in both directions, very accurate reads of up to about 1000 nucleotides of continuous sequence can be obtained. The high accuracy of the technique meant it was fundamental in the effort to sequence the human genome

[9] Sanger sequencing thus uses the same principle as in standard gel electrophoresis explained in section 1.3.1. Automated Sanger sequencing machines can also be used for genotyping genetic markers that exhibit length polymorphisms.

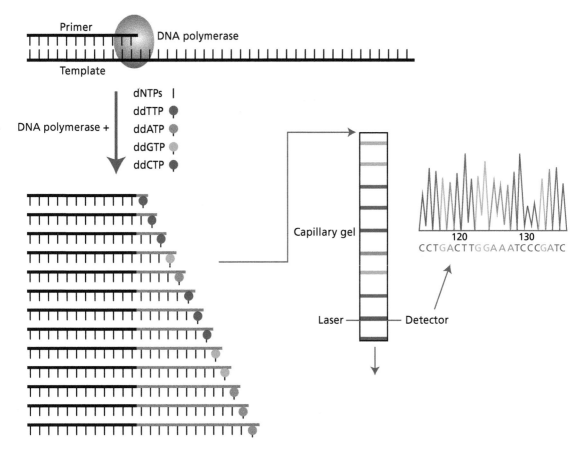

Figure 1.12 Principle of Sanger sequencing. Sanger sequencing is based on selective, chain-terminating, *in vitro* DNA replication. A primer anneals to the targeted single-stranded DNA template based on sequence complementarity. A DNA polymerase assembles free nucleotides (dNTPs) to build a new complementary strand as dictated by the DNA template. Among the free dNTPs, one includes a smaller amount of chain-terminating dideoxynucleotide (ddNTP), labeled with a different fluorescence color for each of the four nucleotide bases A, T, C, and G. DNA extension stops when a ddNTP is incorporated. As a result, a large number of DNA strands of different lengths are generated. These are transported through a capillary gel by electrophoresis. The shortest strands will pass through the capillary gel first followed by strands of successively increasing length. A laser detector reads the fluorescence light signal of the ddNTPs as they pass and the signal is transformed to a chromatogram showing the intensity peaks of each of the four nucleotides and, hence, the complementary DNA sequence of the targeted template.

(Lander et al. 2001). The automation of Sanger sequencing led to the first truly high-throughput sequencing technology—i.e. the ability to sequence multiple reads in parallel. Typical Sanger sequencers contained 96 capillaries, meaning that a single run could produce approximately 100 kbp of sequencing data. Today, due to its high cost per output and relatively slow speed, Sanger sequencing is no longer the major method used to generate sequencing data. However, accuracy is still its greatest asset and it is occasionally used to verify findings from sequences identified using high-throughput technology.

1.3.5 High-throughput sequencing

High-throughput sequencing (HTS) is a colloquial name for several different technologies that have replaced Sanger sequencing as the main source of generating sequence data (e.g. Mardis 2008). Automated Sanger sequencing was a first step towards high-throughput sequencing. True advances in sequencing output, however, came with the rise of so-called next-generation sequencing (NGS), able to sequence millions of reads in parallel (Mardis 2008; Metzker 2010). We note that NGS and HTS are used as interchangeable terms in the literature.[10] In this book, we will use "high-throughput sequencing" as many of the technologies referred to as next generation are already being replaced! A diverse array of HTS technologies have been developed with different core approaches; many of the initial platforms developed in the 2000s are now defunct and no longer in use. The rapid nature of development in this field therefore means that any overview of the available platforms is quickly dated; however, HTS approaches typically require that template DNA is broken into much smaller fragments and that the incorporation and imaging of nucleotides is done in real-time in parallel for many, many DNA strands simultaneously (Rokas & Abbot 2009).

These basic principles of HTS are best illustrated with the Illumina sequencing-by-synthesis technology that is the current dominant platform. The first step requires preparation of libraries—i.e. genomic DNA is fragmented into short sequences (approximately 200 bp in length) and oligonucleotides (short, synthetic DNA sequences) called adapters are ligated to either end. These adapters allow the DNA fragments to be attached to one of millions of short anchors which are complementary oligonucleotides anchored on a glass plate—a flow cell (Shendure & Ji 2008). DNA amplification is then carried out *in situ* on the flow cell using a process called bridge PCR (Shendure & Ji 2008; Metzker 2010) (Figure 1.13). The template, anchored at one end to the flow cell, bends over and attaches its free adapter to another available anchor on the flow cell, again through complementary base pairing. While bent, amplification takes place as in a standard PCR cycle with polymerase enzymes and free nucleotides and, once complete, the DNA is denatured to produce two separate strands anchored close to one another on the flow cell. Repeated cycles of this process allow different templates to be amplified thousands of times, but ensure they remain locally clustered on the cell.

Sequencing then takes place when DNA polymerase incorporates fluorescent-tagged nucleotides to the now amplified template sequence—hence sequencing by synthesis (Figure 1.14). A single base is added each time for each of the strands on the flow cell and the corresponding fluorescent color is detected using laser imaging (Mardis 2008). The fluorescent signal from only one sequence can hardly be detected, but since the sequences are amplified in clusters with the bridge PCR technology a locally colored spot lights up on the flow cell and is detected and registered by a computer. Fluorescent tags are then washed away and the next base is added, detected, and registered, and so forth until the read length limit is reached. Early Illumina machines produced 30–40 bp reads but 100 bp is now standard and it is possible to sequence even larger reads (250–500 bp) depending on the machine used. Because of the nature of genomic data and the error rates of the platform (i.e. miscalling of nucleotides or false positive detections of insertions and deletions), users typically sequence multiple times—resulting in higher coverage of the DNA region of interest.

HTS technologies have been rapidly adopted since their initial introduction, which in turn has led to an explosion in the availability of genomic data. As of writing 15 244 genomes have been sequenced, assembled, published, and archived on public databases[11] and many more sequence projects are underway. Improved platforms and widespread use have also reduced the cost of genome sequencing such that it is now possible to sequence a human genome for around $1000 USD, a stark contrast to the $500 million–$1 billion estimated price tag for the Human Genome Project completed in 2003. Despite this meteoric rise, HTS platforms are not without issues; for example, sequencing error rates for Illumina data are typically higher than Sanger sequencing (Shendure & Ji 2008). Furthermore, the short reads produced by popular platforms such as Illumina can make genome assembly particularly challenging—a topic we will explore in chapter 10 (Metzker 2010). Often computational solutions have been developed to these problems (see next section)

---

[10] The first commercially available NGS technology was released in 2004, making the term "next-generation" somewhat redundant. Technologies such as Illumina are also referred to as "second-generation." However, NGS is still commonly used today to refer to high-throughput technologies other than Sanger sequencing.

[11] Taken from the Genomes Online Database, Dec 2018 - https://gold.jgi.doe.gov/

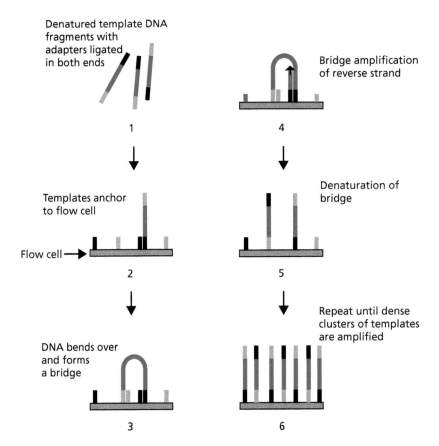

Figure 1.13 Schematic illustration of the principle of bridge PCR used in Illumina sequence by synthesis technology. Genomic template DNA (blue bars) is fragmented into short (≈200 bp) sequence fragments and adapters (black and grey bars) are ligated to both ends of each fragment. 1) The template/adapter-complex is denatured to yield single-stranded DNA. 2) Attached to a glass surface, a flow cell, are anchors, which are oligonucleotides with sequences complementary to the adapters. The template will therefore bind to the flow cell through hybridization between the anchor and adapter sequences. 3) The free adapter at the other end of the template binds to another nearby anchor and a local bridge is formed. 4) Enzymes and free nucleotides now replicate the bridged single-stranded DNA like in a normal PCR to produce double-stranded DNA. 5) The strands are denatured and two anchored single strands have now been produced. 6) The procedure is repeated over several cycles and eventually a local cluster of the same template is amplified on the flow cell. The process occurs in parallel with hundreds of thousands of other templates on the flow cell.

but other more recent HTS approaches also offer promise. Long-read sequence technologies such as the real-time sequencing platform developed by Pacific Biosciences can produce reads of up to 20 Kbp, potentially making genome assembly much more straightforward (Goodwin et al. 2016; English et al. 2012). One of the most exciting recent developments has been Oxford Nanopore's MinION which marks a clear divergence from the amplification and synthesis cycles used in other HTS platforms; DNA is passed through a protein pore and changes in electrical current are measured to identify nucleotide

bases (Goodwin et al. 2016). This has the advantage of being extremely fast and also surprisingly portable—the MinION sequencer is smaller than a smartphone!

Although it is tempting to think so, whole genome (re)sequencing[12] is not always the most effective means to answer evolutionary and population genetic questions. Sequencing is undoubtedly becoming

---

[12] When a reference genome has been sequenced and we then sequence the genomes of other individuals from the same species, we refer to this as resequencing.

Amplified clusters of templates on flow cell

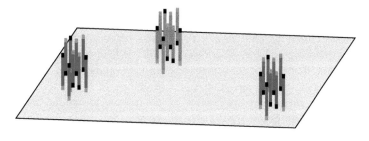

Sequencing   Signal scanning

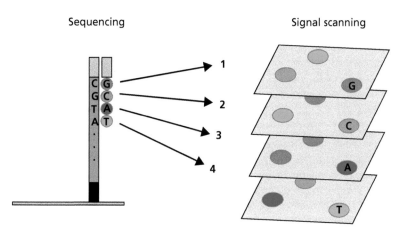

Figure 1.14 Principle of sequence by synthesis used in Illumina technology. Local clusters of bridge amplified templates are anchored on the flow cell (top panel). Focusing on a single template from the blue cluster (bottom left), a primer binds to the adapter and a fluorescent tagged nucleotide is added one at a time as dictated by the template sequence. 1) The fluorescent tag temporarily stops further extension, so when the orange "G" is added to this molecule and the others in the local cluster the signal is scanned and recorded (bottom right). Signals from the other clusters are also scanned in this way, allowing sequencing in parallel. 2) By washing off the fluorescent tags from the first nucleotide, a second tagged nucleotide can now be incorporated (green "C") and scanned, and so forth (3 and 4) until the read length limit is reached.

cheaper but often still prohibitively expensive for the numbers of individuals that are necessary for a population genetics study. Furthermore, whole genomes require an assembled reference genome for the species of interest or a close relative in order to be of much use. Because of this, a suite of protocols has been developed to reduce the proportion of the genome sequenced, collectively referred to as reduced-representation sequencing (Davey et al. 2011). These methods include RNA-seq (see chapter 10) which targets only protein coding regions of the genome (i.e. the exome) and genotyping-by-sequencing approaches. One of the most popular of these is RAD-sequencing (mentioned briefly in section 1.3.1). RAD-seq digests the genome into fragments using restriction enzymes and then sequences at the enzyme cut sites in order to identify single nucleotide polymorphisms (Baird et al. 2008). Various flavors of RAD-seq are available but generally they allow the straightforward genotyping of thousands of markers across the genome, easy multiplexing of multiple individuals, and reliable sequencing coverage (Andrews et al. 2016). Especially important for non-model organisms, it is also possible to analyze reduced-representation data without access to a reference genome.

1.4 Analyzing genetic data

Throughout this chapter, we have learned about many different approaches for generating genetic data. It is important to have a good understanding of such methods, even if you might not actually use them—no practicing scientist should treat technology as a black box. However, learning how to analyze genetic data is even more important. With the advent of high-throughput sequencing and the generation of huge amounts of data, there is an even greater need to have a good understanding of how analysis is conducted. In this section, we turn our attention to this area of modern evolutionary genetics.

1.4.1 Statistics and bioinformatics

Statistical analysis of genetic data derived from high-throughput sequencing is a challenge and with more and more data created with new approaches and technologies, it is essential for biologists to come up with new and innovative ways to store, analyze, and summarize it. Throughout this book, we will introduce you to many of the statistics and methods that are used to analyze such data—thus we will not go into detail of specific statistics or approaches here. Instead we will focus on what it means to work with high-throughput sequencing data. To give you some perspective on the issue, one of the authors (Ravinet) started his PhD in 2008. During his studies, he used nine microsatellite markers to look at population structure in a fish species, which at the time was a reasonable dataset. This meant it was possible to perform analyses on a simple desktop, using simple software with graphical user interfaces. A decade later and the more recent work of both authors regularly uses in excess of 1 million SNPs. To check the quality of a microsatellite locus ten years ago, we might have examined fragment profiles by eye and looked at data tables closely to spot errors or issues. This is simply not possible with orders of magnitude more data. High-throughput datasets therefore require a completely different statistical and methodological approach.

The application of programming languages, statistics, and data visualization for genetic data is referred to as **bioinformatics**. The term is not actually clearly defined but is a general catch-all for the need to use computation to properly investigate high-throughput sequencing. Many laboratories, departments, and research institutes now employ full-time bioinformaticians who specialize in using programming languages and computation tools to solving biological problems. However, you do not

need to be a specialist in computer science to learn how to use bioinformatics in your own research—many evolutionary biologists are self-taught or have taken advantage of community run training courses to spread bioinformatics skills.[13] This book and its online component are also designed as a training aid for biologists wishing to learn more about how to handle, analyze, and interpret such genetic data. It is also important to keep in mind that there is a diverse number of bioinformatic approaches. You might aspire to be a true bioinformatician, using your computing skills and familiarity with programming languages to build analysis pipelines and software packages. You might also aspire to be a biologist who largely makes use of such packages but who has enough knowledge to troubleshoot the problems she encounters. We tend towards the latter, since our specialization lies in the biological sciences and we are not computer programmers. Either way, some familiarity with bioinformatics is now an essential tool for biologists working in evolution and other life science fields.

What does it mean to work with bioinformatics? It is difficult to give a standard answer to this as there are as many uses for bioinformatics as you can dream of. The problems, challenges and tasks you will face will vary hugely among the projects you are working on. However, typically the first thing we are faced with when we work with high-throughput genetic data is a lot of raw data. This is something nearly all of us working in bioinformatics will encounter! A single sequence run on an Illumina sequencer might produce in excess of 100 million 100 base pair reads; each read will be stored in what is essentially a very large text file. There is no way we will have the time (or the patience) to look through such a file and loading it into a graphical program is usually far too memory intensive. Instead we might use the **command line** to interact with it. If you open a laptop or your smartphone and start up an app, you are using a graphical user interface, but if you use the command line, you are using a text-based interface. This might seem incredibly old fashioned but it is extremely powerful and flexible—exactly what you need for bioinformatic analysis. Returning to your sequence

---

[13] We have done both!

reads, you might want to screen them for quality, throwing out all sequences which fail a quality test. You could use one of a wide array of command-line based tools designed exactly for this purpose. Alternatively, you might use a **programming language** to write your own script to do it for you.

A programming language is a means of providing instructions for tasks to a computer. They are referred to as languages because much like a spoken or written human language, they have grammar, syntax, and rules. Popular programming languages and environments for bioinformatics include Python and R. There are many other programming languages also available, although these are the two most commonly used in evolutionary genetics. In the online component to this book, we focus only on R as it is specifically designed for statistical analysis and is widely used in the biological sciences for analysis beyond that of genetic data alone. R is also extremely powerful when it comes to producing high quality data visualization and graphics. Python can also be used for these purposes (and in some fields, such as physics, is the preferred option) but in our opinion is less straightforward for the beginner. Some familiarity with programming is extremely useful for evolutionary genetics. You might only need enough to implement some of the popular open source packages produced by other users to analyze your data. In some cases, you might also need to write custom scripts to carry out specific analyses or tasks. Coding in this sense is extremely powerful as it can make it straightforward to automate and rerun analyses when necessary.

We appreciate it is difficult to give a sense of how bioinformatics works from a short section in a textbook. This is one of the main reasons we have also designed the practical online component to introduce you to working in R. The aim of the online exercises is not to give an exhaustive introduction to all the different tools you might use when handling high-throughput sequencing data—no course, book, or tutorial can do that. Instead, they are designed to reinforce concepts we will encounter in the main text while also demonstrating the power of bioinformatics and programming for evolutionary biology. Throughout the rest of the book, we will discuss certain specific methods and approaches when they are important for developing understanding of the topics of hand. However, our aim with the online section and this brief introduction is instead to help you feel comfortable and familiar with the general approach that we, as evolutionary biologists, use to handle this type of data.

1.4.2 Internet resources

A crucial part of the high-throughput sequencing revolution has been a push towards open, reproducible science. The bulk of this has largely been in the storage of data in online databases which are accessible to anyone. For example, the National Center for Biotechnology Information (NCBI) provides a huge number of resources online, including hosting short-read sequence data, whole genome assemblies, and gene expression data. The European Nucleotide Archive (ENA) provides a similar service, hosting sequence data from many different sequencing projects. The Genomes Online Database (GOLD) also provides a record of all the genome sequencing projects currently underway with information on the organisms targeted and the researchers involved. Other data archives include the Dryad repository, which stores scripts and other (i.e. non-genetic) datasets.

Open science is not just about data access. It is also important to be able to easily and clearly apply analysis pipelines, scripts, and programs used by other researchers. Often this is made very easy through the use of software modules in certain programming languages. For example, the tidyverse package in R or the biopython libraries in Python. These can be loaded from central repositories where users submit pre-built packages that allow different types of analyses. Alternatively, some researchers store custom scripts on code hosting services such as GitHub. Increasingly, scripts and code are also made available online alongside publications, to make the data analyses conducted in a study completely reproducible. Certain journals, such as GigaScience, are now aiming to store all the components of a study—from data to analyses—in order to make reproducibility even more straightforward.

Bioinformatics is still a rapidly developing approach. One of the nice aspects of an arms race between new types of data and new types of analyses is that there are a lot of other researchers trying to make sense of similar challenges and problems. There are a lot of places to turn for help if you need it. For example, StackOverflow is an excellent place to ask programming related questions, as is BioStars. Social media platforms such as Twitter are also good places to ask for help. Often it is possible to get an answer within just a few minutes of posting: researchers are willing and eager to help. Furthermore, many researchers host blogs with tutorials explaining how to conduct a particular analysis or fix a specific problem. We cannot tell you the amount of times that tutorials like this have helped us!

Study questions

1. What do dominant and recessive mean in terms of alleles? How would a partly dominant or partly recessive allele behave?
2. Alleles that cause genetic diseases tend to be recessive. Suggest at least two hypotheses that can explain this pattern.
3. Suggest an experiment that would disprove Lamarck's theory of inheritance of acquired characteristics.
4. In section 1.1.2, Darwin's argument for evolution by natural selection is summarized in four points. Try to reformulate the arguments if Darwin had known about the laws of Mendelian inheritance.
5. Discuss to what extent the model depicted in Figure 1.3 can be used to dismiss the mutation theory of evolution.
6. Discuss whether the adaptive landscape depicted in Figure 1.4 could lead to the formation of two different bird species with different beak lengths.
7. What are the main differences and similarities between Darwin's theory of evolution by natural selection, the modern synthesis of evolution, and the neutral theory of molecular evolution?

Are the three theories compatible with each other?
8. What are the differences between DNA and RNA, and what are the roles of the two molecules in living organisms?
9. Explain the central dogma of molecular biology.
10. Explain these concepts:
    a. Transformation
    b. Transcription
    c. Translation
    d. Genotype
    e. Nucleotide.
11. What are purines and pyrimidines and what is the difference between the two?
12. Some microsatellites are found within the coding region of protein coding genes. Why do we normally only find 3n-repeated microsatellites in such regions?
13. Compare DNA replication occurring during cell division (Figure 1.8a) with the PCR method (Figure 1.11) and Sanger sequencing (Figure 1.12). Discuss whether PCR and Sanger sequencing can be understood as *in vitro* DNA replication. What are the differences between the three processes?
14. Assume that the electrophoresed fragments in Figure 1.10 are PCR products of a microsatellite marker from different individuals. Determine the genotype of each individual.
15. What strategy would you choose to identify the molecular basis for variation in, for example, pea color?
16. A phenotypic trait can be affected by environmental factors, for instance through epigenetic modification. With this in mind, discuss to what extent Mendel's laws and the Mendelian gene concept are still valid.
17. What is the main difference between genetic and epigenetic inheritance?
18. Explain the principle of sequence by synthesis.

CHAPTER 2

Genomes and the origin of genetic variation

Error and chance events, random mutations, are necessary prerequisites for evolution to happen. In a perfect world with no errors (no mutation) there would be no evolution because no genetic variation would be generated that natural selection and/or genetic drift could work upon. In this chapter, we shall explore the processes that create novel genetic variation, which in turn is the raw material for all evolutionary change. We start this chapter, however, by briefly reviewing how DNA is organized into genomes and genes in living organisms.

2.1 Genome structure

2.1.1 Genome organization in bacteria, archaea, and eukarya

Life on Earth can be classified into three main domains: two microbial domains (bacteria and archaea) and eukarya (eukaryotes), which includes protists, plants, fungi, and animals (Woese et al. 1990; Pace 2006). Phylogenetic and biochemical evidence suggests that the three domains share a common ancestor, the last universal common ancestor (LUCA) that lived some 3.8 billion years ago, which then diverged in two branches: bacteria and the line that later split into archaea and eukarya (Figure 2.1).

Thus, eukaryotes and archaea are more closely related to one another than either is to bacteria. Of the microbial domains, bacteria have been studied more intensively than archaea. Many bacteria have attracted research interest because they are important human pathogens. Moreover, bacterial species can often easily be studied because they can be grown in large colonies on petri dishes or in flasks in the lab. More recently, PCR amplification and sequencing of environmental samples, called **metagenomics**, has revealed a large diversity of microbe communities that previously went undetected. Both archaea and bacteria are widespread and numerous in a range of environments: oceans, lakes, soil, hot springs, salt lakes, and even within the bodies of other organisms, such as ourselves.

Genomes of bacteria

In bacteria, the genome is typically organized into a single circular chromosome. However, many additionally carry much smaller, circular DNA molecules called **plasmids** (Figure 2.2). The sizes of bacterial genomes are modest compared to eukaryotes, from less than 150 000 base pairs (0.15 Mb) in certain parasitic lines (McCutcheon & Moran 2012) up to about 14 Mb in certain free ranging bacteria.[1] The bacterial genome is very compact. It contains mainly protein coding genes; usually 80–95% of the genome of a bacterium consists of coding sequence. Accordingly, there is a strong correlation between number of coding genes and genome size in these

[1] In comparison, the human genome size is about 3100 Mb.

Evolutionary Genetics: Concepts, Analysis, and Practice. Glenn-Peter Sætre & Mark Ravinet, Oxford University Press (2019).
© Glenn-Peter Sætre & Mark Ravinet 2019. DOI: 10.1093/oso/9780198830917.001.0001

organisms. In addition, they may carry short intergenic sequences between genes, various control elements involved in orchestrating DNA replication and gene regulation, and other noncoding DNA.

Bacteria normally reproduce asexually through cell division, but they are also able to acquire new genetic information through **horizontal gene transfer** (Figure 2.3). As we saw in chapter 1, some bacteria can incorporate genetic material from their surroundings through their cell membrane, a process called **transformation**. In addition, genetic material can be transferred from one bacterium to another via

bacteriophages (viruses that infect bacteria), a process called **transduction**, or directly from one bacterium to another, a process called **conjugation**.

Genomes of archaea

The genome organization of archaea resembles that of bacteria. Neither of the two microbial groups have any nucleus or other membrane-bound organelles, and both have a single, circular chromosome and, potentially, additional plasmids. Their genome size is also similar to bacteria, ranging from 0.5 Mb in *Nanoarchaeum equitans* (Huber et al. 2002) to 5.7 Mb in *Methanosarcina acetivorans* (Galagan et al. 2002). Due to such superficial similarities, the two microbial groups were traditionally lumped together in one group called prokaryotes (literally: before nucleus). However, such a classification does not reflect the evolutionary relationship between the organisms and we therefore avoid the term. In other respects, archaea are more similar to eukaryotes. Notably, the enzymes involved in transcription of DNA to RNA and translation of RNA to protein are quite similar between archaea and eukaryotes and different from those of bacteria (Langer et al. 1995; Bell & Jackson 1998). Moreover, like eukaryotes and unlike bacteria, protein coding genes in archaea may carry noncoding blocks,

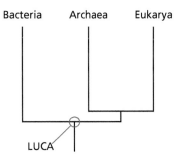

Figure 2.1 Phylogenetic relationship between the three domains of life. Bacteria, Archaea, and Eukarya share a common ancestry in the last universal common ancestor (LUCA) that lived some 3.8 billion years ago.

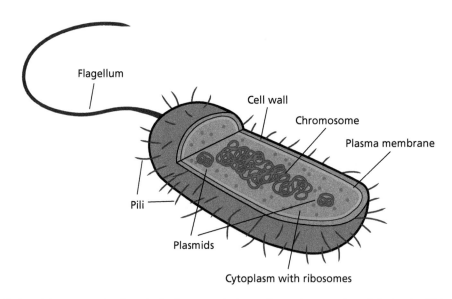

Figure 2.2 The bacterial genome consists of one circular chromosome in the cytoplasm and may contain small circular plasmids in addition.

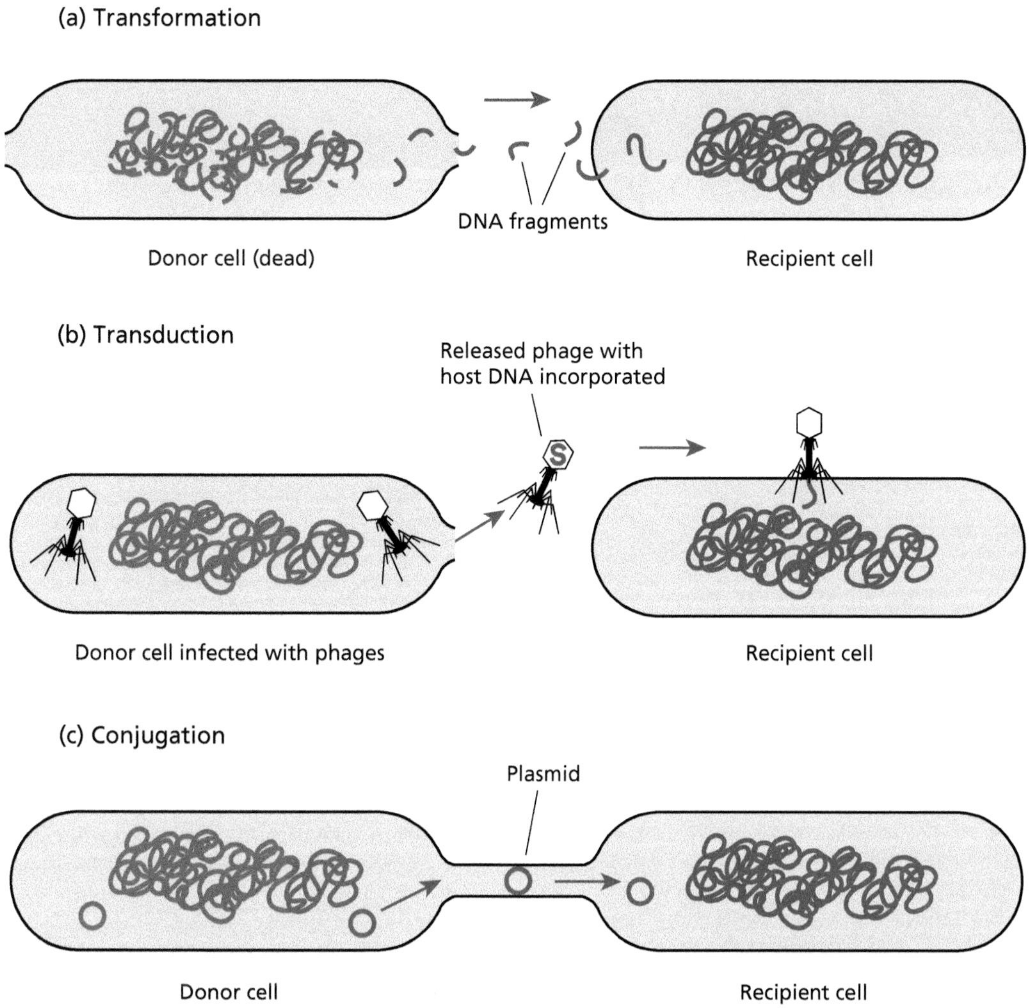

(a) Transformation

Donor cell (dead)   DNA fragments   Recipient cell

(b) Transduction

Released phage with
host DNA incorporated

Donor cell infected with phages   Recipient cell

(c) Conjugation

Plasmid

Donor cell   Recipient cell

Figure 2.3 Means of horizontal gene transfer in bacteria. **a)** Transformation, the uptake of DNA from the surroundings; **b)** transduction, the uptake of foreign DNA via a viral vector and **c)** conjugation, transfer of DNA between bacterial cells that are in direct physical contact.

introns, in between coding sequences (Watanabe et al. 2002), although introns are shorter and rarer in archaea genes than in eukarya ones. Archaea normally reproduce asexually but horizontal gene transfer is also common and widespread. Archaean cells can exchange genetic material in a fashion similar to bacterial conjugation (e.g. Prangishvili et al. 1998). Horizontal gene flow can even occur between archaea and bacteria (Nelson et al. 1999).

Genomes of eukaryotes

The eukaryote genome is mainly contained in linear chromosomes inside the nucleus. Each chromosome consists of one long, linear DNA molecule and associated **histone** proteins that pack the DNA in an orderly but compact way (Figure 2.4). In addition, mitochondria and chloroplasts organelles carry their own, circular DNA. The latter is a testimony to the evolutionary origin of the eukaryote cell (Margulis 1970; Zimorski et al. 2014). The well-supported endosymbiont theory suggests the eukaryote cell originated when large, single-celled archaea evolved an endosymbiotic partnership with engulfed bacteria, which eventually became its organelles. Hence, eukaryotes are the descendants of both archaea and bacteria.

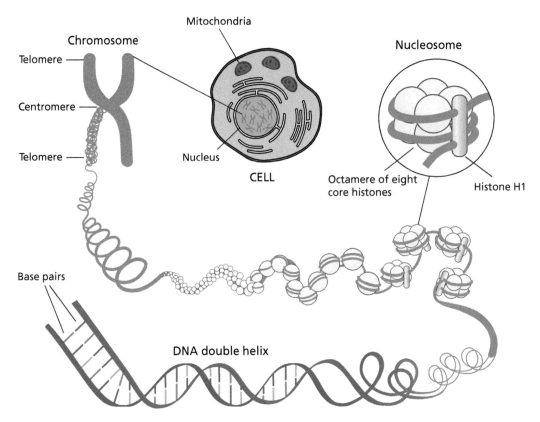

Figure 2.4 The eukaryote genome is organized into linear chromosomes contained in the cell nucleus. The DNA is densely packed by histone molecules consisting of units (nucleosomes) of eight core histones and the histone H1. In addition to the nuclear chromosomes, organelles such as mitochondria and chloroplasts carry their own circular DNA molecules.

One striking feature of eukaryote genomes is the tremendous variation in size between species and, unlike bacteria and archaea, the lack of correlation between genome size and number of genes. The smallest known eukaryote genome belongs to the parasitic microsporidian *Encephalitozoon intestinalis* and is only 2.3 Mb (Corradi et al. 2010), whereas the largest known belongs to the monocot flower plant *Paris japonica* of about 150 000 Mb (Pellicer et al. 2010).

The extraordinary amount of variation in genome size in eukaryotes has puzzled scientists and inspired much research. One would think that more complex organisms would need more genes to handle their increased complexity and hence, larger genomes. Indeed, there is a positive correlation between complexity and gene number, but it is very weak. For instance, it is true that eukaryotes on average have more genes than

microbes, and vertebrates have more genes than invertebrates. However, the increased number of genes in vertebrates is probably at least in part due to genome duplication events (see section 2.4.1) that occurred after the vertebrate line diverged from that of other chordates (tunicates and lancelets). As a comparison, the aforementioned parasitic fungus *Encephalitozoon intestinalis* with the smallest known eukaryote genome contains some 1800 densely packed genes and very little non-genic DNA. Humans have about 11 times as many genes (some 20 000 genes) but a genome size that is more than 1300 times larger (about 3100 Mb). Less than 2% of the human genome consists of protein coding sequence. Indeed, most of the variation in genome size in eukaryotes is due to variation in the amount of noncoding DNA and not due to variation in the amount of coding sequence.

2.1.2 Structure of the eukaryote genome—functional elements

Genomic DNA can be classified in different ways. It is common to distinguish between coding and noncoding DNA. However, the distinction is not straightforward. Some coding sequences are parasitic (viruses and transposable elements) and may be considered foreign elements in the genome. Further, some elements do not code for any protein but are by no means devoid of function. Finally, coding genes may contain large amounts of noncoding DNA. In this subsection, we will consider genes and other functional elements, excluding those that parasitize their host genome, which will be considered in section 2.1.3.

Protein coding genes

Functional DNA includes first and foremost the protein coding genes, responsible for producing all the proteins an organism is built up of, including enzymes and hormones responsible for orchestrating a vast amount of intra- and intercellular biochemical processes. A eukaryote protein coding gene consists of coding sequence that eventually becomes translated into a protein. Usually, the coding sequence is organized into blocks called **exons**, interspersed by noncoding blocks, called **introns** (Figure 2.5). The start codon (ATG) is located at the 5′-end of exon 1 and a stop codon (TAA, TAG, or TGA) ends the **open reading frame** at the 3′-end of the last exon. The open reading frame is the DNA sequence that eventually is translated into protein.

A protein coding gene is first transcribed into a prime RNA (**pre-mRNA**) molecule. The pre-mRNA is then processed into a mature mRNA molecule which is transported out of the nucleus to a ribosome in the cytoplasm where translation takes place.

Immediately upstream of the start codon and immediately downstream of the stop codon there are regions which are transcribed along with the exon and introns, but which are not translated into protein. The **five-prime untranslated region (5′UTR)** contains sequence-specific signals that regulate translation, as well as a sequence that is recognized by the ribosome, allowing the mRNA to temporarily bind to the ribosome so that translation can be initiated. The **three-prime untranslated region**

(3″UTR) contains regulatory sequences that influence gene expression, including binding sites for regulatory proteins and microRNAs (miRNAs). The latter can decrease gene expression, for instance by causing degradation of the transcript.

When a pre-mRNA is being processed into a mature mRNA, special proteins first cut off the 3′-most segment of the pre-mRNA and add a chain of adenine (A) nucleotides to the 3′-end of the transcript. This chain of adenines is called the **poly-A tail** and is important for the export of the mRNA out of the nucleus and for the stability of the mRNA (Figure 2.5).

Introns nearly always start with a GT- and end with an AG-sequence. These are splice signals. After transcription, but before the mature mRNA leaves the nucleus, special RNA–protein complexes called **spliceosomes** recognize these splice signals and catalyze the excision of introns and ligation of the exons. Hence the mature mRNA is free of intron sequences and only contains the open reading frame flanked by the 5′UTR and 3′UTR with the added poly-A tail (Figure 2.5).[2] A single gene may code for several different proteins. This is achieved by controlled **alternative splicing**, in which different exons are spliced out in different resulting proteins (Figure 2.5). Accordingly, eukaryotes can produce many more proteins than the number of genes would imply (see section 2.4.3).

Associated with protein coding genes are a variety of sequence specific motifs that are neither transcribed nor translated but that are important for gene function. The **promoter** is a sequence motif immediately upstream of the gene (Figure 2.5), which functions as an attachment site for RNA polymerase, the protein that initiates and orchestrates transcription (see chapter 1). In addition, eukaryotic genes typically harbor various **regulatory sequences**. These are DNA sequence motifs that function as attachment sites for special proteins, so called **transcription factors**, which either enhance or repress transcription of the gene, called **enhancers** and **silencers** respectively. These transcription factors are of course themselves encoded

[2] Introns can also be found (and excised) in the untranslated regions of the pre-mRNA.

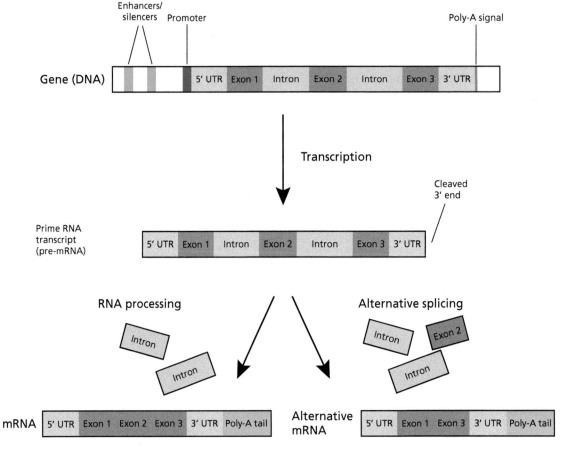

Figure 2.5 The structure of a eukaryote gene. The coding sequence is organized into blocks (exons) interspersed by noncoding introns. The gene is transcribed from DNA into pre-mRNA. Along with the exons and introns the pre-mRNA contains transcribed regions upstream of the start codon, the five-prime untranslated region (5′UTR), and downstream of the stop codon, the three-prime untranslated region (3′ UTR), that contain important signals for translation and gene expression. The immature pre-mRNA is processed into a mature mRNA by excision of introns by means of RNA-protein complexes called spliceosomes, and the addition of a poly-A tail, which stabilizes the mRNA and aids in its transport from the nucleus to a ribosome in the cytoplasm where translation occurs. Alternative splicing of exons enables one gene to code for different proteins. Associated with a protein coding gene are several non-transcribed sequence motifs that act as attachment sites for proteins essential for transcription and gene regulation, including the promoter that binds the RNA polymerase that initiates and controls transcription, and enhancers and/or silencers that bind transcription factors, proteins that can either enhance or suppress transcription.

by specific protein coding genes. Regulatory sequences (enhancers and silencers) can be found upstream and downstream of the gene and even inside introns within the gene.

RNA genes

Some specific DNA sequences are transcribed into functional RNAs, but are never translated into protein. Such sequences are called **RNA genes**. As we have already seen, noncoding RNA molecules have many functions in living organisms. They include tRNAs and rRNAs, which are instrumental in protein synthesis (see chapter 1). Other important classes include the small nuclear RNAs (snRNA) that play critical roles in gene regulation by way of RNA splicing (the RNAs associated with spliceosomes belong to this group) as well as short interfering RNAs (siRNA) and even shorter microRNAs (miRNA), that both play important roles in regulation of gene expression. Special RNA genes throughout the genome specify all these noncoding, functional RNA molecules.

Non-transcribed functional DNA

To the class of functional DNA we may also add sequences that are neither transcribed nor translated but yet serve essential functions in the cell. Important examples include sites of origin of DNA replication, the centromere, and telomeres. During each cell division, the entire genome is replicated. In eukaryotes DNA replication is initiated at multiple sites of **origin of replication** on each chromosome at which replication is initiated at different times during cell division. Each origin of replication binds a protein complex called the pre-replication complex which orchestrates the unwinding of the double-stranded DNA and attracts a DNA polymerase that begins copying DNA bi-directionally (see chapter 1).

The **centromere** is an attachment site for the **kinetochore**, a highly complex protein structure controlling chromosome segregation during cell division (Figure 2.6). In most eukaryotes, the centromere consists of arrays of highly repetitive DNA to which the kinetochore is attached.

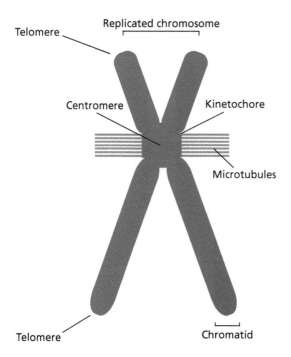

Figure 2.6 A replicated chromosome with kinetochores (red) attached to the centromere of each chromatid. Microtubules attach to the kinetochore and drag the chromatids to opposite sides of the cell during cell division (mitosis and meiosis).

The **telomeres** at the ends of each chromosome tip are caps of highly repetitive DNA that protect the chromosomes from degradation and from fusing with other chromosomes (Figure 2.6).

2.1.3 Structure of the eukaryote genome— parasitic and non-functional DNA

Parasitic DNA

Parasitic DNA elements are so tightly integrated in the genomes of eukaryotes that it is a matter of choice whether to regard them as integral components of the genome or as external entities parasitizing their hosts. Parasitic DNA includes various kinds of **viruses** that most of us would think of as external entities, and intragenomic parasites, so called **transposable elements (TEs)**, that largely lack the ability to infect different individuals, except through inheritance from parent to offspring, but that otherwise share many properties with viruses.

TEs are often referred to as "jumping genes" because they have the ability to change genomic location (McClintock 1950, Ravindran 2012). Such movement, or **transposition**, is caused either by the element being excised and reinserted elsewhere in the genome, so called **conservative transposition**, or by the element making copies of itself and then the new copies being inserted at new places in the genome, so called **replicative transposition** (Figure 2.7). There are two main classes of TEs. **Class I TEs (retrotransposons)** are all replicative and are copied in two stages. First, they are transcribed from DNA to RNA, and then the RNA copy is reverse transcribed back to DNA by means of the protein reverse transcriptase that is often, but not always, encoded by the TE itself. The DNA copy is then integrated at a new position in the genome. Retroviruses, such as HIV (responsible for the serious disease AIDS), constitute the most complex members of Class I TEs (Figure 2.8). They harbor three main genes:

1) The group specific antigen *gag* that encodes the viral matrix, the viral protein shell (capsid) that protects the viral DNA, and various nucleoproteins.

2) The polymerase gene *pol*, that produces proteins responsible for synthesis and integration of viral DNA into the host, including reverse

(a) Conservative transposition

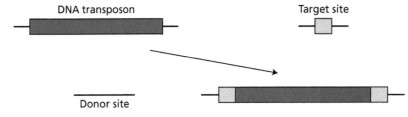

(b) Replicative transposition (Retroposition)

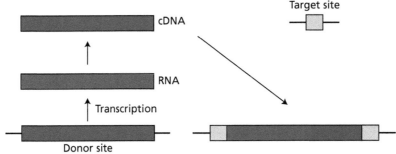

Figure 2.7 Mechanisms of transposition. **a)** Conservative transposition. The transposable element produces a protein called transposase that catalyzes the excision of the element and its integration at a target site elsewhere in the genome. **b)** Replicative transposition. The element is transcribed into RNA, then reverse transcribed into a DNA copy which is then integrated into the genome (Class I TEs). The integration of TEs results in short repetitive sequences in the host DNA flanking the element (light green).

transcriptase (RT) that reverse transcribes RNA to cDNA, endonuclease (EN) that cleaves the host DNA, and integrase, enabling the viral DNA to be integrated into the host genome.

3) The envelope gene *env* that produces the protective viral envelope that enables the virus to bind to the cell surface of a host cell and enter it. The *env* proteins are first and foremost what distinguish retroviruses from non-viral Class I TEs.

Other Class I TEs resemble retroviruses and typically harbor some of the genes that they do. However, the simpler TEs all lack a functional *env* gene and unlike viruses are unable to infect new hosts horizontally: instead they function as intragenomic parasites. The simplest Class I TEs, the short interspersed nuclear repeats (SINEs) even lack the gene for reverse transcriptase and rely on other TEs, such as long interspersed nuclear elements (LINEs), for their propagation in the genome (Figure 2.8).

Class II TEs (DNA transposons) can be both conservative and replicative. A replicative DNA transposon replicates itself to a DNA copy that is then reintegrated elsewhere in the genome. Class II TEs harbor an open reading frame that produces transposase enzymes that catalyze transposition (Figure 2.8) but may also contain other protein coding genes. Transposase acts like a molecular scissor that cuts the element out from its current position, then transfers and pastes it to the target position. As a parallel to the link between Class I TEs and retroviruses, there is an evolutionary relationship between Class II TEs and DNA viruses that attack eukaryotes, which again are related to bacteriophages (Koonin et al. 2015).

Note that transposition is not a universal feature of viruses. We find viruses that transpose as part of their life cycle and others that do not. Likewise, as we have seen, some elements that are able to transpose are viruses whereas others are not, depending on whether or not they have the ability to move horizontally from one host individual to another. Class I TEs and retroviruses are evolutionarily related to one another, as are Class II TEs and DNA

Class I transposable elements

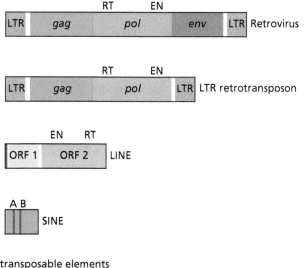

Class II transposable elements

Figure 2.8 Different types of transposable elements. Among class I TEs, retroviruses are the most complex ones, and contain genes that are essential for the virus's life cycle including the genes: group specific antigen (*gag*), polymerase (*pol*), which includes sequences that code for reverse transcriptase (RT) and endonuclease (EN), and envelope (*env*). The genes are flanked by long terminal repeats (LTR). TEs with long terminal repeats resemble retrovirus but lack a functional *env* gene. Long interspersed nuclear elements (LINEs) also carry open reading frames (ORF) that resemble those of retrovirus, including sequences that code for reverse transcriptase and endonuclease. Short interspersed nuclear repeats (SINEs) are non-autonomous elements that lack the reverse transcriptase gene and thus rely on other TEs for their transposition. Box A and B in the depicted SINE constitute promoters. Class II elements all contain the transposase gene which produces enzymes that catalyze transposition and are flanked by inverted repeats (IR). Modified from Finnegan (2012).

viruses. They share genes and sequence similarities and have a common evolutionary ancestry.

A still unresolved question is whether viruses evolved from intragenomic transposable elements (TEs) by escaping the genome or whether intragenomic TEs evolved from parasitic viruses that lost the ability to move horizontally between individual hosts. According to the first hypothesis viruses came from us; selfish genetic elements of the genomes of the organism they once resided in—elements within the genome that gained the ability to escape and spread among us to their own gain and our loss. According to the other hypothesis viruses may themselves be degenerative remnants of once free-living organisms, maybe some ancient bacteria that became so integrated with us in its parasitic way of life that it lost many of the properties that characterize a living organism. Losing its ability to infect new individuals horizontally and instead becoming a

true intragenomic TE parasite would just be the final step of integrating its parasitic way of life into our genomes. Further still, it may be that they are the precursors of life as we know it? Structural similarities and shared proteins between viruses infecting members of all the three main domains of life (bacteria, archaea, and eukarya) suggest that their origin is ancient; maybe even older than the last universal common ancestor (LUCA) that lived some 3.8 billion years ago (Forterre & Prangishvili 2009).

Nonfunctional DNA derived from transposition

Replicative transposition has large-scale effects on genome evolution. An active TE can produce multiple copies of itself at a fast rate and, if not destroyed by mutation—reverse transcription of RNA to cDNA is a highly error prone process—the copies can themselves make additional copies after being integrated into the host genome. Replicative transposition is the main mutation mechanism causing **interspersed repeats** in the genome and is also one of the most potent factors for the growth in genome size. A large proportion of genome size variation among eukaryote species can be attributed to past and present fluctuations in transposition activity. In humans, about 44% of the genome consists of sequences derived from transposition, although most of this sequence does not consist of currently active TEs. Defective copies of TEs (TE pseudogenes) make up a large proportion of the nonfunctional DNA of many organisms. Moreover, old degenerate TE pseudogenes can be difficult to identify as being derived from transposition and may be classified as nonfunctional DNA of unknown origin. An even larger proportion of the nonfunctional DNA found in eukaryote genomes may therefore derive from transposable elements but we are simply unable to properly detect it.

By and large TEs do little good for the organisms they reside in. However, they are not always detrimental; the fitness effects of an inserted sequence can be negligible if it is inserted far away from functional genes, and accidental positive effects have been found in some cases. For instance, retroviruses can sometimes function as vectors for horizontal transfer of a gene from one species to another, where the transferred gene takes up an adaptive function in its new host (e.g. Ciomborowska et al. 2013). However, if a TE is inserted in or close to a coding gene it can impair or destroy the function of that gene. Additionally, it can be costly to replicate all the extra DNA derived from TEs at each cell division. Like viruses can make us ill, so can intragenomic TEs. Indeed, a number of diseases can be linked to TEs (e.g. Belancio et al. 2008; Reiley et al. 2013). From the point of view of the host, there is therefore usually a strong selection pressure to deactivate TEs. A range of host defense mechanisms against TEs has been identified (e.g. Waterhouse et al. 2001; Obbard et al. 2009).

1) The host may silence the TE by epigenetic modification, such as methylation or histone modification. This would stop the element from being transcribed.
2) Even if the TE is transcribed, small RNA molecules can hinder the copies from being inserted into the host genome. These RNA molecules

attach themselves to newly transcribed TE copies and degrade them or hinder their integration.

Origin of introns

Ever since the discovery that eukaryote protein coding genes contain large chunks of noncoding introns in between the coding exons, which are spliced out in the mature mRNA (Chow et al. 1977; Berget et al. 1977), the origin and possible function of introns has been debated. **The introns-early hypothesis** suggests that they are inherited from an early ancestor and that they played an important role in the origin of proteins by facilitating recombination between protein modules (Gilbert 1978; Blake 1979). Organisms such as bacteria would later lose their introns according to this hypothesis, to streamline their genomes. Selection for rapid replication in bacteria would cause non-essential parts of the genome to be removed and thus introns disappeared from these organisms.

The alternative **introns-late hypothesis** suggests that introns first evolved in eukaryotes and that these organisms have continued to gain introns during evolutionary history (Doolittle 1978; Doolittle & Stoltzfus 1993).

Evidence from comparative genomics does not seem to support the introns-early hypothesis. The position of introns often varies between species and exons often do not correspond to functional domains (but see section 2.4.2 on exon shuffling in animals). Koonin (2006) suggested a third hypothesis that combines elements of the two other hypotheses. According to his hypothesis, introns are ancient, virus-like, self-splicing elements that have existed since before the last universal common ancestor (LUCA). These elements invaded eukaryotic genes at an early stage in eukaryote evolution and have been with us since then. According to this hypothesis introns were parasitic DNA, much like present day transposable elements, although they currently do little or no harm to the organisms. Eukaryote features, such as the spliceosome RNA–protein complex (see section 2.1.2) may be host-specific adaptations to cope with introns according to this theory.

Other sources of nonfunctional DNA

Sequences derived from transposition and introns typically make up a large bulk of the noncoding

DNA in eukaryotes. In the human genome about 44% of the genome is derived from transposition and about 26% is intron sequence. However, these are not the only sources of apparently nonfunctional DNA. Much like TEs can mutate to become non-functional pseudogenes unable to transpose, so can genes of the host itself become inactivated. Such pseudogenes often originate after gene duplication events (see section 2.3.3) in which one (or more) of the copies accumulate mutations that inactivate the gene. The human genome contains some 13 000 pseudogenes. Finally, some proportion of the genome of a eukaryote is nonfunctional but of unknown origin. They may be degenerate pseudogenes or duplicated chunks of DNA that have diverged to the extent that they bear no sequence similarity to other elements in the genome.

2.2 Mutations

2.2.1 Random mutations

A **mutation** is usually understood as a permanent change in the nucleotide sequence in the genome of a cell or an organism, but is also commonly used to describe the phenotypic effect a given DNA alteration may have. Mutations are often said to be random. However, they are random only in one important sense; they occur without foresight. They do not arise to solve a problem an organism or species is facing. They just happen. This may seem like an obvious point, but it is a common misunderstanding of the evolutionary process. Once in a while a mutation will happen to be beneficial. Then, natural selection will act to increase its frequency in a population. More often, mutations are harmful and are quickly removed by selection. Finally, a mutation may have little or no effect on fitness at all. In such cases its fate is left to random chance events. It may disappear altogether from a population or increase to higher frequency and eventually become fixed—in short, it evolves by genetic drift. In other ways mutations are not random. Certain agents such as radiation and various chemicals can increase the mutation rate. Moreover, different sites in the genome have different probabilities of mutating. For instance, repeated DNA sequences are more mutable (they have a higher mutation rate) than single copy DNA (see section 2.2.3).

A mutation can be caused by an error in DNA replication or in DNA repair during cell division. Mutations that occur in a somatic cell will not be passed on to the offspring and are thus of little concern in evolution, although they can have consequences for the individual carrying the mutated cell. For instance, cancer can result from somatic mutations in genes affecting cell growth (Greenman et al. 2007). In evolutionary genetics, however, we are interested in genetic variation that is passed on across generations and thus, **germline mutations**—those occurring during meiosis when gametes are being produced—are our primary concern.

2.2.2 Point mutations

In its simplest form a mutation is the **substitution** of one nucleotide, say an A, with another, say a G, at a specific site in the genome; a **point mutation**. In chapter 1 we saw that the four nucleotides are of two main types: purines (A and G), which are double-ringed molecules, and pyrimidines (C and T), which are single-ringed molecules. Perhaps not surprisingly, given their different size and shape, when errors occur during DNA replication, it is more common that a purine is replaced with another purine (A <–> G) or a pyrimidine with a pyrimidine (C <–> T), than that a purine is replaced with a pyrimidine or vice versa. The former, more common type of point mutation is a **transition**, and the latter is a **transversion**.

A point mutation may or may not affect the phenotype of an organism depending on where in the genome it occurs. Mutations in noncoding regions such as introns will rarely have any phenotypic effect. However, many mutations in coding regions also have no effect on the phenotype. There are 64 ways that the four different nucleotides can be combined into triplets (codons) but only 20 different amino acids (see Table 1.1). The genetic code is redundant; many codons specify the same amino acid. For instance, both AAA and AAG specify the amino acid lysine. A transition mutation at the third position (A <–> G) will thus not cause any change in the amino acid sequence of the protein. Mutations that do not alter the amino acid sequence are called **silent** or **synonymous mutations**. In contrast, substituting an A with a G at the first position of the AAA codon, to GAA, will result in lysine being

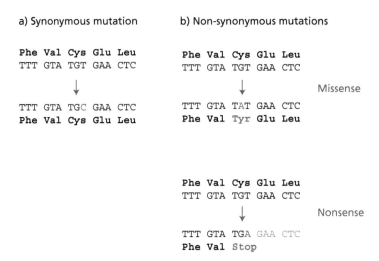

Figure 2.9 Classification of genic point mutations. Synonymous mutations do not change the amino acid sequence. Non-synonymous mutations change the amino acid sequence of a gene (red), either by changing which amino acid a codon specifies (missense) or by changing an amino acid specifying codon into a stop codon (nonsense).

replaced with glutamic acid. A similar substitution at the second position, from AAA to AGA, will result in lysine being substituted with arginine. Mutations that cause an amino acid to be replaced with another are called **missense** and is one of two classes of **replacement** or **non-synonymous mutations**, the other being **nonsense** mutations in which an amino acid specifying codon (say AGA which codes for arginine) mutates into a stop codon (say TGA) yielding a shortened and usually non-functional protein (Figure 2.9). Non-synonymous mutations alter the resulting protein and may affect its function. They are therefore likely targets of natural selection.

2.2.3 Replication slippage

Replication slippage is a form of mutation that causes short insertions or deletions in the genome during DNA replication and is thought to be the main mutation mechanism for microsatellites; that is, short (2–5 nucleotides) **tandem repeats** (e.g.…CACACACACA…). Such tandem repeated sequences are relatively unstable and can easily form secondary structures on single stranded DNA, such as hairpin loops. If, during replication, the template and growing strand temporarily dissociates, these hairpin loops can cause a mismatch between the template and the strand that is being

synthesized when they realign. This results in one (or more) extra repeat units being added, or that fewer units are being synthesized, depending on whether the template or the growing strand harbors the hairpin loop (Figure 2.10). DNA-repair enzymes will recognize the resulting mismatch in length between the template and the new strand and either insert or delete a repeat unit to the complementary strand.

Because of their instability and susceptibility to replication slippage, short tandem repeats—more commonly referred to as microsatellites—are useful genetic markers. The high mutation rate of microsatellites typically means a large number of alleles (different number of repeat units) segregate in a population. This makes them practical for several purposes. For one, it is highly unlikely that two genetically unrelated individuals will have exactly the same genotype at a number of different microsatellite loci. Multi-locus microsatellite genotypes are like molecular fingerprints—unique to the individual. Such DNA fingerprinting can be used to find genetic relatedness between individuals, including identifying the genetic father in a paternity case, or to link a suspect to a crime scene, to name but a few of their applications.

Replication slippage is also one of the mutation mechanisms that can explain insertions and deletions (indels) in the genome. Insertions and deletions can have serious consequences if they occur

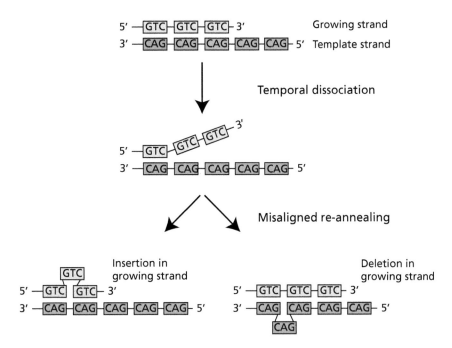

Figure 2.10 Replication slippage is an error in DNA replication that typically affects short tandem repeats (microsatellites). Top: DNA replication of a tandemly repeated sequence motif. Center: double-stranded DNA temporarily dissociates during replication. Bottom: Secondary hairpin structures form, resulting in misalignment when the strands reanneal. Consequently, an extra repeat unit will be synthesized (bottom left) or replication will miss a repeat unit (bottom right), depending on whether the hairpin loop forms at the growing strand or the template strand.

```
Phe Val Cys Glu Leu Gly Ser Asn Ala Pro
TTT GTA TGT GCA CTC GGA TCC AAT GCA CCC


        Insertion of T

TTT GTT ATG TGC ACT CGG ATC CAA TGC ACC C
Phe Val Met Cys Thr Arg Ile Gln Cys Thr
```

Figure 2.11 A frame-shift mutation is the insertion or deletion of a DNA segment into a coding sequence that distorts the reading frame. Here an additional T is inserted at position 6, causing a different chain of amino acids (marked in red).

within coding sequences. A **frame-shift mutation** is an insertion or deletion that distorts the reading frame of a gene. The reading frame consists of triplets of nucleotides (codons). Therefore, the insertion or deletion of a number of nucleotides that is not divisible by three will result in a completely different protein (Figure 2.11). The resulting arbitrary chain of amino acids has never been tested by natural selection and will most likely be harmful to the organism, resulting in selection against it.

Indels can vary substantially in size but even very small ones can have harmful effects. A classic example is the trinucleotide (i.e. 3 nucleotide) insertion-deletion associated with Huntington's disease in humans, which results in uncontrollable jerking movements, impairment of cognition, and ultimately death. The disease is caused by the insertion of an extra CAG repeat-unit in the coding sequence for the *huntingtin* gene, which results in an abnormal protein that causes death of brain cells (O'Donnovan 1993). Such trinucleotide insertions within genes are not universally harmful, however (Kashi & King 2006). For instance, some of the genes involved in controlling circadian rhythms (such as alternations between awakeness and sleep during the day or the timing of reproduction and migration during the year) harbor length polymorphisms in trinucleotide tandem repeats within a coding sequence. The number of repeats present affects the timing and length of the circadian rhythm, allowing for evolutionary fine-tuning of the biological clock to the local environment (Kyriacou et al. 2008).

2.3 Recombination and associated mutations

2.3.1 Recombination and reshuffling of alleles

During meiosis, the alleles that an individual has inherited from its mother and father are reshuffled. This is the process of **recombination**. For instance, at a given gene on chromosome 1 an allele A_1, inherited from the individual's father, may make it to a given gamete. However, at another gene on chromosome 2 it is the allele B_2, inherited from the individual's mother, that ends up in that same gamete, resulting in a recombinant A_1B_2 allelic combination in the gamete. Also within chromosomes alleles at different loci are reshuffled due to **crossing over** between homologous chromosomes (Figure 2.12). As a result, the allele combination at different loci, or **haplotype**, in the gametes can be different from those carried by the individual itself. For instance, an individual may have the haplotype A_1B_1 for locus A and B on the copy of chromosome 1 it inherited from its mother and A_2B_2 on the homologous chromosome inherited from its father. Crossing over between locus A and B during meiosis may result in gametes with the recombinant haplotypes A_1B_2 and A_2B_1. Recombination is therefore one of the reasons why offspring are not exact intermediates between their parents; the offspring will carry allele combinations that neither of its parents carried. In addition to reshuffling alleles, recombination can generate novel mutations.

2.3.2 Gene conversion

During prophase I of the meiotic cycle, homologous chromosomes pair up and align according to sequence similarity. This is the phase in which crossing over and other interesting recombination events occur. **Gene conversion** is one such event in which, upon recombination, the allele on one of the chromosomes dictates the sequence of the homologous allele at the other chromosome in heterozygotes (Figure 2.13). DNA-repair enzymes will recognize a mismatch in base-pairing at a heterozygous site of recombining chromosomes and replace the variant at one of the chromosomes so that it becomes similar to the other. That is, the site will mutate so that the resulting gametes will carry the same allele. Gene conversion may or may not be symmetric. In some cases, it is more likely that, say, a G nucleotide will convert a T at the other chromosome than vice versa (biased gene conversion) but other times changes in either direction are equally likely (unbiased gene conversion).

2.3.3 Unequal crossing over

Misalignment of homologous chromosomes during prophase I of meiosis can lead to **unequal crossing over**. Following an unequal crossing over event one of the resulting chromosomes will get an insertion and the other a deletion (Figure 2.14). Thus, unequal crossing over is a second mutation mechanism that causes insertions and deletions (in addition to replication slippage). Homologous chromosomes align

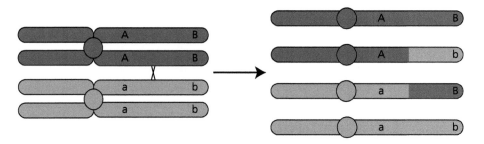

Figure 2.12 During prophase I of meiosis, homologous chromosomes are paired and aligned according to sequence similarity. During this phase crossing over takes place which results in recombinant chromosomes that differ from the parental chromosomes. Here crossing over takes place between two genes in the central chromatids yielding the recombinant haplotypes Ab and aB in the resulting chromosomes.

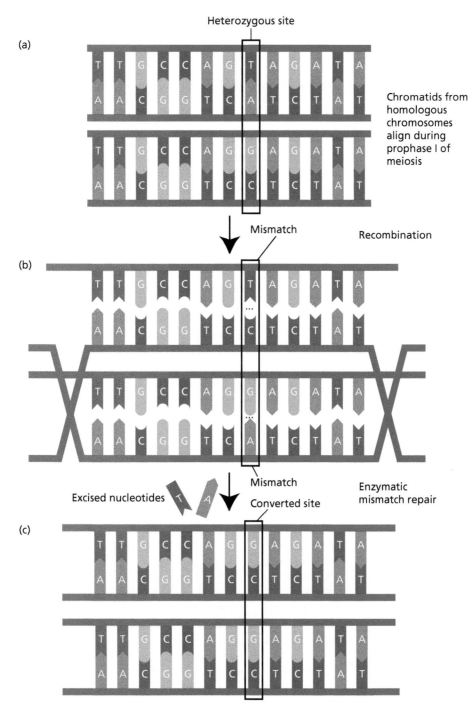

Figure 2.13 Gene conversion. **a)** Chromatids from homologous chromosomes align during meiosis. **b)** During recombination, a heteroduplex is formed. That is, the strands from the homolog marked in blue are paired with the strands from that marked in red. Accordingly, there will be a mismatch in base pairing at a heterozygous site. **c)** DNA repair enzymes recognize the mismatch and replace nucleotides (T and A with G and C) so that complementarity is regained. Consequently, one of the homologs (red) is converted by the other (blue).

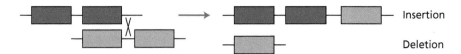

Figure 2.14 Unequal crossing can happen when homologous chromosomes are misaligned during meiosis. The resulting chromosomes in the haploid gametes will possess an insertion or a deletion.

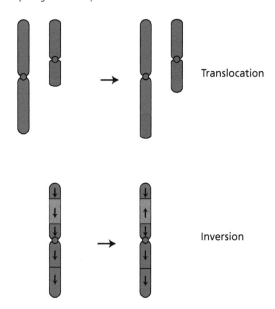

Figure 2.15 Unequal crossing over can cause chromosome rearrangements including translocation, the exchange of chromosome parts between non-homologous chromosomes (upper panel) and inversions, the 180° flip of a segment of a chromosome (lower panel).

due to sequence similarity, but misalignment may occur when the chromosomes carry non-homologous repeated sequences. Such repeated sequences are common in the genome and consequently unequal crossing is a relatively frequent phenomenon. One important source of repetitive sequence is transposable elements. One TE can produce multiple copies, or **paralogues**, that become widely distributed across the genome. Hence, unequal crossing over following misalignment can sometimes cause large insertions and deletions.

Due to the presence of dispersed TE repeats, misalignment may even occur between non-homologous chromosomes, resulting in **translocation**, the exchange of parts between chromosomes. Other large-scale genomic mutations, such as **inversions**, the 180° rotation of a segment of a chromosome, can also happen through unequal crossing over (Figure 2.15).

Many large-scale mutations caused by unequal crossing over are undoubtedly harmful to the organism and would be selected against. Indeed, several known human diseases are caused by unequal crossing over. For instance, non-homologous recombination of *Alu*-elements (a highly numerous human SINE) in or near important genes has been associated with a wide range of diseases, including acute myelogenous leukemia, von Hippel-Lindau syndrome (an inherited disorder characterized by the formation of cysts and tumors in different parts of the body), and family breast cancer (Belancio et al. 2009). Yet evidently such mutations sometimes do become fixed. Genomes of different species regularly differ in ways that can only be explained by unequal crossing over having taken place. Thus, mutations derived from unequal crossing over are sometimes beneficial (or harmless) and become fixed by natural selection (or genetic drift).

Unequal crossing over can lead to **gene duplication**. One example is the opsin genes for long-wave (OPN1LW) and medium-wave (OPN1MW) cone photopigments (red and green respectively) in Old World monkeys, apes, and humans (Ibbotson et al. 1992; Dulai et al. 1999). In these animals, the long-wave and one or more copies of the medium-wave opsin genes are located adjacent to one another on the X chromosome. These opsin genes have very similar DNA-sequences; they are paralogues. An unequal crossing over event must have taken place at some time in our evolutionary past, resulting in a gene duplication that gave us two opsin genes. The two genes have subsequently diverged to become sensitive to slightly different wavelengths of light, enabling trichromatic color vision (as all primates also carry an opsin gene for short-wave (blue) cone photopigment, OPN1SW).

Tandem repetition of similar, but not identical, opsin genes is a characteristic that we share with all apes and monkeys from the Old World, but not with New World monkeys, which only have one opsin gene on their X chromosome. Accordingly, the duplication event must have occurred after the two primate lineages diverged some 35 million years ago. Interestingly, however, howler monkeys have acquired trichromatic vision by an independent gene duplication event (Jacobs et al. 1996) (Figure 2.16). In other New World monkeys, alleles sensitive to different wave lengths are segregating at their single, X-linked opsin gene. Thus, females that are heterozygous for different

wavelength opsins have trichromatic color vision, whereas males, which only have one X chromosome, and homozygous females are red–green color blind.

A duplicated gene does not always result in new genes with different functions, however. Sometimes, one of the copies mutates in such a way that it loses its functionality—it mutates into a **pseudogene**. There are many mutations that can turn a

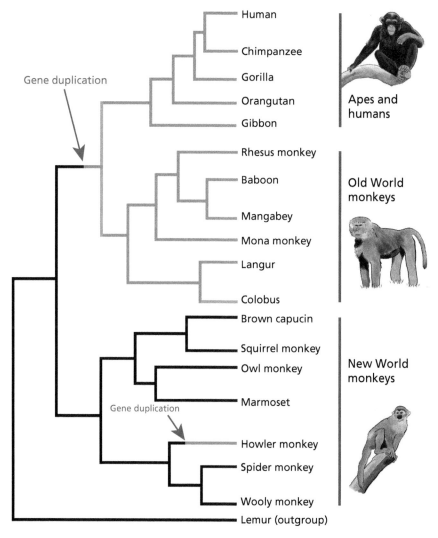

Figure 2.16 Phylogenetic relationship between primate species and events of gene duplication in opsin genes. The species with green branches possess two different opsin genes located adjacent to each other on the X chromosome, whereas species with black branches have only one opsin gene. Gene duplication has occurred independently in the common ancestor of apes, humans, and the Old World monkeys, and in the howler monkey. The duplicated genes have diverged to produce photopigments sensitive to different wavelengths of light, enabling color vision. Modified from Gilad et al. (2004).

functional gene into a pseudogene. For instance, a mutation in the promoter region can prevent RNA polymerase from attaching, thus preventing the gene from being transcribed.

Whereas replication slippage is the main mutation mechanism for short, tandem repeated sequences such as microsatellites, unequal crossing over explains the occurrence of longer tandem repeats. A duplicated locus is a likely target for further episodes of misalignment and unequal crossing over. Consequently, long chains of tandem repeated paralogues might accumulate. Multiple paralogue genes can be beneficial if the organism requires large amounts of the gene product the gene produces. Histones are highly conservative proteins that package and order DNA into chromosomes in eukaryotes (cf. Figure 2.4). Thus, at each cell division, an organism must produce large amounts of these proteins. The **gene family** encoding histones has been identified from the human genome sequence (Warzluff et al. 2002). The largest cluster of tandem-repeated histone genes, HIST1, on human chromosome 6, contains no less than 55 histone genes. Many histone genes produce exactly the same protein. Correct packaging of DNA depends on the exact secondary structure of the histone proteins. Hence, natural selection will weed out mutant histones genes that produce proteins with suboptimal secondary structure, thus maintaining very similar DNA sequences among the paralogues.

2.3.4 Concerted evolution

Stabilizing natural selection is, however, not the only factor that can maintain sequence similarity among tandem repeated paralogues. A peculiar phenomenon is that the sequences of such paralogues are often surprisingly similar within a species, although different species may be quite divergent at the same loci. If the paralogues had evolved independently from each other, and if two species have inherited the same loci from a common ancestor, one would expect the paralogues within each species to be equally different as those between the species. Some mechanism(s) are apparently causing the paralogues within each species to not evolve independently but to remain similar. This

phenomenon is called **concerted evolution**. As an example, although the genes for red and green opsins have diverged at a functionally important site to yield trichromatic color vision, within any one species, including us humans, the two genes have surprisingly similar DNA sequences, even in their noncoding introns. The duplication event occurred some 35 million years ago, so there has been plenty of time for the paralogues to diverge. Concerted evolution is evidently taking place to maintain sequence similarity within each evolving species.

Two main mutation mechanisms can explain concerted evolution and both should be familiar to us by now: (1) gene conversion and (2) unequal crossing over. Biased gene conversion can cause a mutation to spread among tandem repeated paralogues when homologous chromosomes are misaligned during meiosis (Figure 2.17). For instance, paralogue 1 may contain a new mutation at a given position. If it is misaligned with paralogue 2 during meiosis the mutation can be converted to that paralogue. Further events of misalignment can result in the mutation being spread to all the paralogues in a repeat. In a different species, a mutation at a different position might spread this way. This will mean that while the paralogues diverge between species, within species they will remain similar. Biased gene conversion can likewise hinder new mutations from spreading by converting the newly mutated paralogue to become similar to the ancestral ones.

Understanding how unequal crossing over can cause concerted evolution is potentially a little more difficult. Remember that unequal crossing over is equally likely to cause duplication as the deletion of a repeat unit. Thus, over evolutionary time a tandem repeated motif may shrink and grow in copy number. For instance, a new mutation in paralogue number 8 would disappear if that copy was deleted. Subsequently, a new paralogue number 8 may emerge that lacks this mutation. More generally, repeated cycles of shrinking and growth in copy number over evolutionary time would ensure that the paralogues that are present at any one point in time would have a relatively recent common ancestry. They are derived from relatively recent duplication events and would not have had time to accumulate many mutations.

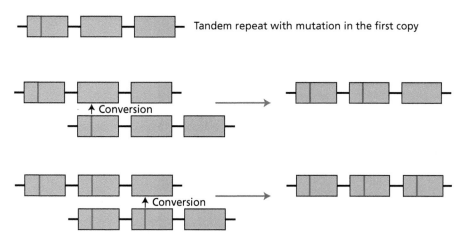

Figure 2.17 Concerted evolution by gene conversion. Gene conversion between paralogs homogenizes their DNA sequence. Here a novel mutation in one of the paralogs (red) spreads to the others by repeated events of gene conversion.

2.4 Origin of new genes and protein functions

2.4.1 Gene and genome duplication

The human genome contains some 20 000 different protein coding genes that have originated at some time in the past. We share most of these genes with closely related species such as the chimpanzee and the gorilla (albeit slightly different versions of those genes), but as we compare more and more distantly related species gene content becomes more and more different and our simplest relatives, such as bacteria, have much fewer genes than us. Clearly, new genes do arise from time to time across the tree of life.

It is straightforward to imagine how a new gene could arise relatively easily by chance from noncoding DNA. An arbitrary mutation generating a start codon (ATG) would produce an open reading frame immediately. Downstream of the new start codon, a stop codon (TAA, TAG, or TGA) will eventually be present, stopping the reading frame. Despite the ease of this process, a large number of random mutations would typically be needed to transform this open reading frame into a functional gene, including a promoter sequence that can attach an RNA-polymerase and various sequence specific regulatory sequences that can attach transcription factors. Moreover, even if all the necessary mutations should occur, the new gene would code an arbitrary polypeptide that would most likely have little or no benefit for the organism it has arisen in. Nevertheless, some rare examples of *de novo* origin of genes from noncoding sequences have been suggested, even in our own species (Knowles & McLysagth 2009).

However, most new genes probably arise by duplication events and subsequent divergence (Ohno 1970). The genes for long- and medium-wave photopigments in primates provide one example. Two copies of a functional gene are not necessarily any worse in terms of fitness than one copy, and in certain cases the copies may diverge and take up new functions. These functions may differ only slightly at first, but over long evolutionary time spans they can become very divergent. Myoglobin is the oxygen carrying protein in muscle tissue. It is evolutionarily related to hemoglobin and shares sequence similarities. Hemoglobin genes probably originated when a myoglobin gene was duplicated in an early chordate ancestor some 500 million years ago, after which one of the copies took up the oxygen carrying function in the blood, leaving the other gene for muscle tissue function (Holmquist et al. 1976). Even within the hemoglobin gene family, there have been more recent duplication events, allowing the paralogues to perform slightly different functions. In humans, a different but related set of genes code the active hemoglobin protein in fetuses compared to adults. The fetal hemoglobin has a higher oxygen affinity than the adult one. At the age of about 6 months the genes for fetal hemoglobin are epigenetically turned off and the adult genes turned on (Xu et al. 2013).

As we have seen, gene duplication and the evolution of new genes with divergent functions can result from unequal crossing over. This is the most likely explanation for multiple photopigment genes in primates, for example (Figure 2.16). Replicative TEs can also sometimes cause gene duplication, when a host gene is replicated and transposed along with the transposable element (e.g. Morgante et al. 2005). As in the case of unequal crossing over, the duplicated genes may diverge and take up new functions.

Sometimes an error during meiosis occurs, resulting in unreduced, diploid gametes (2n). Fusion of diploid gametes will result in tetraploid offspring (4n). Such **polyploidy** is very common in plants but several animal examples are also known. Polyploidy can occur within a species (**autopolyploidy**), but often it occurs in connection with hybridization between species (**allopolyploidy**). A tetraploid hybrid will often be instantaneously reproductively isolated from its diploid parent species. This

is because the resulting triploid (3n), hybrid offspring would produce **aneuploid** gametes and be sterile or have low fertility. During meiosis, each daughter cell of a triploid may receive one copy of a certain chromosome and two copies of another (Figure 2.18). Such aneuploid gametes would thus have an imbalance in chromosome number making them nonfunctional or unable to produce viable offspring.

The duplication of whole genomes, and by definition all the genes they contain, provides an opportunity for the diversification and divergence of gene function among copies. In reality however, many of the paralogues appear to mutate into pseudogenes or become deleted (Wolfe 2001).

2.4.2 Exon shuffling

New gene functions can originate through **exon shuffling**. This is the process in which one or more exons from one gene become inserted into a different

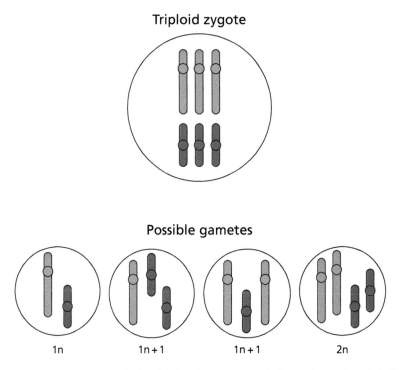

Figure 2.18 Hybridization between a tetraploid and either of its diploid parent species will often result in sterile, triploid offspring. Triploid zygotes can produce many possible gametes, because each gamete can get either one or two copies of each chromosome, in many possible combinations. Even in the simple case of only two different chromosomes, only one out of four possible gametes will have the normal haploid complement of chromosomes. Aneuploid gametes are usually inviable.

gene (Figure 2.19) or in which exons within a gene become duplicated. The primary molecular mechanism for exon shuffling is thought to be recombination within introns (Moran et al. 1999). In eukaryotes introns are typically much longer than exons. Accordingly, crossing over is much more likely to occur within introns than within exons. Moreover, introns frequently harbor repetitive sequences derived from transposable elements. Thus, recombination between paralogous sequences (unequal crossing over) can result in exchange of exons between genes and exon duplication within genes.

The number of genes that have clearly evolved by exon shuffling is much higher in animals than in other eukaryotes (protists, plants, and fungi). Moreover, the majority of the resulting modular proteins have functions that can be linked to multi-cellularity (Patthy 1999). They are not found within cells but in body fluids or on cell membranes where they function in regulating cell–cell communication. The taxonomic distribution and function of genes that have evolved by exon shuffling may suggest that this mode of evolution was important for the radiation of the animal kingdom that occurred during the onset of the Cambrian period some 540 million years ago (Patthy 1999).

2.4.3 Protein moonlighting, RNA editing, and alternative splicing

The number of protein functions in humans greatly exceeds the number of genes in the genome. Our genomes house some 20 000 genes but they produce more than 100 000 different proteins. This means that the same genes often produce several different proteins. Moreover, the same protein may serve different functions depending on where it is expressed. The latter phenomenon is called **protein moonlighting**. The transparent eye lenses of vertebrates are made up of structural proteins called crystalin. In birds and crocodiles one of these structural proteins, ε-crystalin, also functions as an enzyme in the digestive system, where it metabolizes lactate (Wistow et al. 1987).

The same gene can produce different proteins by means of post-transcriptional modification of the mRNA, so called **RNA editing**. In humans, the *APOB* gene codes for two different proteins (Chen et al. 1987). Apolipoprotein B-100 is produced in liver tissue and is a protein that transports cholesterol and other lipids in the blood. In the intestine, however, the same gene produces an edited protein, apolipoprotein B-48, that aids in the absorption of lipids from digested food. In intestine tissue the

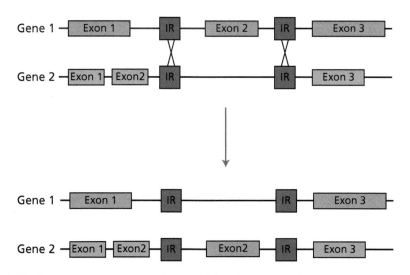

Figure 2.19 Exon shuffling between two genes. Interspersed repeats (IR) derived from transposable elements located in introns can cause misalignment of non-homologous genes. Here, double crossing over results in exon 2 being deleted from gene 1 and inserted into gene 2.

enzyme cytidine deaminase catalyzes a change in the mRNA, editing a codon for the amino acid glutamine (CAA) to a stop codon (UAA), resulting in a shortened protein with a different function.

Finally, the same gene may produce different proteins by means of alternative splicing of the exons (see Figure 2.5). This is an extremely common phenomenon. According to Johnson et al. (2003) at least 74% of human multi-exon genes are alternatively spliced to yield two or more different proteins per gene. Again, this means that the genome is able to produce a much wider array of proteins than the number of genes it contains.

2.5 External sources of genetic variation

2.5.1 Introgression

Ultimately, mutations are the source of all genetic variation. Yet, a given species can sometimes receive new genetic variation from a different species. Hybridization between different species is relatively frequent. Often, it will have no genetic consequences as the hybrid would be sterile or inviable and leave no descendants. Sometimes, however, a hybrid may succeed in backcrossing with one of its parent species. In such cases one or more genes from one species can be transferred to the other species. Such gene flow across a species barrier is called **introgression**. Genes from one species are often incompatible with genes from another species. However, incompatible genes should be weeded out by selection in early generation backcrosses following the initial introgression event. However, not all genes are incompatible across species. Sometimes, a foreign gene could be beneficial in the new genetic background. Indeed, one difference between genetic variation originating from mutation and introgression is that mutations are random changes whereas introgressed variants have been molded by natural selection, albeit in a different genomic background. Therefore, introgressed alleles may have a higher probability of being beneficial and contribute to adaptation than a random mutation. There are many examples in which introgressed alleles appears to have contributed to adaptation. *Heliconius* is a group of butterflies that live in Central and South America (Figure 2.20). These are toxic butterflies

Figure 2.20 Müllerian mimicry in *Heliconius* butterflies. Four varieties of the unpalatable species *H. erato* (left) and *H. melpomene* (right). Each pair comes from the same geographic region and the two species gain protection from predations from resembling each other. Photo courtesy of Chris Jiggins and the MacGuire Centre. The photographs are from the Neukirchen collection.

that signal their unpalatability to potential predators by possessing striking wing patterns in red, black, white, yellow, and orange colors. Moreover, different *Heliconius* species mimic the color patterns of each other to gain extra protection (e.g. Benson 1972; so called Müllerian mimicry, after Müller 1878). For instance, a rare poisonous species is better off mimicking the warning signal of a more common species than to produce a different warning signal of its own. This is because it is more likely that predators would already have learned to avoid the more common signal. In several cases, adaptive mimicry has been made possible by introgression. For instance, the gene for a red wing band is shared among several mimetic *Heliconius* species and has likely been transferred between these species through introgressive hybridization (Pardo-Diaz et al. 2012).

2.5.2 Horizontal gene transfer

Genes can sometimes be transferred between distantly related organisms without reproduction taking place between them. Horizontal gene transfer is very common in bacteria and archaea, which possess specific mechanisms for transfer and uptake of DNA. However, examples of horizontal transfer are also known from eukaryotes. Intimate relationships between bacteria and their animal hosts have sometimes resulted in uptake of bacterial DNA (Dunning Hotopp 2011). Viruses can also be vectors for horizontal gene transfer. Gladyshev et al. (2008) investigated the genomes of bdelloid rotifers and found many genes of bacterial, fungal, and plant origin along with mobile genetic elements in the genomes of these tiny animals. Some of the transferred genes were indeed intact and were transcribed in the rotifers.

Study questions

1. Which are the main differences between the genomes of bacteria, archaea, and eukarya?
2. Which factor(s) may explain the large variation in genome size between eukaryote species?
3. What is the evidence that transposable elements parasitize host genomes? Make a search on Google Scholar with the keywords "transposable elements adaptation". What are the main beneficial effects of transposable elements that have been found according the main hits you receive? Discuss to what extent these beneficial effects contradict the hypothesis that transposable elements are parasitic entities.
4. Which are the main evolutionary consequences of transposable elements?
5. Would you characterize an event of conservative or replicative transposition a mutation? Why or why not?
6. Transposable elements share many properties with viruses and appear to share a common evolutionary origin. Make arguments for and against the likelihood of each of the following three hypotheses: 1) Transposable elements evolved from viruses; 2) Viruses evolved from transposable elements; 3) Viruses and transposable elements are descendants of self-replicating molecules that predate the origin of the living cell.

Discuss what kind of evidence that could be obtained which would support one hypothesis over the others.
7. Which mechanisms can explain the occurrence of dispersed and tandem repeats in the eukaryote genome?
8. In what ways could we say that mutations are random?
9. These are the first five codons (after the start codon ATG) in the chicken Melanocortin 1 receptor gene, which is involved in the synthesis of melanin pigment: TCG ATG CTG GCC CCC. Use the genetic code in Table 1.1 from chapter 1 to address the following questions.
   a. Find the corresponding mRNA sequence and the resulting amino acid sequence.
   b. For each of the nucleotide sites, how many point mutations would cause a non-synonymous (replacement) mutation and how many would be synonymous (silent)?
   c. Which of the non-synonymous would be missense and which would be nonsense?
   d. Imagine that an A-nucleotide was inserted between the G at position 3 and the A at position 4. What would be the resulting amino acid sequence? What do we call such a mutation?
   e. For each position, determine how many transitions and how many transversions would be possible? How many of the transitions and how many of the transversions would result in a non-synonymous mutation?
10. Suppose that a novel frame-shift mutation occurred during meiosis in a gene in an individual magpie that lived 500 years ago. What do you think would be the consequence of that mutation for:
    a. Magpies as a species?
    b. The individual in which the mutation occurred?
    c. The gene that mutated?
    d. The gene copy that mutated?
11. Explain in brief the following concepts and suggest mechanisms that may account for each of them:
    a. Translocation
    b. Insertion
    c. Inversion
    d. Concerted evolution
    e. Exon shuffling.

12. List all the different types of mutations that have been covered in this chapter and explain the mechanism(s) by which they occur.

13. How can new genes arise in evolution and which are the most important mechanisms for their origin?

14. Explain the concepts of protein moonlighting, RNA editing, and alternative splicing and their role in explaining how relatively few genes in e.g. the human genome can explain the much higher number of proteins and protein functions.

15. What characterizes introgression and horizontal gene transfer? What are the similarities and differences between these two processes and, respectively, gene flow and mutation?

CHAPTER 3

Changes in allele and genotype frequency

Evolution is the change in heritable traits of populations over successive generations. At the molecular level this translates into changes in their genetic composition. One important goal in evolutionary genetics is to reconstruct the evolutionary past of contemporary populations. Where did they come from? How many individuals existed? How are the different populations related to each other? Which evolutionary processes have affected them? The past has shaped the genetic variation we observe today. A general theoretical investigation of how different demographic and evolutionary processes affect genetic variation within and between populations provides us with tools to reconstruct evolutionary history and begin to answer these questions. This is the fundamental purpose of population genetics. It is important to keep in mind that population genetic theory is abstract, and the models we use are approximations of reality. Reality is complex and, by comparison, population genetic models are simple, making certain necessary assumptions. Nonetheless, they are means of developing an expectation for the patterns we might observe in empirical data. In this chapter, we shall first investigate the relationship between allele and genotype frequencies in a hypothetical population that is not subjected to any evolutionary forces and then, one by one, introduce demographic and evolutionary factors to investigate in what ways they affect allele and/or genotype frequencies. We start with the simplest case, by investigating a single locus with two segregating alleles.

3.1 The Hardy–Weinberg model

3.1.1 Assumptions

At a given locus, we can denote the frequency of an allele A_1 as p. It is simply the number of A_1 alleles in a population, or population sample, divided by the total number of alleles in that population or sample. For example, at a locus with two alleles we may genotype 100 diploid individuals from a population and find 9 individuals with the genotype A_1A_1, 42 with the genotype A_1A_2 and 49 with the genotype A_2A_2. We thus have a total of $2 \cdot 9 + 42 = 60$ A_1 alleles out of a total of 200 alleles (since each individual carries two alleles). Hence $p = 60/200 = 0.3$. We may denote the frequency of the alternative allele A_2 as q. Since there are only two alleles in this example $q = 1 - p$. In our example here $q = 1 - 0.3 = 0.7$. We denote the frequency of the genotypes as g with subscripts that specify the alleles they include. In our numerical example the frequencies of the three genotypes A_1A_1, A_1A_2, and A_2A_2 are respectively:

$$g_{11} = 9/100 = 0.09$$

$$g_{12} = 42/100 = 0.42$$

$$g_{22} = 49/100 = 0.49.$$

In a given population, the allele and/or genotype frequency at a locus may change due to several factors:

1. The genotype frequencies may change due to *non-random mating*. For instance, genetic relatives

Evolutionary Genetics: Concepts, Analysis, and Practice. Glenn-Peter Sætre & Mark Ravinet, Oxford University Press (2019).
© Glenn-Peter Sætre & Mark Ravinet 2019. DOI: 10.1093/oso/9780198830917.001.0001

may tend to mate with each other (inbreeding) or genetic relatives may tend to actively avoid mating with each other (outbreeding).

2. The allele frequencies may change at random, that is, they are subjected to *genetic drift*.
3. Certain genotypes may have phenotypic effects that enhance the individual's probability of surviving or reproducing relative to individuals with other genotypes, in which case there is *natural selection* that can affect both allele and genotype frequencies.
4. An allele may mutate into a different allele due to an error during cell division. If such a *mutation* occurs during meiosis, it can be passed on to the next generation.
5. Finally, the genetic composition of a population may change due to influx of alleles from a different population, that is, *gene flow*.

In order to investigate how the five processes above affect the genetic composition of a population in quantitative terms it is useful to have a null model: a description of the expected relationship between allele and genotype frequencies if none of these factors are operating. The classical null model for the one-locus case is called the Hardy–Weinberg model, after its independent demonstration by the English mathematician Godfrey H. Hardy (1908) and the German physician Wilhelm Weinberg (1908). The Hardy–Weinberg model is a simple, mathematical relationship between allele and genotype frequencies in an ideal population, not subjected to any evolutionary processes. The assumptions of the Hardy–Weinberg model are as follows:

1) Individuals in the population mate randomly.
2) The population is infinitely large.
3) No selection occurs.
4) No mutation occurs.
5) No gene flow occurs.

Imagine such a population and focus on one diploid locus with two segregating alleles, A_1 and A_2. The frequency of A_1 is p and that of A_2 is q $(= 1 - p)$. The genotypes of the next generation are made by randomly combining gametes from the large population into pairs. Since the population is infinitely large, so is the number of pairs of gametes—

meaning the Hardy–Weinberg model is a binomial process (see section 3.3.1 for a deeper explanation of this). By assuming random mating and that no other factors are changing allele frequencies from one generation to the next, the expected genotype frequencies simply translate into the probabilities of combining any pair of alleles into a genotype. Mathematically, random mating is analog to collecting all the gametes from a population, transferring them into a big container, stirring well, and blindly picking sperm and egg pairs for fertilization to make the next generation.[1] In our imaginary container, p of the sperm carry A_1 and q carry A_2. Likewise, p of the eggs carry A_1 and q carry A_2. Furthermore, since each individual produces a large number of gametes, picking one pair of gametes for fertilization (say $A_1 \times A_1$) does not alter the probability of picking the same (or a different) pair of gametes the next time (or any time), that is, the probabilities are independent. In other words, the probability of picking a pair of gametes for fertilization translates into the expected frequency of the resulting genotype the next generation.

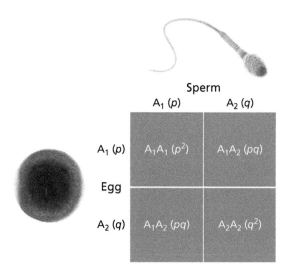

Figure 3.1 The expected genotype frequencies at Hardy–Weinberg equilibrium. Under ideal conditions (the Hardy–Weinberg assumptions) the frequency of each genotype after one generation of random mating in a population equals the probability of combining the corresponding gametes into zygotes.

[1] In probability theoretic terms random sampling of alleles at fertilization is equivalent to sampling with replacement.

The probability of making a homozygote A_1A_1 is the probability of combining one A_1- carrying sperm (p) with one A_1-carrying egg (p), that is, $p \times p = p^2$ (Figure 3.1). With our assumptions, the probability of making a particular genotype doesn't alter the probability of making other genotypes so the frequency of the genotype A_1A_1 in the next generation will also be p^2, and, importantly, it will remain p^2 in future generations as long as our five assumptions hold. The equivalent argument holds for the A_2A_2 genotype; the probability of making any A_2A_2 genotype equals the probability of combining an A_2 sperm (q) with an A_2 egg (q), that is, $q \times q = q^2$. The frequency of A_2A_2 in the next generation is q^2 and it will never change under our ideal assumptions. There are two ways of making the heterozygote genotype A_1A_2. An A_1 sperm can fertilize an A_2 egg (probability pq) or an A_2 sperm may fertilize an A_1 egg (probability pq). Therefore, the combined probability of making a heterozygote, and hence the frequency of heterozygotes the next (and future) generation(s), will be $2pq$.

The expected genotype frequencies under the Hardy–Weinberg model,

$$E(g_{11}) = p^2$$
$$E(g_{12}) = 2pq$$
$$E(g_{22}) = q^2$$

in the two-allele case, is referred to as the **Hardy–Weinberg expectation (HWE)**. It can easily be expanded to multiple alleles. Denoting allele i as A_i and its frequency p_i, the Hardy–Weinberg expectations of genotypes are simply the binomial frequencies:

$$E(g_{ii}) = p_i^2$$
$$E(g_{ij}) = 2p_ip_j \quad \text{(for i} \neq \text{j)} \quad (3.1)$$

3.1.2 Testing for deviations from the Hardy–Weinberg expectation

In our numerical example above we have $p = 0.3$ and $q = 0.7$. Let us check if the observed genotype frequencies are in accordance with the Hardy–Weinberg expectation. The expected genotype frequencies are:

$$E(g_{11}) = 0.3^2 = 0.09$$
$$E(g_{12}) = 2 \cdot 0.3 \cdot 0.7 = 0.42$$
$$E(g_{22}) = 0.7^2 = 0.49$$

which is exactly what we observed; there is no deviation from the Hardy–Weinberg expectation in this particular case.

Let us take a second example. We genotype 100 individuals from another population of the same species and obtain the following genotype frequencies:

$$g_{11} = 12/100 = 0.12$$
$$g_{12} = 42/100 = 0.42$$
$$g_{22} = 46/100 = 0.46$$

The allele frequencies are $p = 66/200 = 0.33$ and $q = 1 - p = 0.67$. The expected genotype frequencies from Hardy–Weinberg are:

$$E(g_{11}) = 0.33^2 = 0.1089$$
$$E(g_{12}) = 2 \cdot 0.33 \cdot 0.67 = 0.4422$$
$$E(g_{22}) = 0.67^2 = 0.4489$$

The observed and expected frequencies are compared in Table 3.1.

The observed genotype frequencies deviate slightly from the expected, but not very much. Should we expect that some assumption of the Hardy–Weinberg model is violated in this population or is it statistically likely that the 100 individuals we happened to sample are drawn from a population in Hardy–Weinberg equilibrium? In statistical hypothesis testing we calculate the probability that our sample is drawn from a population specified by the null hypothesis, which in this case is the Hardy–Weinberg model. Only if this probability is sufficiently low, usually, by convention, lower than $P = 0.05$, do we reject the null hypothesis. For categorical data, such as genotype frequencies, we can use a chi-square test to investigate the goodness of fit between the observed and expected genotype frequencies. The chi-square test is defined as:

$$\chi^2 = \sum_{i=1}^{k} \frac{(E_i - O_i)^2}{E_i} \quad (3.2)$$

That is, the sum of squared differences between expected and observed values, divided by the

Table 3.1 Observed and expected genotype frequencies in numerical example 2.

	N	g_{ii}	$E(g_{ii})$
A_1A_1	12	0.12	0.1089
A_1A_2	42	0.42	0.4422
A_2A_2	46	0.46	0.4489
Total	100	1	1

expected values. In our case, the expected values are the number of individuals of each of the genotypes we would expect under Hardy–Weinberg equilibrium, given our sample size. Since our sample size is 100 individuals we find the expected number of individuals by multiplying $E(g_{ij})$ with 100. We can then calculate chi-square as:

$$\chi^2 = \frac{(10.89 - 12)^2}{10.89} + \frac{(44.22 - 42)^2}{44.22}$$
$$+ \frac{(44.89 - 46)^2}{44.89} = 0.252$$

The probability P that the 100 individuals are drawn from a population in Hardy–Weinberg equilibrium can be found by comparing our chi-square value (0.252) with the critical values of the chi-square distribution with 2 degrees of freedom (df = 2). Performing the chi-square test by hand is an excellent way to learn and understand how it is carried out. However, it is more convenient and practical to use a statistical software function, such as one implemented in the R programming environment, to do the calculation. In our case $P = 0.8816$, which is much higher than the conventional significance limit of $P < 0.05$, so we do not reject the null hypothesis of Hardy–Weinberg equilibrium.[2] Below we shall explore how violations of the different Hardy–Weinberg assumptions affect the genetic composition of a population, starting with the assumption of random mating.

---

[2] In real populations, the Hardy–Weinberg assumptions are never truly met. For one, no population is infinitely large. Hence, although we sometimes do not reject the null model it does not imply that we hold it as universally true. The Hardy–Weinberg is useful, not because it is realistic, but because violations of the different assumptions can cause deviations from expectation at different levels, for instance within populations or in between-population comparisons.

3.2 Non-random mating

3.2.1 Inbreeding and outbreeding

Random mating is a critical assumption of the Hardy–Weinberg model. Clearly, if the alleles were not combined randomly at fertilization but according to some other rule, the genotype frequencies in the next generation are likely to deviate from the Hardy–Weinberg expectation.

Inbreeding is one form of non-random mating and it occurs when individuals are more likely to mate with close genetic relatives than with random members of the population. The most extreme form of inbreeding is self-fertilization, which is common in many hermaphroditic plant species (including the garden peas studied by Mendel) and some animals. Let us consider a population similar to our last set of examples in which all the Hardy–Weinberg assumptions hold except that the individuals reproduce by self-fertilization. We assume that the population initially is in Hardy–Weinberg equilibrium at a locus with two alleles, A_1 and A_2, but that self-fertilization is then introduced. Clearly, all A_1A_1 individuals that self-fertilize will produce A_1A_1 offspring and all A_2A_2 individuals will produce A_2A_2 offspring (Figure 3.2). In contrast, with Mendelian segregation, heterozygote A_1A_2 individuals that self-fertilize will produce offspring in which on average 25 percent are A_1A_1, 50 percent A_1A_2 and 25 percent A_2A_2. In other words, only half the offspring of heterozygote self-fertilizers will be heterozygous. For each generation in a self-fertilizing population there will therefore be a 50 percent reduction in the frequency of heterozygotes. If self-fertilization continues, the population will eventually end up with only homozygotes. Note that the allele frequencies would not change in an imaginary infinitely large population (i.e. if all the other Hardy–Weinberg assumptions hold). The alleles are simply redistributed from heterozygous to homozygous genotypes. Less extreme forms of inbreeding, such as mating between cousins, also qualitatively yields the same result. A reduction in the frequency of heterozygotes over time occurs but it is slower than the extreme case of self-fertilization.

Outbreeding is the opposite of inbreeding and occurs when individuals are less likely to mate with

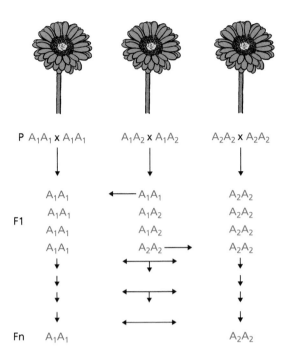

Figure 3.2 Loss of heterozygosity due to inbreeding. Self-fertilization in a hermaphroditic organism causes a 50% reduction in the frequency of heterozygotes per generation because, on average, only half the offspring of heterozygote parents would be heterozygous, whereas all the offspring of homozygotes would be homozygous.

genetic relatives than expected from random mating. Genetic relatives tend to share alleles inherited from a recent common ancestor whereas non-relatives are likely to possess different alleles. Accordingly, mating between non-relatives will more often result in heterozygous offspring than mating between relatives. Outbreeding leads to more heterozygotes than expected from the Hardy–Weinberg model because when mating is random we would expect some accidental mating between genetic relatives.

One consequence of inbreeding is that rare, deleterious recessive alleles are more likely to be combined in homozygote genotypes. This leads to low viability and/or fertility of individuals and hence decreased mean fitness of the population, a so-called **inbreeding depression**. Natural selection would tend to favor traits that reduce the probability of maladaptive inbreeding. Such inbreeding avoidance can result in outbreeding (heterozygote excess). A large number of species possess traits that reduce

the probability of inbreeding. Many flowering plants possess a self-incompatibility mechanism that ensures that self-fertilization does not take place. Self-/nonself-recognition appears to be caused by a multi-allelic locus in many flowering plant species (Takayama & Isogai 2005). Pollen development is inhibited when the same allele is expressed by both pollen and pistil. In many animal species sex-biased natal dispersal reduces the probability of inbreeding (Pusey 1987). Members of one of the sexes stay close to where they were born (usually females in mammals and males in birds) whereas the other sex disperses. Accordingly, mating will tend to take place between genetically unrelated individuals.

Given the high costs of inbreeding one may wonder why it still is common in many organisms. Many hermaphroditic flowering plant species are obligate self-pollinators. One advantage with self-pollination is that it assures reproduction when outcrossing is difficult. For instance, in cold climates where insect pollinators are rare or absent, self-pollination could be the best reproductive strategy despite its costs (Darwin 1876). The blue-eyed mary (*Collinsia verna*) has been shown to switch adaptively between outcrossing and self-pollination depending on the frequency of pollinator visits (Kalisz et al. 2004) (Figure 3.3).

Figure 3.3 The blue eyed mary (*Collinsia verna*) is an outcrossing flowering plant that actively switches to self-pollination when insect pollinators are rare in the environment, according to a study by Kalisz et al. (2004). Photo courtesy of James St. John.

3.2.2 Assortative and disassortative mating

Non-random mating is not restricted to inbreeding and outbreeding. **Assortative mating** is preferential mating between individuals with the same genotype or phenotype. The Gouldian finch (*Erythrura gouldiae*) occurs in two color morphs, a redheaded and a blackheaded morph (Figure 3.4). Both sexes exhibit a preference for mates with the same head color as themselves, that is, they mate assortatively (Pryke 2010). Head color shows sex-limited inheritance: females always inherit the head color of their father. This suggests that the gene for head color is located on a sex chromosome. Furthermore, head color is consistent with being determined by one locus with two alleles, in which the allele for red color Z^R is dominant and that for black color Z^r is recessive. In birds, it is the female that has different

Figure 3.4 Assortative mating based on head color in the Gouldian finch (*Erythrura gouldiae*). A blackheaded male (left) is receiving interest from a blackheaded female whereas the redheaded male (right) is ignored. Photo courtesy of Sarah Pryke and Simon Griffith.

sex chromosomes (ZW) whereas males have two Z chromosomes. Hence, females inherit the Z chromosome from their father and the W from their mother, whereas males inherit one Z chromosome from each parent. Genomic data confirms the locus for head color is indeed located on the Z chromosome as suggested by its pattern of inheritance (Toomey et al. 2018; Kim et al. 2018). In males, the genotypes Z^RZ^R and Z^RZ^r at a Z-linked locus near the *Follistatin* gene have red heads whereas birds carrying the Z^rZ^r genotype have black heads. The locus does not have a W-homolog so females are either Z^RW (red head) or Z^rW (black head). Due to the pattern of assortative mating the frequency of heterozygous males is much lower in natural populations of the Gouldian finch than expected if random mating were occurring.

Disassortative mating is preferential mating between individuals with different genotypes and / or phenotypes. The Tanganyikan cichlid *Perissodus microlepis* provides one example (Takahashi & Hori 2008). This cichlid fish feeds by biting off scales of larger fish species. *P. microlepis* occurs in two morphs; in one morph the mouth opens to the right, causing the left side of the head to face forward (lefty), whereas in the other morph the mouth opens to the left, causing the right side of the head to face forward (righty). Takahashi and Hori (2008) found that mating between lefties and righties occurred at a much higher frequency than expected from random mating; that is, mating is disassortative.

The molecular genetic basis for the laterality polymorphism has not been identified, but the pattern of inheritance of the phenotypes is very interesting. Lefty and righty individuals coexist at approximately equal frequencies in Lake Tanganyika and no intermediates are found. Hori et al. (2007) compared the laterality between parents and their young and inferred that the trait is most likely caused by one Mendelian locus with two alleles in which the lefty L-allele is dominant over the righty l-allele. Interestingly, however, lefty–lefty pairs produce a 2:1 ratio of lefty:righty young, although ratios of 3:1 (Ll × Ll) or 1:0 (LL × Ll) would be expected under simple Mendelian genetics (see chapter 1). Further, lefty–righty pairs produce a 1:1 ratio of lefty:righty young, whereas the Mendelian expectation is 1:0 (LL × ll) or 1:1 (Ll × ll). These patterns are consistent with the lefty homozygote LL genotype being lethal (Figure 3.5). If so, the population consists

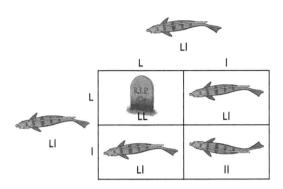

Figure 3.5 Inheritance of mouth/skull asymmetry in the Lake Tanganyikan scale eating cichlid *Perissodus microlepis*. Crossing between lefty fishes results in 2:1 lefty and righty offspring. The pattern of inheritance is consistent with the trait being coded by a single locus with two alleles in which L (lefty) is dominant and l (righty) is recessive, but in which the homozygous LL genotype is lethal.

entirely of Ll lefties and ll righties. Mating between lefties and righties would thus always be between Ll × ll genotypes, which would result in an equal proportion of lefty and righty offspring. Mating between two lefties would always be between Ll x Ll, which would yield a 2:1 ratio of lefty:righty provided that the LL homozygote is lethal. Disassortative mating may therefore contribute to the maintenance of the lateral dimorphism in this system.

3.2.3 Inbreeding/outbreeding versus assortative/disassortative mating

Inbreeding and assortative mating have similar effects on genotype frequencies: they both decrease the proportion of heterozygous individuals relative to the Hardy–Weinberg expectation. Likewise, outbreeding and disassortative mating both increase the proportion of heterozygotes relative to the Hardy–Weinberg expectation. The difference between these pairs of factors is the scale at which they operate in the genome. Inbreeding and outbreeding have genome-wide effects whereas assortative and disassortative mating have only local, locus-specific effects. In an inbred population, heterozygosity is reduced at all polymorphic loci. In contrast, assortative mating only reduces heterozygosity at the locus/loci that determines the mating preference, such as the locus for head color in the Gouldian Finch. It is recombination that allows loci elsewhere in the genome to segregate independently and to escape

the effects of reduced heterozygosity (although loci in close physical linkage with the mating preference locus are affected as recombination events are rarer in this case).

3.2.4 F_{IS}—a coefficient for measuring within-population deviations from HWE

Non-random mating is a powerful factor in causing within-population deviations from the Hardy–Weinberg expectation (HWE). Certain forms of selection can also have such local effects on genotype frequencies (see chapter 4). Wright (1922) introduced the coefficient F_{IS} to quantify local deviations from HWE. The F_{IS}-index measures the difference between expected (HWE) and observed heterozygosity:

$$F_{IS} = \frac{H_S - H_I}{H_S} \qquad (3.3)$$

H_S is the expected frequency of heterozygotes at HWE ($H_S = E(g_{12}) = 2pq$ in the two-allele case) and H_I is the observed frequency of heterozygotes ($H_I = g_{12}$ in the two-allele case). F_{IS} is 0 at HWE, positive when $H_S > H_I$ (heterozygote deficit) and negative when $H_S < H_I$ (heterozygote excess). F_{IS} is 1 when there are no heterozygotes in the population despite the presence of multiple alleles, as would be the case in a completely inbred population. F_{IS} would also be 1 at a mate preference locus if mating were 100% assortative. However, in this case, the different morphs would belong to different species, since they would be reproductively isolated from each other (see chapter 8).

Taking the time to calculate an example will help make the F_{IS}-index clearer. Assume that Hori and co-workers have deduced genotypic variation correctly in the *P. microlepis* cichlid: we assume that the homozygote genotype LL is lethal and hence absent, and that the frequency of the two morphs is equal (0.5), which is close to what Hori and co-workers observed (Hori et al. 2007; Takahashi & Hori 2008). Then half the individuals in the population have the genotype Ll ($g_{Ll} = 0.5$) and the rest are homozygous ll ($g_{ll} = 0.5$). The frequency of the L-allele must therefore be $p = 0.25$ and that of the l-allele $q = 0.75$ (make sure you understand why). $H_S = E(g_{Ll}) = 2pq = 2 \cdot 0.25 \cdot 0.75 = 0.375$, whereas $H_I = g_{Ll} = 0.5$. Then $F_{IS} = (0.375 - 0.5)/0.375 = -0.333$. The way to interpret the

F_{IS}-value is that the frequency of heterozygotes is 33.33 percent higher than expected from the Hardy–Weinberg model, which is quite a lot. If we had sampled 100 individuals from the population and found these proportions (50 Ll and 50 ll), the probability that our sample had been drawn from a population in Hardy–Weinberg equilibrium would be:

$$\chi^2 = \frac{(6.25-0)^2}{6.25} + \frac{(37.5-50)^2}{37.5} + \frac{(56.25-50)^2}{56.25} = 11.11$$

and $P=0.0039$ (df=2).[3] The probability that our 100 individuals are drawn from an HWE population is only 0.39 percent, which is much lower than the conventional 5 percent limit. There is a *significant* deviation from the null expectation, a significant excess of heterozygotes, and we can confidently reject HWE. This example underlies the value of the HWE model as a means of testing for the contribution of demographic and evolutionary processes to the patterns of allelic frequencies we observe.

Both disassortative mating and selection (lethal LL) contribute to heterozygote excess in the *P. microlepis* cichlid example. In addition, there is a rare-morph advantage in this system (Hori 1993), so-called negative frequency-dependent selection (see chapter 4). Apparently, lefties and righties tend to attack their prey from opposite sides. If one of the morphs becomes very common, prey species may become more alert towards attacks from that morph's preferred side of attack. This would give the rarer morph a selective advantage (more successful attacks and hence more food and more offspring) and it would increase in frequency. Therefore, in addition to disassortative mating, negative frequency-dependent selection contributes to the maintenance of the two morphs and the occurrence of heterozygote excess as it allows the L-allele to remain at a high frequency despite lethality of the LL genotype.

3.3 Genetic drift—evolution in finite populations

3.3.1 Within-population effects of genetic drift

The second assumption in the Hardy–Weinberg model is that the population size is infinitely large,

which obviously is never the case in nature. Populations are finite and this has evolutionary consequences. In a finite population, random events have non-negligible effects on allele frequencies. Whether an individual survives to adulthood and how many descendants it leaves can have many non-deterministic causes. For instance, by chance an individual with the genotype A_2A_2 might be at the wrong place at the wrong time and die whereas an A_1A_1 individual is lucky and stays clear. The subsequent increase in frequency of the A_1 allele in that population would then be entirely due to random chance. In a large population one or a few accidental deaths would have negligible effect, but the smaller the population the larger the impact. Evolutionary change (change in allele frequency) due to random events is referred to as evolution by **genetic drift**. One important source of genetic drift is the random sampling of alleles that takes place at fertilization. Just like some parents by chance produce an excess of daughters or sons, they may produce an excess of a certain genotype. A pair of heterozygote parents $A_1A_2 \times A_1A_2$ may for instance produce offspring that are all A_1A_1. Let us consider random sampling in a very small population.

First, we assume that all the Hardy–Weinberg assumptions hold except that the population size is finite and small. Assume that we have a population of $N=8$ individuals and consider a locus with two alleles. For simplicity, we will assume that both alleles are equally common $p = q = 0.5$ and that the population is in Hardy–Weinberg equilibrium; that is, individuals mate randomly, there is no selection, mutation, or gene flow. Thus, there are 2 each of the homozygote genotypes A_1A_1 and A_2A_2, and 4 heterozygotes A_1A_2.

Now, we let the individuals mate randomly to produce the next generation; they do so to replacement, meaning another 8 individuals are produced. Think back to the start of the chapter, where we imagined mixing all the gametes of all individuals in a large container and then we randomly picked egg–sperm pairs. Inside the container, the allele frequencies in the gametes will reflect that of the population, but there will be many more gametes than individuals. In our example here, the two allele frequencies, p and q are identical so the probability of choosing either allele from our container is 0.5. In this particular example, we could therefore flip a

---

[3] The *P*-value was calculated using R.

coin to simulate randomly picking gametes from a large sample. Let heads represent A_1 and tails A_2. We suggest that you try a coin-flipping simulation of your own and note the results you get. This is the series we got from 16 coin flips ordered into pairs to represent genotypes at fertilization: HT TT HT HT TH TT TH HT. Translating them into genotypes yields 0 A_1A_1 homozygotes, 6 A_1A_2 heterozygotes and 2 A_2A_2 homozygotes. The frequency of A_1 after one simulated generation of random sampling is $p' = 6/16 = 0.375$ and that of A_2 is $q' = 0.625$. One generation of random sampling has led to a quite dramatic change in allele frequency, from $p = 0.5$ to $p'=0.375$. What about the genotype frequencies? If we calculate the chi-square test as in equation 3.2 we get $\chi^2 = 2.88$, (df = 2) and $P = 0.237$. The deviation from Hardy–Weinberg expectation is not significant. This is a general finding: the odds are favorable that in your own trial, you too found that $P \neq 0.5$ and the genotype frequencies in the new generation did not differ significantly from HWE. Random sampling of alleles in finite populations leads to *changes in allele frequencies*. However, as long as mating is random, no systematic change in genotype frequencies relative to HWE is expected. In our coin-toss trial, we happened to get a few more heterozygotes than expected under HWE, although not significantly so. You may have gotten a few more homozygotes than expected, but again most likely not significantly so.

The coin-flipping, allele-picking experiment above is an example of a binomial sampling process. Let us assume that a large number of people did the same experiment, or that we repeated it a large number of times. How would the different outcomes of allele frequencies after one generation of random sampling be distributed? Sometimes p' would be higher than the initial p, sometimes lower, and sometimes p' would be exactly equal to p. The frequency distribution of possible outcomes would be a bell-curve with mean $= p$ (Figure 3.6). The expected frequency of A_1 after one generation of random sampling is $E(p') = p$, but with some variation around the mean. The variance of such a binomial distribution is:

$$V = \frac{p(1-p)}{2N} \qquad (3.4)$$

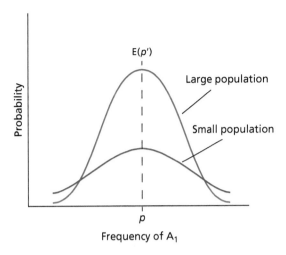

Figure 3.6 Probability distribution of allele frequencies after one generation of random sampling. The expected frequency E(p') equals the allele frequency before random sampling (p) but with a variance of V = [p(1 – p)]/2N. The probability distribution of allele frequencies the next generation is therefore wider in a small than in a large population.

It is this variance term that constitutes the source of evolution by genetic drift. Variance is inversely proportional to N, the population size. The smaller the population size the greater the variance (Figure 3.6) and thus, the greater the average change in allele frequency per generation. Conversely, when N is large the variance is small, and as N approaches infinity, V approaches zero. Genetic drift will not occur in an infinitely large population.

What will happen over time? In each generation, random sampling of alleles takes place. Sometimes an allele A_1 by chance increases in frequency from one generation to the next, other times it decreases. As time goes by and generation follows generation the frequency of A_1 will go up and down in a random dance. Will the dance ever stop? Yes. Sooner or later one thing or the other will occur and bring evolution by genetic drift to a halt: either A_1 will become fixed in the population or it becomes lost. When A_1 (or the alternative allele) becomes fixed no more evolution by genetic drift can take place because there is no more genetic variation present at this locus. Drift can only continue if a new mutation occurs or new genetic variation is brought into the population through gene flow.

The probability that a given allele becomes fixed or lost depends on the starting frequency of that allele. In the two-allele case and when A_1 and A_2 are equally common, it is equally likely that either allele is the one that becomes fixed. If A_1 is more common than A_2 it is more likely that A_1 becomes fixed (although by chance A_2 might be the "lucky" allele). The probability that an allele becomes fixed at some time in the future equals its current frequency. When A_1 is twice as common as A_2 it is twice as likely that A_1 becomes fixed rather than A_2. What is the probability that a new mutation eventually will become fixed by genetic drift? The frequency of a new mutation would be $1/(2N)$ since each of the N individuals in the population has two alleles at a locus and by definition, the new mutation is only present as a single allele. The probability that a new mutation becomes fixed is therefore $1/(2N)$. Fixation of new mutations by drift is therefore much more likely when the population size is small.

The time it takes for an allele to become fixed or lost by genetic drift is likewise proportional to population size. Evolution by genetic drift is faster the smaller the population. This can easily be seen

in simulations of genetic drift in populations of different sizes (Figure 3.7).

3.3.2 Effects of genetic drift in a structured metapopulation

Most populations are structured in one way or another. A flowering plant species may occur in habitable patches of meadow separated by inhabitable land such as marshes, woods, or agricultural land. Crosspollination is likely to occur between individual plants within the same meadow but less likely between individuals in different meadows, although occasionally a pollinator such as a bumblebee may fly from one meadow to the next, causing fertilization among individuals otherwise separated by geographical barriers. In this sense, you can begin to picture how population structure can affect the exchange of genes among populations. The plants in each given meadow constitute a population and the network of (more or less) geographically separated populations constitutes a **metapopulation**. The populations may be more or less genetically

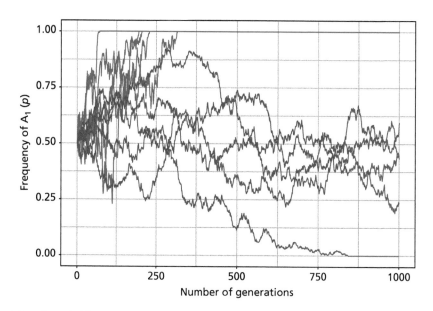

Figure 3.7 Simulations of genetic drift in a small (red lines, $N = 100$) and a large (blue lines, $N = 1000$) population. The amount of change in allele frequency per generation is greater and the expected time before one of the alleles goes to fixation is shorter the smaller the population size. In all simulations, the starting frequency of A_1 is $p = 0.5$.

isolated from one another and this isolation affects the genetic composition of the metapopulation.

It can be useful to make simple, arbitrary thought experiments to enforce clear thinking about processes which in reality are likely to be very complex. Consider first an infinitely large, unstructured population in which all the Hardy–Weinberg assumptions are met. We focus on a locus with two alleles A_1 and A_2 with frequencies p and q, so the genotype frequencies are:

$$g_{11} = p^2$$

$$g_{12} = 2pq$$

$$g_{22} = q^2$$

Then, we split this population into an infinite number of finite populations (Figure 3.8). Assume that the populations are completely isolated from one another so that there is no exchange of genes between them (no gene flow); this means that, except for the fact that population sizes are finite within each population, all the other Hardy–Weinberg assumptions are met. What will happen? Since genetic drift is operating independently in each of the populations A_1 will eventually become

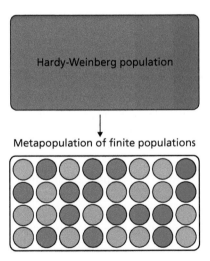

Figure 3.8 Illustration of a thought experiment. A population in which all the Hardy–Weinberg assumptions hold (above) is transformed into a metapopulation consisting of an infinite number of isolated, finite populations. Alleles segregating in the ancestral population will go to fixation by genetic drift in the finite composite populations of the metapopulation. See text for further explanation.

fixed in some of them and A_2 in the others. We just learned that the probability that an allele is eventually fixed by genetic drift equals its current frequency. Hence, assuming that the initial allele frequencies in each of the populations are p and q, A_1 will become fixed in p of the populations and A_2 will be fixed in q of them. In the (infinitely large) metapopulation *as a whole*, the frequency of the two alleles would not have changed, but because of the local fixation events in each of the isolated populations, there will eventually be no heterozygote individuals left in the metapopulation.

We have learned that, within a population, genetic drift changes allele frequencies but has no systematic effect on genotype frequencies relative to HWE. In contrast, non-random mating changes genotype frequencies away from HWE, but has, in itself, no effect on allele frequencies. Now we see that, in an infinitely large but structured metapopulation, genetic drift has an effect that resembles non-random mating (inbreeding); there is no change in allele frequency in the metapopulation but a gradual loss of heterozygosity occurs due to local fixations.

Is there any connection between inbreeding and population structuring? Well, yes there is. Although individuals mate at random within each local population, mating is non-random within the metapopulation. Within each of the finite, isolated populations the same alleles, inherited from recent common ancestors, are redistributed among the individuals. Over time, they will therefore become more closely related to each other, genetically speaking, than to random members of the metapopulation. Population structuring can crucially be seen as a subtle form of inbreeding since mating is non-random in the metapopulation, albeit random at a local scale.

3.3.3 F_{ST}—the fixation index

The expected loss of heterozygosity relative to HWE in a metapopulation can be used to measure genetic divergence between populations. F_{ST}, the fixation index, can be defined[4] as:

---

[4] The F_{ST} definition in equation 3.5 is not the only way to define the statistic, but it is the easiest to calculate by hand and to grasp intuitively. Refined formulas take into account biases caused by differences in sample size and more.

$$F_{ST} = \frac{H_T - H_S}{H_T} \qquad (3.5)$$

It is structurally identical to F_{IS} but includes other estimates of heterozygosity. We commonly compute F_{ST} between population pairs although it is possible to use more populations. For simplicity, here we focus on an example with two populations only. H_T is the expected heterozygosity according to the Hardy–Weinberg model, assuming random mating in the total metapopulation (the two populations combined), whereas H_S is the average expected heterozygosity assuming random mating only within the populations. Let us take a numerical example. We genotype 100 individuals of a fish species from two populations and obtain the following genotypes (Table 3.2).

In Dauphin Lake, the frequency of A_1 is

$$p_1 = (2\cdot4 + 32) / 200 = 0.2(q_1 = 0.8), \text{ and}$$
in Lake St Martin

$$p_2 = (2\cdot16 + 48) / 200 = 0.4(q_2 = 0.6).$$

The frequency of A_1 in the total metapopulation is thus

$$p_T = (p_1 + p_2) / 2 = (0.2 + 0.4) / 2 = 0.3,$$
and that of A_2 is

$$q_T = 1 - p_T = 0.7.$$

We can now calculate the expected heterozygosity in the metapopulation:

$$H_T = 2p_T q_T = 2\cdot0.3\cdot0.7 = 0.42.$$

In Dauphin Lake, the expected heterozygosity under HWE is

$$H_{S1} = 2p_1 q_1 = 2\cdot0.2\cdot0.8 = 0.32,$$
and in Lake St Martin it is

$$H_{S2} = 2p_2 q_2 = 2\cdot0.4\cdot0.6 = 0.48.$$

Table 3.2 Number of individual fishes of each genotype sampled from two different lake populations.

Population	A_1A_1	A_1A_2	A_2A_2
Dauphin Lake (1)	4	32	64
Lake St Martin (2)	16	48	36

The average expected heterozygosity when mating is random within the two populations is thus:

$$H_S = (H_{S1} + H_{S2}) / 2 = (0.32 + 0.48) / 2 = 0.4.$$

Now we can calculate the fixation index:

$$F_{ST} = (H_T - H_S) / H_T = (0.42 - 0.4) / 0.42 = 1 / 21 \approx 0.048$$

F_{ST} ranges from 0 when two populations have equal allele frequencies (i.e. there is no difference in expected heterozygosity at the sub or total population levels), to 1 when the two populations are fixed for different alleles. An F_{ST} of 0.048 is modest: a 4.8 percent reduction in heterozygosity in the total population due to population structuring. Another way to interpret the statistic is as the proportion of genetic variance that is partitioned among the subpopulations. So, in our simple example, 4.8 percent of the genetic variance differs among lakes. F_{ST} is a much-used statistic in evolutionary genetics. It can seem a little difficult to grasp conceptually at first, but it has many practical applications. For example, it is used as a measure of genetic differentiation between populations and species in studies of population structure, and to identify genes or genomic regions subjected to divergent natural selection (see section 3.4 and chapter 7).

3.3.4 Empirical studies of population structure

The example above of an infinitely large, structured metapopulation is of course highly arbitrary. In many cases, it can be useful to find out how real populations are structured, for instance for conservation purposes or as part of a research program on a given organism. Some organisms may be structured into relatively discrete, easily identified populations, with limited gene flow occurring between them. Organisms inhabiting islands provide one example. In such cases calculation of pairwise F_{ST} between local populations can provide valuable information on population structure and evolutionary history. Other organisms have a more continuous distribution. Such organisms can be expected to exhibit a pattern of **isolation by distance**. Individuals are likely to mate with individuals in their geographic vicinity but less likely to mate with individuals far away. Even without any geographical barrier this will lead to a pattern of increasing genetic differentiation with geographic distance. Isolation by distance can be analyzed by

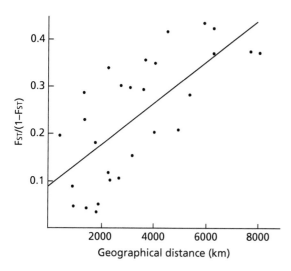

Figure 3.9 Isolation by distance in a ring around the Himalayas and the Tibetan plateau in the greenish warbler (*Phylloscopus trochiloides*). Genetic differentiation $F_{ST}/(1-F_{ST})$ at 2334 SNP markers increases with geographic distance around the ring. Redrawn from Alcaide et al. (2014).

plotting genetic distance, such as $F_{ST}/(1-F_{ST})$ among population pairs, as a function of the geographic distance between those pairs. As an example, Alcaide et al. (2014) studied pair-wise F_{ST} between populations of the greenish warbler (*Phylloscopus trochiloides*) around the Himalayas and the Tibetan plateau using biallelic SNP-markers. Genetic distance increases with geographical distance between the populations (Figure 3.9). Interestingly, however, the greenish warblers are actually a form of "ring" species. In Siberia two genetically distinct greenish warblers meet: *P. t. viridanus* and the two-barred warbler *P. plumbeitarsus*. These two are regarded as separate species by taxonomists because there are strong barriers to gene exchange between them and thus very little gene flow occurs. However, the two forms are connected through a ring of interbreeding populations around the Himalayas and the Tibetan plateau. Moving south from western Siberia and around the Tibetan plateau the populations become progressively more genetically dissimilar from *P. t. viridanus* and more similar to *P. plumbeitarsus* (Figure 3.10). A likely interpretation is that an ancestral southern population (*P. t. trochiloides*) has expanded northwards in two

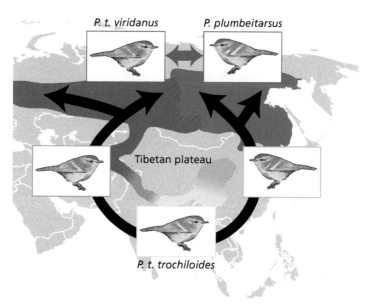

Figure 3.10 Evolutionary scenario that can explain the pattern of genetic variation in the greenish warbler species complex (the greenish warbler *Phylloscopus trochiloides* and the two-barred warbler *P. plumbeitarsus*). An ancestral southern population *P. t. trochiloides* (yellow) expanded northwards on either side of the Tibetan plateau and the two lineages gradually diverged from each other into *P. t. viridanus* (blue) and *P. plumbeitarsus* (red) as they expanded northwards. Eventually, the two lineages experienced secondary contact in Siberia. The map shows the current distribution of the birds and the likely routes of expansion with black arrows. Modified from Irwin et al. (2005).

directions—west and east of the Tibetan plateau. These expanding populations gradually developed isolation by distance from the mother population. Then, when the expanding warblers finally met north of the Tibetan plateau in Siberia, they had differentiated to the extent that barriers to gene exchange had evolved between them.

Sometimes delimitation of populations is difficult and geography can be a poor predictor of actual populations. One way to circumvent this problem is to statistically infer population delimitation based on the individual's genotypes. Statistical software such as STRUCTURE (Pritchard et al. 2000) identifies likely populations from multilocus genotype data and assigns individuals to these populations based on statistical inference according to criteria that maximize Hardy–Weinberg equilibrium and linkage equilibrium (see

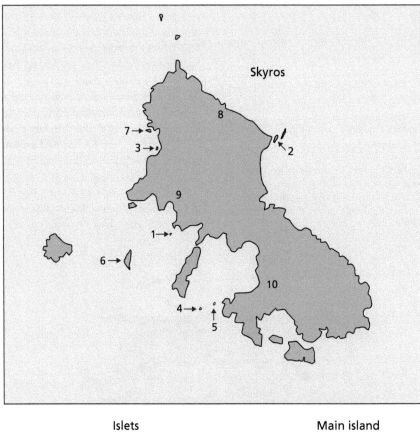

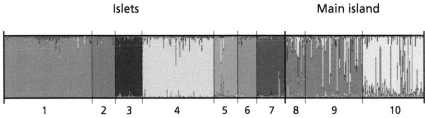

Figure 3.11 Genetic structure in the Skyros wall lizard *Podarcis gaigeae*. Seven populations on islets surrounding the island of Skyros show clear genetic structure and are assigned as distinct populations in a STRUCTURE analysis, whereas the three populations on the main island are genetically much more homogeneous and only show weak structuring. Each colored bar represents the probability that an individual is assigned to population 1 to 9 ($K = 9$). Modified from Runemark et al. 2011.

chapter 6) within, but not between, inferred populations. As an example, Runemark et al. (2011) studied the population structure of the Skyros wall lizard (*Podarcis gaigeae*), which is endemic to the island of Skyros of Greece and adjacent islets. A STRUCTURE analysis based on several microsatellite markers showed that each of the islets harbors a well-defined population whereas sampling sites on the main island showed much less clear structuring (Figure 3.11). On the main island the analysis suggested that two populations are present, but these seem to grade into each other, consistent with a pattern of isolation by distance. Open water between the islets and the main island most likely reduce gene exchange, causing the populations to diverge by genetic drift, whereas the absence of geographic barriers on the main island promotes gene exchange and causes these populations to diverge more slowly. We will return to analysis of population structure in more detail in chapter 9.

3.4 Other deviations from the Hardy–Weinberg model

3.4.1 Selection and the Hardy–Weinberg model

Natural selection is a highly important process that can have a range of effects on genotype and allele frequencies, causing strong deviation from HWE. Given the importance of this process to evolutionary genetics, we have devoted two chapters to the analysis of the topic (chapters 4 and 5). A few important points can, however, be made here. In many cases natural selection will favor one allele over the other(s). Genotypes containing the favored allele will on average leave more descendants than alternative genotypes that lack it. This has the effect that the favored allele goes to fixation in the population. Fixation of alleles by natural selection will typically be much faster than when evolution is only caused by random events (i.e. genetic drift). Given that beneficial, adaptive mutations are likely to be rare (see chapter 2) and that the effects of selection can be rapid, in many cases we are unlikely to be able to directly observe the deviations from HWE at loci that natural selection causes. Instead, natural selection at a locus will manifest itself as a factor that causes populations to diverge from one

another at a faster rate at the locus under selection compared to loci only affected by genetic drift. Selection typically has a local effect in the genome. That is, it will affect specific genes (and loci physically linked to them) rather than the genome as a whole. This is because of recombination; crossing over becomes progressively more likely to occur between sites in the genome that are physically distant—i.e. further away from each other on a chromosome (see chapters 2 and 6). Consequently, by comparing population differentiation at a large number of loci, or across the entire genome, selection can leave footprints in the form of local peaks of differentiation in the genome, such as a locally much higher F_{ST}-value around the locus targeted by selection compared to the genomic average. We shall explore this effect of selection further in chapter 7.

Some forms of natural selection will, however, yield within-population deviations from HWE and can thus be detected by using the F_{IS} coefficient. The peculiar case of the laterally dimorphic *P. microlepis* cichlids of Lake Tanganyika mentioned in section 3.2.2 provides one example. Whenever heterozygotes have higher fitness than the homozygotes there would be an excess of heterozygotes relative to HWE (negative F_{IS}). Similarly, when heterozygotes are selected against and both homozygotes favored (disruptive selection) there would be heterozygote deficit at the locus under selection (positive F_{IS}).

3.4.2 Mutation and the Hardy–Weinberg model

Mutations are relatively infrequent. For instance, in the human genome the rate of mutation has been estimated to be about 1.2×10^{-8} per nucleotide site per generation (Scally & Durbin 2012). Such mutation events will therefore have negligible effects on within-population genotype frequencies (non-significant deviations from HWE). The importance of mutation is rather that it creates new genetic variation that selection and genetic drift can act on. In the absence of mutation, genetic drift and selection would rapidly deplete natural populations of genetic variation. We shall explore the interplay between mutation and genetic drift further in Chapter 7.

3.4.3 Gene flow and the Hardy–Weinberg model

Gene flow is the result of interbreeding between resident individuals and immigrants from a different population. There are some paradoxes concerning gene flow and the Hardy–Weinberg model. First, "no gene flow" is one of the basic assumptions of the Hardy–Weinberg model. Yet, as long as mating between immigrants and residents is random, gene flow does usually not cause much within-population deviation from HWE. Second, gene flow acts towards bringing a structured metapopulation into Hardy–Weinberg equilibrium and it counteracts isolation by distance. We shall explore the within-population effects of gene flow and its homogenizing effect between populations in a moment. But first, why is the exclusion of gene flow one of the basic assumptions of the Hardy–Weinberg null model? This is because in an ideal Hardy–Weinberg population, the genotype frequency in the next generation would be the same as in the previous generation and it will remain the same in subsequent generations. Much like mutation, gene flow can introduce novel genetic variation into natural populations that natural selection and genetic drift can then act upon to cause evolutionary change. It is mainly in this sense that gene flow violates the ideal null model of Hardy–Weinberg and therefore needs to be excluded.

Let us first convince ourselves that gene flow usually does not cause much deviation from HWE within a population. The Hardy–Weinberg expectation within a population is a simple mathematical relationship between allele and genotype frequency. Let us assume that population 1 initially was fixed for the A_1-allele, but started to receive a very few immigrants from population 2 each generation some time ago. We assume that the other population is fixed for A_2. Assume further that mating is random between residents and immigrants. At a given time, we may observe that, say, 5 percent of the alleles in population 1 are derived from rare immigrants from population 2, that is, $p = 0.95$ and $q = 0.05$. If mating is random we expect the genotype frequencies:

$$E(g_{11}) = p^2 = 0.95^2 = 0.9025$$

$$E(g_{12}) = 2pq = 2 \cdot 0.95 \cdot 0.05 = 0.095$$

$$E(g_{22}) = q^2 = 0.05^2 = 0.0025$$

As long as mating between immigrants and residents is random we should not observe any systematic deviation from these expectations. However, some small deviations may occur, for instance if we included some recently arrived A_2A_2 immigrants in our sample (a slight heterozygote deficit); but this deviation would be marginal as long as the number of immigrants per generation is small relative to the size of the resident population. Things change however, if mating is assortative between residents and immigrants, which is sometimes the case in nature. When this occurs, a deviation from HWE (heterozygote deficit) is expected but this would be due to non-random mating and not to migration per se.

One general effect of gene flow is that it opposes differentiation by genetic drift. Let us return to our infinitely large metapopulation, structured into an infinite number of finite populations, from section 3.3.2 (Figure 3.12). Again, we focus on one locus with two alleles A_1 and A_2 with respective frequencies of p and q. In the absence of gene flow between the populations genetic drift will eventually lead to fixation of A_1 in p of the subpopulations and of A_2 in q of them. If on the other hand gene flow between the populations is sufficiently high, mating is essentially random in the entire metapopulation; all the populations would have the same allele frequencies (p and q) and F_{ST} in the metapopulation would be zero. Between these two extremes we could imagine that

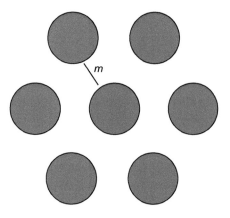

Figure 3.12 Illustration of Sewall Wright's model of gene flow between populations in a metapopulation. A number m of migrants is exchanged between each population each generation. See text for further explanation.

CHANGES IN ALLELE AND GENOTYPE FREQUENCY

the diversifying effect of genetic drift and the homogenizing effect of gene flow would cancel each other out so that an equilibrium between the two forces is established. In other words, as alleles are lost from subpopulations due to drift, migration from one of the other populations will reintroduce them. Assuming that all the populations are of equal size N and that any population is equally likely to receive a migrant m from any of the other populations we have the following approximate relationship between F_{ST} and migration at drift/migration equilibrium:

$$\hat{F}_{ST} \approx \frac{1}{1+4Nm} \qquad (3.6)$$

The derivation of this formula involves some algebra and approximations that we choose to omit here for simplicity. This is a famous theoretical result derived by Sewall Wright, but it must be interpreted with care. It is based on some very specific assumptions and it should therefore not be used directly as a predictor of the amount of gene flow occurring between natural populations. For instance, F_{ST} between populations may be low even in the absence of current gene flow if they became isolated only very recently. The populations may simply not be at drift/migration equilibrium. Moreover, real populations may vary in size and to the extent that they receive migrants from or contribute migrants to the metapopulation. Finally, selection may be occurring that would affect F_{ST}.

3.5 The gene pool

In every population, a number of alleles at a number of different genes are segregating over generations. This is the **gene pool**. Alleles can be lost from the gene pool by selection and/or genetic drift, and alleles can be added to it through mutation and gene flow. Breeding individuals constitute the link between generations. Over time the composition of the gene pool changes and not necessarily in parallel across loci.

3.5.1 Hardy–Weinberg and real populations

So far, we have explored the impact of violating each of the Hardy–Weinberg assumptions in turn. Non-random mating has an immediate effect on geno-

type frequencies relative to HWE, genetic drift affects allele frequencies and causes populations to diverge, selection can have a range of different effects on genotype and allele frequencies depending on its nature, whereas mutation and gene flow are sources of novel variation that selection and drift can act upon to yield evolutionary change. In real populations however, these factors do not operate in isolation. Take for example the scale-eating cichlid *Perissodus microlepis* in Lake Tanganyika. Both non-random mating (disassortative) and selection (negative frequency-dependent as well as selection against the LL genotype) appear to affect the allele frequencies of the dexterity locus in this species. Furthermore, in a finite population a locus affected by natural selection will simultaneously be affected by genetic drift. A favorable mutation may accidentally be lost by genetic drift when it is still rare. In a small population, genetic drift can be such a powerful force that it overrides selection. By chance, an allele can drift to fixation even if it is selected against on average. Selection is therefore a much more effective force the larger the population size.

The effect of the various demographic and evolutionary forces will also vary along the genome. As we have seen, assortative mating and selection can have locus-specific effects. Due to recombination, the impact of these forces will rapidly diminish as we move away from the targeted site(s). In this way, it is useful to think of recombination as a means for loci to escape the effects of these processes and to segregate independently of nearby genome regions. In contrast, inbreeding and outbreeding are mating rules that will have genome-wide effects.

3.5.2 Effective population size N_e

The total number of individuals in a population can be a poor predictor of the size of the gene pool. This is in part because individuals vary in reproductive output and some individuals don't reproduce at all. The alleles carried by non-reproducing individuals do not contribute to the gene pool of the population. Accordingly, genetically speaking, the population size is effectively smaller than the census size. Moreover, the number of alleles segregating in a population can vary between loci.

First, not all loci in the genome are diploid. Many organisms have different sex chromosomes. In mammals, the X-chromosome carries many more genes than the Y-chromosome and genes at the two different sex chromosomes generally do not recombine; they evolve independently. A male only has one X chromosome. Therefore, there are only three copies of an X-linked locus segregating in a population per four copies of an **autosomal** locus, where both sexes have two copies. Likewise, since only males carry Y-chromosomes there is only one copy of a Y-linked locus per four copies of an autosomal locus. Finally, organelles such as mitochondria and chloroplasts are maternally inherited. Sons inherit mitochondria (and in plants, chloroplasts) from their mother, but they do not transfer their organelles to their own offspring. Like in Y-linked loci there is therefore one copy of an organelle-DNA locus per four copies of an autosomal one. Because of the lower number of copies segregating at sex-linked and organelle loci compared to autosomal ones, genetic drift will be faster at the former loci (e.g. Charlesworth et al. 1987).

Second, a locus under directional selection would in effect have fewer alleles segregating than a locus not affected by selection averaged over generations. This is because alleles are being selectively removed each generation at a locus experiencing selection. Again, because of recombination, the effect of selection will weaken with physical distance (in the genome) from the locus targeted by selection.

Effective population size is an important estimator of the size of the gene pool (Wright 1938). We denote the effective population size as N_e and it is defined as the number of breeding individuals in an idealized population that would show the same amount of genetic drift as the population under consideration (Wright 1938). Here, an idealized population is one in which mating is random, all individuals contribute equally to reproduction, the sex ratio is equal, the population size is constant, no selection occurs, and generations are non-overlapping. In nature, effective population size N_e is nearly always smaller than the census size N. Deviations from the assumptions of the idealized population are the reason that the effective population size and the census size are not equivalent. Note that the term *effective population size* is somewhat unfortunate. Because the number of copies

segregating differs between autosomal, sex-linked, and organelle loci, and between loci affected by selection or not, the effective population size varies between different loci. We can therefore think of effective population size as a statistical measure of how many copies of a locus actually are segregating in a population (contributing to the gene pool), expressed in terms of number of genetically contributing individuals. The effective number of gene copies in circulation would thus be $2N_e$ for an autosomal gene, since each individual carries two gene copies.

We can use the expected variance caused by random sampling of alleles (genetic drift) to define N_e. From equation 3.4 we have that the variance in allele frequency after one generation of random sampling in an idealized population is $V_{ideal} = \frac{p(1-p)}{2N}$. Let V_{actual} denote the same, typically larger, variance of the actual population under consideration (remember that variance increases as population size decreases). The effective population size N_e would be the size of an idealized population in which $V_{ideal} = V_{actual}$. We can therefore find the effective population size by substituting V_{actual} for V_{ideal} and solve for N, which gives:

$$N_e = \frac{p(1-p)}{2V_{actual}} \qquad (3.7)$$

For example, let us say we are studying lab populations of *Drosophila*. In each of 100 flasks we keep 6 males and 6 females, so the census size is $N=12$ in each. In each flask, we start with allele frequencies of, say $p = q = 0.5$ at a given locus. We let the flies reproduce freely and genotype the allele frequencies of their offspring to calculate the variance of the resulting allele frequency after one generation p' and obtain $V_{actual} = 0.015625$. Then, according to equation 3.6 we can calculate the effective population size of these *Drosophila* flies as $N_e = [0.5(1–0.5)]/(2·0.015625) = 8$. Maybe some of the flies in the various flasks failed to reproduce, maybe some pairs produced significantly more offspring than others, or maybe some other assumption of the idealized population was violated. In any case, the hypothetical experiment illustrates the general finding that N_e is usually smaller than N. In chapter 9 we shall see that we can estimate effective population size from data on genetic variation.

Study questions

1. The Hardy–Weinberg model hardly applies to any natural population. Why is it nevertheless useful?

2. Review the similarity and difference between inbreeding and assortative mating and outbreeding and disassortative mating. What are the genetic consequences of these forms of nonrandom mating?

3. Explain the following concepts:
   a. Population structure.
   b. Isolation by distance.
   c. Outcrossing vs outbreeding.
   d. Genetic drift vs gene flow.

4. You genotype a species of grasshoppers along a north–south transect across the European Alps. Near Munich, Germany, north of the Alps, you sample 120 individuals; near Innsbruck, Austria, within the Alps, you sample 122 individuals; near Verona, Italy, south of the Alps, you sample 118 individuals. You find the following number of each genotype.

	A_1A_1	A_1A_2	A_2A_2
Munich	6	33	81
Innsbruck	20	59	43
Verona	65	39	14

   a. Calculate the frequencies of the two alleles in each population.
   b. Are there any deviations from Hardy–Weinberg expectation in any of the populations?
   c. Calculate F_{IS} for each of the populations. How do the results fit your findings in (b)? How would you interpret the results?
   d. What are you measuring with F_{IS}?
   e. As you found in (a) the allele frequencies differ between the three populations. Calculate F_{ST} between Munich–Innsbruck, Munich–Verona, and Innsbruck–Verona.
   f. What do the F_{ST} results tell you about the relationship between the three populations? How do they fit the geography/topography? Remember that the Alps were covered in glaciers during the last ice age and that large parts of Northern Europe were uninhabitable for animals adapted to temperate climate.

5. You have genotyped 500 individuals from population A and B at a locus with two alleles. You find that $p = 0.1$ in population A and $p = 0.8$ in population B. Calculate F_{ST} between the two populations.

6. Discuss how F_{IS} and/or F_{ST} could be affected in the following circumstances:
   a. In a zone in which two subspecies recently have come together. The two subspecies interbreed freely (random mating) but hybrids have lower survival than "pure" individuals.
   b. An island population that receives migrants from a continental population. Assume that there is random mating between residents and immigrants.
   c. A mainland population in which locus A causes assortative mating, so that A_1A_1 and A_1A_2 tend to mate with each other (A_1 is dominant), as do A_2A_2 individuals. Contrast locus A with the unlinked locus B also with two alleles that neither affects assortative mating nor fitness (it is neutral). Assume further that a geographically isolated population exists on an island (no current gene flow between the mainland and island) where the two loci behave in exactly the same way: A affects assortative mating, B is neutral.

7. Discuss whether these claims are true or false, and why:
   a. In a small population, genetic drift will lead to a reduction in the frequency of heterozygotes relative to the Hardy–Weinberg expectation.
   b. A small population can experience inbreeding depressions even if mating is random in that population.
   c. A population that has been inbred a long time will suffer more from inbreeding depression than one that has been inbred only a short time.
   d. Gene flow would normally yield an excess of heterozygotes relative to the Hardy–Weinberg expectation.
   e. When an allele has gone to fixation in a population there are no heterozygotes left, yet there is no deviation from Hardy–Weinberg expectation in this case.

f. Given that a novel neutral mutation becomes fixed in a population it will on average take twice as long time in a population of 200 individuals as in one of 100 individuals.

g. If the frequency of an allele is p at a given moment in time it will on average be p also after one generation of random sampling. Accordingly, on average genetic drift does not occur.

CHAPTER 4

The theory of natural selection

Natural selection is the scientific explanation for the evolution of adaptations. Wonders of the living world, such as the anatomy and physiology that grants the cheetah its unchallenged running speed; the seductive colors and scents of a flower that are irresistible to its pollinators; and the accuracy and sophistication of sense organs such as the human eye are the ultimate results of this one creative force in evolution. In this chapter, we shall investigate some simple models of natural selection to explore its power in causing evolutionary change. Natural selection is all about differential fitness of biological entities. Hence, we start by defining this key concept.

4.1 Fitness

4.1.1 Defining fitness

> "I have two kids; my brother-in-law has four. Does he have higher fitness than me?"

Surprisingly few, even those acquainted with the concept of fitness, will answer this question both correctly and for the right reason. The answer is most certainly no. Darwinian fitness is not an individual attribute, or, at least, it is not useful to define it in that way in most cases. We need a definition of fitness that contrasts natural selection with genetic drift. As we saw in chapter 3 an allele can increase in frequency by chance. Likewise, chance and circumstance may determine whether my brother-in-law or I leave more descendants; that is, chance and circumstance may play a role in determining the relative success of our respective alleles. Difference

in reproductive output between individuals *per se* does not bring us any closer to understanding the evolution of adaptations. Rather, it is the *characteristics* of individuals or other biological entities (e.g. genes or family groups) that matters, in the sense that these characteristics can have causal effects on survival and/or fecundity. In population genetics, we commonly assign fitness to genotypes; more precisely, we assign fitness to the phenotypic effect these genotypes have. Individuals with the genotype A_1A_1 will have a higher fitness than those with A_2A_2 if they produce more surviving offspring on average *because* the phenotypic characteristics their genotype specify affect survival and/or fecundity in a causal way. An example of this is a flowering plant with two alleles that code for different flower color pigments. If the color of A_1A_1 individuals is more seductive and thus attracts more pollinators on average than that of the A_2A_2 individuals, the A_1A_1 individuals will produce on average more offspring. However, a randomly chosen A_1A_1 may still attract fewer pollinators than a randomly chosen A_2A_2 individual. You can see then why posing a question about fitness between individuals is the wrong way to think about the concept—fitness is determined by whether individuals exhibiting a characteristic on *average* produce more offspring, due to a causal relationship between that character and survival/fecundity.

Fitness is, however, not just restricted to discrete entities such as genotypes—it can vary with continuous traits too. In such cases it is more useful to assign a **fitness function** to the phenotype that describes the relationship between reproductive

Evolutionary Genetics: Concepts, Analysis, and Practice. Glenn-Peter Sætre & Mark Ravinet, Oxford University Press (2019).
© Glenn-Peter Sætre & Mark Ravinet 2019. DOI: 10.1093/oso/9780198830917.001.0001

output and the continuously varying trait. In many fish species, for instance, there is a positive relationship between female body size and fecundity (e.g. Beldade et al. 2012). We can classify forms of natural selection according to the shape of the fitness function relative to such trait values (Figure 4.1). **Stabilizing selection** occurs when extreme trait values at both ends of the distribution are selected against and intermediate trait values are favored, **disruptive selection** occurs when intermediate trait values are selected against and trait values at both ends of the distribution are favored, and **directional selection** occurs when trait values that are higher (or lower) than the current population mean are favored.

Phenotypic traits are determined by both genetic and environmental factors. We will not restrict our definition of fitness to only genetic variation. In Figure 4.1, the fitness function varies with phenotype and hence incorporates both phenotypic and genetic contributions. However, it is important to remember that fitness differences between phenotypes will only result in evolutionary change if this phenotypic variation has a heritable component.

Sometimes fitness differences between genotypes or phenotypes can best be described as differences in survival, other times as differences in fecundity. Survival, say from hatching to adulthood, and fecundity are examples of **fitness components**. Total fitness of a biological entity (e.g. individuals with genotype A_1A_1) is its *per capita* growth rate R, or **reproductive success**. This is basically the expected number of fertile surviving offspring that a given entity produces, which can be estimated by measuring the mean

reproductive success of a reasonably large sample of members of the entity. We can assign fitness to different entities that share some phenotypic property, including genes, individuals, family groups, and populations. The majority of biological adaptations appear to be individual adaptations. That is, they increase the reproductive success of the individuals that carry them. Natural selection shapes the characteristics of individuals, and secondarily those of other biological entities, such as groups or populations, to make them good at surviving and reproducing under prevailing conditions (Maynard Smith 1989).

4.1.2 Absolute, relative, and marginal fitness

Absolute fitness concerns the expected reproductive success of individual members of a biological entity (e.g. a specific genotype). However, in population genetics our primary concern is evolutionary change and equilibria. It is therefore often more convenient to operate with **relative fitness**: how well one genotype fares relative to other genotypes in the population and relative to the population mean. We will use w_{ij} to denote the relative fitness of genotype A_iA_j. This is a measure of the fitness of that genotype relative to the genotype in the population with highest fitness, which we set equal to 1. For example, if the genotypes A_1A_1 and A_1A_2 both on average produce 4 fertile, surviving offspring, whereas A_2A_2 produces 3, then $w_{11} = w_{12} = 4/4 = 1$ and $w_{22} = 3/4 = 0.75$. Another important parameter is the **mean population fitness**, which we denote $\bar{w}$. Let us say that at a given time 70 percent of the population is A_1A_1, 20 percent is A_1A_2, and 10

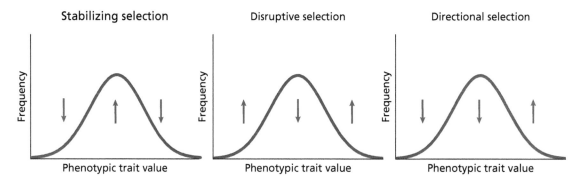

Figure 4.1 Three forms of natural selection. Under stabilizing selection, intermediate trait values are favored and trait values above and below the current population mean are selected against. Under disruptive selection, trait values above and below the current population mean are favored and intermediates are selected against. Under directional selection, trait values above (or below) the current population mean are favored. Arrows indicate selection for (up) or against (down) a trait value.

percent is A_2A_2 and their respective relative fitness are as above (1, 1, and 0.75 respectively), then $\bar{w} = 0.7 \cdot 1 + 0.2 \cdot 1 + 0.1 \cdot 0.75 = 0.975$.

Alleles can certainly contribute to fitness, but because selection acts on phenotypes specified by genotypes we cannot directly assign fitness to alleles. As we saw in the example above, the effect on fitness from either allele depends on which genotype they are in. A_2 was associated with high fitness in heterozygotes (1), but lower fitness in homozygotes (0.75). We will use the term **marginal fitness** for an allele i, denoted as w_i^* (Rice 2004) to signify the average fitness of that allele in the population. The marginal fitness of an allele depends both on the fitness of the various genotypes it is contained in and the frequency of those genotypes. With the genotypic fitnesses given above, the marginal fitness of A_2 would be high when the frequency of heterozygotes in the population is high ($w_{12} = 1$) compared to the frequency of A_2A_2 homozygotes ($w_{22} = 0.75$), but lower when most carriers of A_2 are homozygote. Let p and q be the allele frequencies of A_1 and A_2 respectively. Assuming random mating, the marginal fitness of the two alleles is:

$$w_1^* = pw_{11} + qw_{12}$$
$$w_2^* = pw_{12} + qw_{22}$$

$$(4.1)$$

In other words, the marginal fitness of allele A_1 is the probability that it is together with another A_1 allele (which equals the frequency of that allele (p) when we assume random mating) multiplied by the fitness of the resulting genotype (w_{11}), plus the probability that it is rather together with an A_2 allele (q) multiplied by the fitness of that genotype (w_{12}).

4.2 One-locus model of viability selection

4.2.1 The model

Let us consider a locus with two alleles A_1 and A_2 with respective frequencies p and q ($q = 1 - p$). We assume that the three genotypes A_1A_1, A_1A_2, and A_2A_2 have different **viability**. This simply means that individuals with different genotypes have different probabilities of surviving from birth to adulthood. Besides allowing the genotypes to differ in fitness we shall assume that other Hardy–Weinberg conditions are met (infinite population size, random mating etc.; see chapter 3). Further, we assume

Table 4.1 The one-locus viability selection model.

Genotype	Frequency of zygotes	Fitness
A_1A_1	p^2	w_{11}
A_1A_2	$2pq$	w_{12}
A_2A_2	q^2	w_{22}

that the genotypic fitnesses are constant and can be represented as relative fitness with a number, denoted w_{11}, w_{12}, and w_{22} as above. We assume a population with discrete, non-overlapping generations. Thus, the life cycle includes zygotes that are in Hardy–Weinberg equilibrium followed by a phase of differential survival (fitness), after which the surviving adults mate at random to produce the next generation of zygotes (Table 4.1).

The one-locus viability selection model describes the frequency of A_1 after one generation of selection (p_{t+1}), and depends on its frequency before selection (p), the marginal fitness of A_1 (w_1^*), and to what extent this marginal fitness is higher or lower than the population average ($\bar{w}$). We get the mean population fitness by summing the product of the frequency of each of the three genotypes before viability selection (the Hardy–Weinberg frequencies) with their respective fitness, thus:

$$\bar{w} = p^2w_{11} + 2pqw_{12} + q^2w_{22}$$

$$(4.2)$$

The general one-locus viability selection model can thus be formulated as:

$$p_{t+1} = \frac{pw_1^*}{\bar{w}}$$

$$(4.3)$$

If on average the allele A_1 performs better in terms of fitness than the population mean, it will increase in frequency. You can gain considerable insight by playing with different values of the model parameters, such as the start frequencies of the alleles, and the strength of selection (the differences in genotypic fitness). R and similar programming languages allow you to plot p_{t+1} over successive generations and draw graphs to demonstrate how the frequencies of the alleles change through time. In fact, we have a similar example in the practical course that accompanies this book. Below, however, we shall explore some general properties of the model. Note, we strongly recommend that you work out the algebra between successive steps in the formulas given below to make sure that you understand the operations.

4.2.2 Finding all possible equilibria using Δp

At an equilibrium there is by definition no change in allele frequency from one generation to the next. In other words, at equilibrium $p_{t+1} = p$. Let us introduce the dynamic parameter Δp, which we define as $p_{t+1} - p$, that is, the change in allele frequency per generation. At an equilibrium $\Delta p = 0$. Using equation 4.3 for p_{t+1}, we can deduce that:

$$\Delta p = \frac{pw_1^*}{\bar{w}} - p = \frac{p(w_1^* - \bar{w})}{\bar{w}} \quad (4.4)$$

This formula (equation 4.4) is not particularly easy to interpret, so we shall manipulate the terms to make the equation more informative. One problem is that we have both marginal fitness and mean population fitness in the nominator. However, if we compare the formula for mean population fitness (equation 4.2) with those for marginal fitness (equation 4.1), we see many similarities—i.e. the weighting of genotype frequencies by relative fitness estimates. With a little bit of algebra, we can express one with the other. Notice the little trick we used with the second component of equation 4.2, $2pqw_{12}$, below:

$$\bar{w} = (p^2w_{11} + pqw_{12}) + (pqw_{12} + q^2w_{22})$$
$$\bar{w} = p(pw_{11} + qw_{12}) + q(pw_{12} + qw_{22})$$
$$\bar{w} = pw_1^* + qw_2^* \quad (4.5)$$

Hence, equation 4.4 can be expressed as:

$$\Delta p = \frac{p(w_1^* - pw_1^* - qw_2^*)}{\bar{w}} = \frac{p[w_1^*(1-p) - qw_2^*]}{\bar{w}}$$
$$= \frac{pq(w_1^* - w_2^*)}{\bar{w}} \quad (4.6)$$

From equation 4.6 it is easy to see that $\Delta p = 0$ when either:

$$p = 0 \quad \text{or}$$
$$q = 0 \quad \text{or}$$
$$w_1^* = w_2^*$$

This shows that the model can have a maximum of three equilibria: an internal equilibrium where the two alleles on average do equally well in terms of fitness, and two equilibria in which either of the alleles is fixed and the other is absent. We now need a procedure to determine which of these equilibria are stable or unstable.

4.2.3 Invasion fitness analysis

A useful tool when analyzing selection models, including our one-locus viability selection model, is to compare the marginal fitness of a rare mutant (invasion fitness) with mean population fitness when the population is almost fixed for the alternative allele (resident fitness) to determine if the external equilibriums ($p = 0$ and $p = 1$) are stable or unstable. In other words, is the rare mutant able to invade and establish itself in a population? Let us take a straightforward example. Assume that the genotypic fitnesses are as follows:

$$w_{11} = 0.7$$
$$w_{12} = 1$$
$$w_{22} = 0.8$$

In this case, the heterozygote has a higher fitness than either homozygote (heterozygote advantage). Assume also that the resident population is fixed for A_1 except for some extremely rare carriers of the mutation A_2. From equation 4.2, resident fitness (mean population fitness) is $\bar{w} = p^2w_{11} + 2pqw_{12} + q^2w_{22}$. When $q \approx 0$ and $p \approx 1$ we have resident fitness $\bar{w} \approx w_{11} = 0.7$. The invasion fitness of A_2 equals its marginal fitness, $w_2^* = pw_{12} + qw_{22}$. When $p \approx 1$ and $q \approx 0$ we have invasion fitness $w_2^* \approx w_{12} = 1$. Thus, the invasion fitness is higher than the resident fitness and A_2 will increase in frequency in the population. Clearly then, $p = 1$ ($q = 0$) is not a stable equilibrium in this case; it can be invaded by individuals carrying A_2. Likewise, if the population is fixed for A_2, resident fitness would be 0.8 and invasion fitness 1, so $p = 0$ ($q = 1$) is also an unstable equilibrium as it can be invaded by individuals carrying A_1. A simple conclusion from this example of heterozygote advantage is that both alleles have to be present for a stable equilibrium to exist, and on average they must do equally well in terms of fitness; in other words, $w_1^* = w_2^*$. Solving this equation with respect to p using the genotypic fitnesses shown above and that $q = 1 - p$ yields:

$$w_1^* = w_2^*$$

$$pw_{11} + qw_{12} = pw_{12} + qw_{22}$$

$$0.7p + q = p + 0.8q$$

$$\hat{p} = 0.4$$

where the hat over p indicates that this is an equilibrium frequency ($\hat{p}$). Below (in section 4.2.6) we shall see that this internal equilibrium (i.e. where both alleles do equally well) is indeed stable in the case of heterozygote advantage.

4.2.4 Adaptive landscapes

In chapter 1, we introduced Sewall Wright's metaphor of adaptive landscapes. We shall formalize this metaphor to investigate the dynamic properties of the one-locus viability selection model. We will use plots of mean population fitness, $\bar{w}$, against the frequency of A_1, p. Mean population fitness is a function of one variable (p) because $q = 1 - p$. So, what is the shape of the graph of this function? By inspecting the equation for mean population fitness $\bar{w}(p) = p^2 w_{11} + 2pq w_{12} + q^2 w_{22}$ we see that it is a quadratic function. Hence, the graph will be a parabola with either a maximum or a minimum depending on the sign of the function. Note, however, that p is only defined between 0 and 1 (i.e. an absence of A_1 to only A_1 present in the population). This means that, depending on the values of the fitnesses, the maximum or minimum of the parabola may fall inside or outside the biologically relevant values of p. The graph of mean population fitness against p will therefore always have at least one maximum: $p = 0$ or $p = 1$, or an internal maximum. Finally, we can have two maxima ($p = 0$ and $p = 1$) of mean population fitness when we have an internal minimum.

We find the slope of a graph by differentiating the function with respect to the variable. Because our function is a sum of different components we use the addition rule $(u + v)' = u' + v'$. It is easier to differentiate if we replace q with $1 - p$ and multiply into the resulting parentheses:

$$\bar{w}(p) = p^2 w_{11} + 2p(1-p)w_{12} + (1-p)^2 w_{22}$$

$$\bar{w}(p) = p^2 w_{11} + 2pw_{12} - 2p^2 w_{12} + w_{22} - 2pw_{22} + p^2 w_{22}$$

Then we differentiate:

$$\frac{d\bar{w}}{dp} = 2pw_{11} + 2w_{12} - 4pw_{12} - 2w_{22} + 2pw_{22}$$

We can simplify this equation by expressing the derivative of mean population fitness using the marginal fitness of the two alleles as given in equation 4.1:

$$\frac{d\bar{w}}{dp} = \left(2pw_{11} + 2w_{12} - 2pw_{12}\right) - \left(2pw_{12} + 2w_{22} - 2pw_{22}\right)$$

$$\frac{d\bar{w}}{dp} = 2\left[\left[pw_{11} + w_{12}(1-p)\right] - \left[pw_{12} + w_{22}(1-p)\right]\right]$$

$$\frac{d\bar{w}}{dp} = 2\left(w_1^* - w_2^*\right) \tag{4.7}$$

The graph of mean population fitness against p has a maximum or minimum when $\dfrac{d\bar{w}}{dp} = 0$, or when $2(w_1^* - w_2^*) = 0$, that is, when $w_1^* = w_2^*$. In other words, the internal maximums or minimums in the adaptive landscape are the equilibriums we envisaged in the previous subsection.

We will use this result to reformulate the expression for Δp. Substituting equation 4.7 into equation 4.6 yields:

$$\Delta p = \frac{pq}{2} \cdot \frac{1}{\bar{w}} \cdot \frac{d\bar{w}}{dp} \tag{4.8}$$

Equation 4.8 is one of Wright's (1937) equations for an adaptive landscape and it has some interesting properties. First, consider the first component $pq/2$. This is the genotypic variance of a locus with two alleles. Let's break this down; by assuming random mating, producing an offspring genotype involves randomly picking two alleles from a large sample of gametes, a binomial process, as we explained in chapter 3. For each individual we can pick 0, 1, or 2 A_1 alleles, with a resulting A_1 allele frequency in each of the three genotypes of 0, 0.5, and 1. The variance in a binomial sampling process with these three values is simply $pq/2$. Δp, the change in allele frequency per generation, is therefore partly determined by genotypic variance. Evolutionary change

(Δp) would tend to be faster the higher the variance and this is highest when the two alleles are equally common.

Second, consider the last component, $d\bar{w}/dp$ — the slope of the mean population fitness. Because the variance term $pq/2$ by definition is always positive or zero, the direction of change in allele frequency is determined solely by the sign of this slope. If the slope is positive, A_1 will increase in frequency, whereas if it is negative A_1 will decrease in frequency. When the slope is zero there is no change in allele frequency and so equilibrium has been reached. When fitness is *constant*, selection acts to maximize mean population fitness. In the next two subsections, we will use this purely mathematical insight to predict the evolutionary trajectories of different adaptive landscapes.

4.2.5 Directional selection

It may be a little difficult to grasp some of the concepts in section 4.2.4 without a graphical representation of the adaptive landscape function. In Figure 4.2 we have plotted mean population fitness ($\bar{w}$) against the frequency of A_1 (p) for three different examples of genotypic fitnesses. In all three cases A_1A_1 has the highest genotypic fitness (set to 1) and A_2A_2 has the lowest fitness (set to 0.2). In all cases, we have directional selection favoring A_1A_1. The blue line shows the case when A_1 is dominant ($w_{11} = w_{12}$), the red line shows additive inheritance (the heterozygote has an exactly intermediate fitness), and the green line shows the case where A_1 is recessive ($w_{12} = w_{22}$). If we try to visualize the slope $d\bar{w}/dp$ as we move from $p = 0$ to $p = 1$ we see that it is always positive. According to equation 4.8, A_1 will increase in frequency until it is fixed in all three cases. You may confirm this with an invasion fitness analysis. Examining the differences between the slopes gives us some important insight into how adaptation can proceed. The slopes on the green line (where the favored allele is recessive) are very shallow at low values of p. Thus, the increase will be slow when A_1 is rare. Likewise, the slopes on the blue line (the favored allele is dominant) are very shallow at high values of p, so A_1 will increase very slowly as it approaches fixation. By contrast, the slope remains

unchanged in the case of additive inheritance (red line). In this instance, according to equation 4.8, Δp will only depend on the amount of genotypic variance present ($pq/2$); the increase in allele frequency will be fastest when $p = q = 0.5$.

When we compare this reasoning with plots of the trajectories of change in frequency of A_1 against time, we see that the shapes of the three lines indeed match the predictions (Figure 4.3). What does this mean biologically? When an allele is rare and mating is random, most carriers of the allele will be heterozygous. For instance, at an allele frequency of 1 percent, no less than 99.49 percent of the carriers of the allele would be heterozygous. A beneficial recessive allele (green line) would therefore increase slowly when at low frequencies because there are so few homozygotes with high fitness around for selection to favor. Most carriers of A_1 would be heterozygous and have low fitness. Likewise, a beneficial dominant allele (blue line) will increase slowly in frequency when it is very common because there

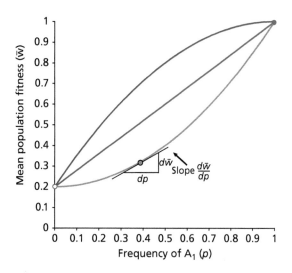

Figure 4.2 Plots of mean population fitness against allele frequency for a locus with two alleles in three cases of directional selection. In all three cases A_1A_1 has the highest genotypic fitness ($w_{11} = 1$) and A_2A_2 the lowest ($w_{22} = 0.2$). The blue curve shows the case when A_1 is dominant ($w_{12} = w_{11}$); the red curve shows a case of additive inheritance ($w_{12} = 0.6$); the green curve shows the case when A_1 is recessive ($w_{12} = w_{22}$). The slope of mean population fitness ($d\bar{w}/dp$) is always positive. Accordingly, there is no internal equilibrium and evolution will proceed until A_1 is fixed; $p = 0$ represents an unstable equilibrium indicated by a red open dot and $p = 1$ represents a stable equilibrium indicated by a blue dot.

are so few A_2A_2 individuals around for selection to act against. In this scenario, most carriers of A_2 would be heterozygous and thus have high fitness. Dominance hides rare, recessive alleles from exposure to selection.

The three evolutionary trajectories in Figure 4.3 share some features. They are all S-shaped, meaning that the transition from low to high frequency of the favored allele is fast (although not necessarily to fixation, as we saw in the previous paragraph). This has some practical implications. We are rarely around to witness evolution by directional selection in action. The exceptions would be cases of sudden environmental change, provided that relevant allelic variation is present for selection to act on. A famous example of directional selection caught in the act is the rise (and fall) in frequency of the melanistic form of the peppered moth (*Biston betularia*) in England. This coincided with a conspicuous and rapid change in the background color of their environment (Cook 2003). The light form of *B. betularia* is well camouflaged on lichen-clad branches and tree trunks (Figure 4.4) but with industrial pollution the lichens disappeared, leaving the surfaces naked and soot-covered. In this new environment, the dark morph had an advantage, almost certainly related to reduced predation risk from visually hunting predators such as birds, and increased dramatically in frequency at the expense of the light morph. With

more recent decreases in pollution soot has disappeared and lichens have reappeared on branches and tree trunks, and the frequency of the two morphs is approaching preindustrial values in many areas. Melanism in the peppered moth is caused by a single mutation in the *cortex* gene that appears to have arisen very recently (van't Hof et al. 2016). Other genes also affect color variation in this species, but the single locus selection model is actually a good approximation of the conspicuous and recent evolutionary changes that have taken place in this species as the mutant allele causing melanisms was targeted by directional selection. Likewise, rapid increases in antibiotic resistance in bacteria (Baquero & Blázquez 1997) and pesticide resistance in insects (Mallet 1989) are associated with dramatic environmental changes. Human activities have created artificially toxic environments for these organisms; resistant genotypes, if present, have a huge selective advantage and will rise to high frequencies very rapidly.

More commonly we only observe the end result of directional selection. We can infer past episodes of selection by comparing the phenotypes and genotypes of populations that reside in different environments and selection regimes. The rock pocket mouse (*Chaetodipus intermedius*) of southern Arizona, USA, occurs in two color morphs (Figure 4.5); a dark-colored morph that inhabits

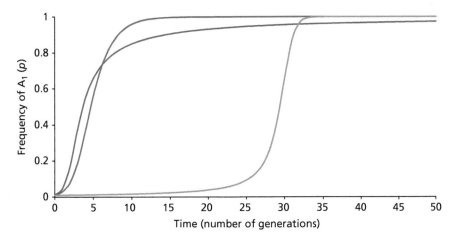

Figure 4.3 Change in frequency of an advantageous allele A_1 over time (positive, directional selection). In all cases the start frequency of A_1 is 0.001. In all three cases A_1A_1 has the highest genotypic fitness ($w_{11} = 1$) and A_2A_2 the lowest ($w_{22} = 0.2$). The blue curve shows the case when A_1 is dominant ($w_{12} = w_{11}$); the red curve shows a case of incomplete dominance ($w_{12} = 0.6$); the green curve shows the case when A_1 is recessive ($w_{12} = w_{22}$). See main text for interpretations.

Figure 4.4 Directional selection caught in the act. Top left panel: The light morph of the peppered moth (*Biston betularia*). Top right panel: The melanistic morph of the same species. Bottom panel: contrasting camouflage on light and dark surfaces of light and melanistic moths. The light morph is well camouflaged on light surfaces such as branches and tree trunks covered by lichens. With increased industrial pollution during the twentieth century the lichens disappeared and surfaces became sooty and dark. In parallel, the dark morph, which is better camouflaged in the new environment, experienced a dramatic increase in frequency. In recent years, reduced pollution has caused soot to disappear and lichens to re-appear; in many places, the light morph is approaching preindustrial frequencies. A single, recent mutation in the *cortex* gene gave rise to the melanistic form (van't Hof et al. 2016). The conspicuous and recent evolutionary changes in this species can therefore to a large extent be explained by changes in frequency of the mutant allele caused by selection for crypsis. Top left and right panel: photo courtesy of Ben Sale. Bottom panel: photo courtesy of Ian Reddig.

dark, volcanic lava and a light-colored morph that inhabits nearby light granite rocks (Nachman et al. 2003). The color difference is caused by allelic variation at the Melanocortin 1 receptor gene (*MC1R*), which is involved in the synthesis of melanin pigment. The allele for dark coat is dominant: *DD* and *Dd* genotypes are dark, whereas *dd* are light. The frequency of the two alleles, and hence coat color, strongly correlates with habitat type, whereas other genetic markers unrelated to color show no such pattern (Hoekstra et al. 2004). As with the peppered moth, strong habitat-dependent selection, presumably

for camouflage protection against visually hunting predators, can explain this pattern. The absence of geographic structure in other genetic markers in the rock pocket mouse suggests either that the adaptation is very recent or, perhaps more likely, that there is quite extensive gene flow going on between the populations. However, due to ongoing strong selection, the color alleles largely fail to move between populations. When directional selection drives populations in different directions, like in the case of the rock pocket mice, we refer to it as **divergent selection** (also called diversifying selection).

Figure 4.5 Divergent selection on coat colors in rock pocket mice (*Chaetodipus intermedius*). Normally, the dark morph resides on dark lava rocks (top left) whereas the light morph resides on light granite rock (bottom left). Either morph is poorly camouflaged in the opposite environment (right panels) and is selected against. Accordingly, coat colors diverge in the two environments.

4.2.6 Overdominance and underdominance

Overdominance means that the phenotypic value and/or fitness of the heterozygote exceeds those of either homozygote. In terms of fitness, heterozygote advantage, where the heterozygote has higher fitness than either homozygote, is an example of overdominance. Figure 4.6 shows the adaptive landscape of one such case, when:

$$w_{11} = 0.2$$

$$w_{12} = 1$$

$$w_{22} = 0.4$$

As we saw in section 4.2.3, an invasion fitness analysis tells us that the external equilibria $p = 0$ and $p = 1$ are unstable in the case of heterozygote advantage. We will now use adaptive landscape analysis to demonstrate that the internal equilibrium (i.e. where $w_1^* = w_2^*$) is stable in this case. If, as above, we superimpose slopes of $d\bar{w}/dp$ onto the graph in Figure 4.6 they are positive on the left side of the graph, leveling off to zero at the peak, and then turning negative. In terms of allele frequency, if we

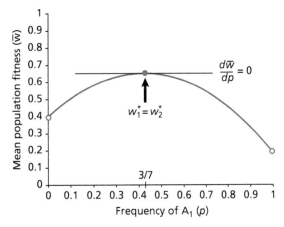

Figure 4.6 Plot of mean population fitness against allele frequency for a locus with two alleles in a case of heterozygote advantage. The genotypic fitnesses are $w_{11} = 0.2$, $w_{12} = 1$, and $w_{22} = 0.4$. The slope of mean population fitness is zero at $\hat{p} = 3/7$ where it is at its maximum and where the marginal fitness of the two alleles is equal. This represents a stable equilibrium (blue dot). The two unstable equilibriums are indicated by red open dots.

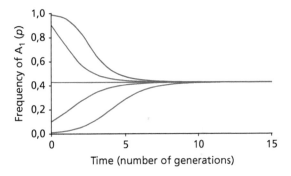

Figure 4.7 Change in frequency of allele A_1 in a case of heterozygote advantage with different start frequencies. The genotypic fitnesses are $w_{11} = 0.2$, $w_{12} = 1$, and $w_{22} = 0.4$. In all cases the allele frequencies converge on the stable equilibrium of $\hat{p} = 3/7$.

start with a low frequency of A_1 (positive slope), the allele will increase in frequency until it reaches the point where the slope is zero (equilibrium). If we start with a high frequency of A_1 (negative slope), the allele decreases in frequency until the slope is zero. The peak in the adaptive landscape is an attractor, a stable equilibrium. From equation 4.6, we find this equilibrium frequency by solving the equation $w_1^* = w_2^*$ with respect to p and find that $\hat{p} = \frac{3}{7}$. Trajectories with different initial frequencies of A_1 are plotted in Figure 4.7.

Heterozygote advantage is an example of **balancing selection**, a form of selection that maintains polymorphisms. A well-established example is that of the hemoglobin gene involved in sickle cell anemia in human populations (e.g. Williams et al. 2005). In areas with high incidence of malaria, individuals that are heterozygous for the sickle cell and the normal allele gain considerable protection against malaria and do not get sickle cell anemia. Individuals that are homozygous for the sickle cell allele can develop severe anemia associated with increased mortality. Individuals homozygous for the normal allele are susceptible to malaria, which can be a deadly disease. The frequency of the sickle cell allele is kept high by balancing selection in regions of the world where malaria is common.

Underdominance means that the heterozygote has a lower phenotypic value and/or fitness than either homozygote and thus represents the mirror image of overdominance. In Figures 4.8 and 4.9 we have plotted similar kinds of graphs as above but for the case of heterozygote disadvantage with:

$$w_{11} = 0.9$$

$$w_{12} = 0.3$$

$$w_{22} = 1$$

By superimposing the slopes of $d\bar{w}/dp$ onto the graph (Figure 4.8), you can clearly see that the fitness minimum of $\bar{w}(p)$ represents a repellent, an unstable equilibrium, and that both $p = 0$ and $q = 0$ are stable equilibriums (you may confirm this by using invasion fitness analysis). As before, the internal equilibrium is found by solving the equation $w_1^* = w_2^*$ with respect to p ($\hat{p} = \frac{7}{13}$). A population that has this exact frequency of p will remain at this equilibrium frequency, but any perturbation away from the equilibrium would drive the population towards fixation of either A_1 or A_2 (Figure 4.9).

Heterozygote disadvantage is an example of disruptive selection, in which selection acts within a population favoring either of the two extremes. One could imagine disruptive selection as a starting point for *in situ* (sympatric) speciation, because intermediates are selected against and either extreme favored. Although this is possible, there is an inherent instability in a system under disruptive selection. As shown in the analysis above, with the assumptions we have made, the outcome is most likely fixation of one of the alleles. However, we shall soon see that other selective forces can act to stabilize a system under disruptive selection.

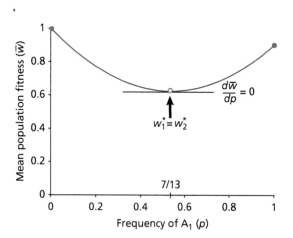

Figure 4.8 Plot of mean population fitness against allele frequency for a locus with two alleles in a case of heterozygote disadvantage. The genotypic fitnesses are $w_{11} = 0.9$, $w_{12} = 0.3$, and $w_{22} = 1$. The slope of mean population fitness is zero at $\hat{p} = 7/13$ where it is at its minimum and where the marginal fitness of the two alleles is equal. This represents an unstable equilibrium (red open dot). The two stable equilibriums are indicated by blue dots.

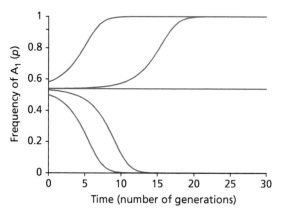

Figure 4.9 Change in frequency of allele A_1 in a case of heterozygote disadvantage with different start frequencies. The genotypic fitnesses are $w_{11} = 0.9$, $w_{12} = 0.3$, and $w_{22} = 1$. Allele frequencies diverge from the unstable equilibrium of $\hat{p} = 7/13$ except when the start frequency is exactly 7/13. Start frequencies of A_1 below the unstable equilibrium results in fixation of the alternative allele A_2, whereas start frequencies above the unstable equilibrium results in fixation of A_1. Hence there are two (locally) stable equilibriums at $\hat{p} = 0$ and $\hat{p} = 1$ respectively.

4.3 Frequency-dependent selection

4.3.1 Negative and positive frequency-dependent selection

In the analysis of viability selection in section 4.2 we assumed that genotypic fitnesses are constant. In nature, often the fitness of a phenotype or genotype actually depends on how common or rare it is in the population; that is, **frequency-dependent selection**. There are positive and negative forms of frequency-dependent selection. The European elder-flowered orchid (*Dactylorhiza sambucina*) provides one example of negative frequency-dependent selection (Figure 4.10).

This orchid occurs in two color morphs, one with purple and one with cream-colored flowers. As in other flowering plants the colorful flowers attract pollinators, such as bumblebees, but in this orchid the insects are not rewarded with any nectar. Gigord et al. (2001) demonstrated that the color polymorphism is maintained by negative frequency-dependent selection. They experimentally varied the relative frequency of the color morphs in the natural habitat of the species and showed that the rarer color morph was more frequently visited by pollinators. Presumably the lack of reward caused the pollinators to learn to associate the color with a negative experience, leading them to avoid that color and approach the other. There would be fewer opportunities to learn

to avoid the rarer morph. Accordingly, either morph would have a reproductive advantage when rare. At equilibrium both types are maintained at intermediate frequencies. Like heterozygote advantage, negative frequency-dependent selection tends to stabilize a system. It is an example of balancing selection that acts to maintain polymorphisms (Figure 4.11).

Positive frequency-dependent selection occurs when a morph or genotype has low fitness when rare and high fitness when common. Warning colors are a good example of this. A predator will learn to associate a striking color pattern, such as the black and yellow stripes of a wasp, with something unpleasant and will avoid attacking again. When the warning color is common in the population, most predators will have experienced strikingly colored and unpleasant prey and therefore have learned to avoid them.

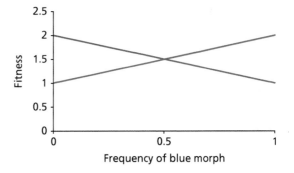

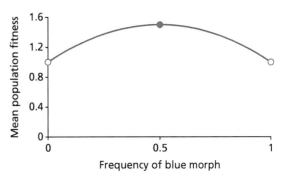

Figure 4.11 Negative frequency-dependent selection. a) Either morph (or genotype) has high fitness when rare and low fitness when common—line color denotes morph; b) mean population fitness is at a maximum at intermediate frequencies in this example, where the two morphs have equal fitness. The stable equilibrium is indicated with a blue dot, the two unstable equilibriums are indicated with red open dots. In general, negative frequency-dependent selection has a stabilizing effect on a system.

Figure 4.10 The rewardless elder-flowered orchid (*Dactylorhiza sambucina*) occurs in two color morphs that are kept at intermediate frequencies by negative frequency-dependent selection. Presumably either morph has an advantage when rare because pollinators in this case have little opportunity to learn to associate the color with a lack of nectar reward. Photo courtesy of Klavs Nielsen.

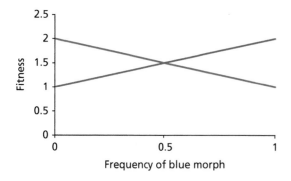

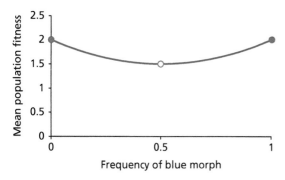

Figure 4.12 Positive frequency-dependent selection. Either morph (or genotype) has low fitness when rare and high fitness when common (top panel). Mean population fitness is at a minimum at intermediate frequencies of the morphs in this example, where the two morphs have equal fitness (lower panel). The unstable equilibrium is indicated with a red open dot, the two stable equilibriums are indicated with blue dots. In general, positive frequency-dependent selection has a destabilizing effect on a system.

As a result, the prey species is well protected. But on the contrary, a striking color is conspicuous, so a naïve predator may be more likely to attack such a prey. When the warning signal is rare most predators would not have experienced any prey with striking colors and would readily attack. The level of protection gained from a warning color therefore increases the more common that color is in the population. Like heterozygote disadvantage, positive frequency-dependent selection tends to destabilize a system (Figure 4.12).

4.3.2 Relationship between balancing, disruptive, and frequency-dependent selection

When fitness is constant the polymorphic equilibrium is stable in the case of heterozygote advantage

(balancing selection) and unstable in the case of heterozygote disadvantage (disruptive selection). Likewise, we have seen that a polymorphic equilibrium can be stable under negative frequency-dependent selection but unstable under positive frequency-dependent selection (Figures 4.11 and 4.12). What is the outcome when heterozygote disadvantage and negative frequency-dependent selection, or heterozygote advantage and positive frequency-dependent selection, act simultaneously in a system?

In Figure 4.13 we have three genotypes that all are under negative frequency-dependent selection. The genotypic frequency-dependent fitness functions in this figure are dependent on p, the frequency of allele A_1 and are summarized in Table 4.2.

As the genotypic fitnesses are symmetric, we have a polymorphic equilibrium at $\hat{p} = 0.5$ where $w_1^* = w_2^*$. Here, the two homozygotes have equal fitness ($w_{11} = w_{22} = 2 - 0.5^2 = 1.75$) and the heterozygote has lower fitness than either of the homozygotes ($w_{12} = 2 - 2 \cdot 0.5 \cdot 0.5 = 1.5$). However, is the equilibrium stable or unstable? There is heterozygote disadvantage (disruptive selection) at the polymorphic equilibrium that would tend to destabilize the system but negative frequency-dependent selection that would tend to stabilize it.

Udovic (1980) introduced the two indices S and M to quantify these opposing selective pressures and showed that the relationship between them determines if the polymorphic equilibrium is stable or unstable. To portray these simply we will first introduce another set of indices that measure the fitness differences between both homozygotes and the heterozygote:

$$a = w_{11}(p) - w_{12}(p)$$
$$b = w_{22}(p) - w_{12}(p) \tag{4.9}$$

Table 4.2 Numerical example in which a polymorphic equilibrium is affected both by disruptive selection and negative frequency-dependent selection.

Genotype	Frequency-dependent fitness functions
A_1A_1	$w_{11}(p) = 2 - p^2$
A_1A_2	$w_{12}(p) = 2 - 2pq$
A_2A_2	$w_{22}(p) = 2 - q^2$

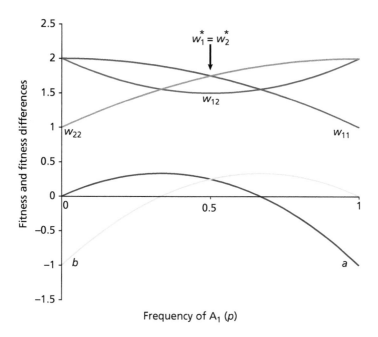

Figure 4.13 Negative frequency-dependent selection at a locus with two alleles. The genotypic fitness functions are set to $w_{11}(p) = 2 - p^2$ (dark blue curve), $w_{12}(p) = 2 - 2pq$ (red curve), and $w_{22}(p) = 2 - q^2$ (green curve). At the polymorphic equilibrium where the two alleles have equal marginal fitness, $w_1^* = w_2^*$, the heterozygote has lower fitness than either homozygote. Hence, at this equilibrium, disruptive selection acts to destabilize the system whereas negative frequency-dependent selection acts to stabilize it. The relative strength of the two forces determines whether the internal equilibrium is stable or unstable. The figure further shows a, the fitness difference between w_{11} and w_{12} (purple curve) and b, the fitness difference between w_{22} and w_{12} (light blue curve). These two indices are used to calculate the relative strength of the two opposing forces and hence the stability of the polymorphic equilibrium. The equilibrium is stable in this particular case. See main text for further explanations.

In our example, we have $a = (2 - p^2) - (2 - 2pq)$ and $b = (2 - q^2) - (2 - 2pq)$. The index S, which measures the degree of heterozygote advantage or disadvantage at the polymorphic equilibrium, is defined as:

$$S = \hat{a} + \hat{b} \qquad (4.10)$$

In our example (Table 4.2) we have $S = (1.75 - 1.5) + (1.75 - 1.5) = 0.5$. S is therefore the cumulative effect of the difference in fitness between the heterozygote and the two homozygotes. It will always be positive when there is heterozygote disadvantage and negative when there is heterozygote advantage.

The index M measures the degree (and direction) of frequency-dependent selection at the polymorphic equilibrium and is defined as:

$$M = \hat{p}\frac{da}{dp}(\hat{p}) - \hat{q}\frac{db}{dp}(\hat{p}) \qquad (4.11)$$

In words, M is the difference between the equilibrium frequencies of A_1 and A_2 multiplied, respectively, by the slope of the indices a and b at that equilibrium. In our numerical example, we have:

$$a = w_{11}(p) - w_{12}(p) = (2 - p^2) - (2 - 2pq) = -3p^2 + 2p$$
$$b = w_{22}(p) - w_{12}(p) = (2 - q^2) - (2 - 2pq) = -3p^2 + 4p - 1$$

Then we differentiate:

$$\frac{da}{dp} = -6p + 2$$
$$\frac{db}{dp} = -6p + 4$$

and calculate M as:

$$M = \hat{p}\frac{da}{dp}(\hat{p}) - \hat{q}\frac{db}{dp}(\hat{p}) = 0.5(-6\cdot0.5+2)$$
$$-0.5(-6\cdot0.5+4) = -1$$

The index M is negative when there is negative frequency-dependent selection, positive when there is positive frequency-dependent selection, and zero when selection is frequency independent. Udovic (1980) showed that a polymorphic equilibrium under frequency-dependent selection is stable when:

$$S + M < 0 \qquad (4.12)$$

In our numerical example, we have $S = 0.5$ and $M = -1$. Thus $S + M = 0.5 - 1 = -0.5$. The inequality (equation 4.11) is satisfied so the equilibrium is stable. There is sufficiently strong negative frequency-dependent selection (negative M) to stabilize the destabilizing effect of heterozygote disadvantage (disruptive selection; positive S) in this example.

When we observe phenotypic traits in nature that appear to be under disruptive selection we would predict that negative frequency-dependent selection stabilizes the system according to the theoretical analyses above. Yang et al. (2010) found evidence for disruptive selection on egg coloration in the ashy-throated parrotbill (*Paradoxornis alphonsianus*) and its avian brood parasite, the common cuckoo (*Cuculus canorus*). The ashy-throated parrotbill, and the cuckoo parasitizing it, lay eggs that are either blue, pale blue, or white; pale-blue eggs are intermediate between blue and white ones in both chroma and brightness. Likely, egg color is determined by one bi-allelic locus in which homozygotes for the blue allele have blue eggs, homozygotes for the white allele have white eggs, and heterozygotes have pale-blue eggs. The parrotbill will reject a foreign egg that differs significantly in color from its own eggs, an adaptation against brood parasitism.

Yang and co-workers conducted experiments in which they laid an artificial egg in the nests of parrotbills. They found that the greater the contrast between the parrotbills' own eggs and the foreign egg, the more likely the birds were to reject the foreign egg (Figure 4.14). Thus, birds with white eggs are very good at rejecting blue foreign eggs and birds with blue eggs are very good at rejecting white foreign eggs. However, the low contrast between pale-blue eggs and any foreign egg, whether blue, pale blue, or white, would make birds laying such eggs more inclined to accept a foreign egg. This is evidence for disruptive selection against intermediate pale-blue colored eggs in the ashy-throated parrotbill. Indeed, field surveys show that birds laying either blue or white eggs are much more common than those laying pale-blue eggs.

From the cuckoo's point of view, it is advantageous if the host does not recognize the cuckoo egg as foreign. Experiments showed that blue foreign eggs were generally accepted among blue host eggs and white eggs were accepted among white host eggs. By contrast, a significant proportion of pale-blue foreign eggs were rejected among both blue and white host eggs. This suggests that there is also disruptive selection on egg color in the cuckoo, because pale-blue cuckoo eggs are only well camouflaged in the rare pale-blue egg host nests.

Clearly, negative frequency-dependent selection stabilizes this system. In a hypothetical parrotbill population where all individuals lay blue eggs, cuckoos with blue eggs would have an advantage and rise to a similarly high frequency. A rare mutant parrotbill with white eggs would now have a huge selective advantage because any blue cuckoo egg would be easily visible and hence rejected. Likewise, in a parrotbill population where all lay white eggs, rare individuals laying blue eggs would be at an advantage. The net result is a stable polymorphism where blue and white eggs are kept at intermediate frequencies and, as we have seen, in which the intermediate pale-blue egg color is selected against in both the host and the parasite at the polymorphic equilibrium. Evolution of egg color in ashy-throated parrotbill and its brood parasite is an example of **coevolution,** in which an adaptation in one of the parties induces a counter-adaptation in the other party.

4.3.3 Evolutionary game theory

As evident from section 4.3.2., modeling frequency-dependent selection is more difficult than when fitness is constant. The genotypic fitnesses $w_{11}(p)$, $w_{12}(p)$, and $w_{22}(p)$ are no longer constants but are instead functions of p, and not necessarily simple functions either. We may therefore benefit from making some simplifying assumptions when modeling such selection. Game theory is a useful tool for simplifying genetics to study the complexity of selection. Developed in the early twentieth century, game

Figure 4.14 Disruptive selection on egg colors in ashy-throated parrotbills (*Paradoxornis alphonsianus*) that are parasitized by the common cuckoo (*Cuculus canorus*) according to a study by Yang et al. (2010). Birds that lay intermediate pale-blue eggs would be more prone to accept a differently colored cuckoo egg than those that lay blue or white eggs, because of the reduced contrast in color.

theory was originally the mathematical study of zero-sum games—i.e. the success of one participant results in the loss for all others—in an economic framework. John Maynard Smith was fundamental in applying game theory to evolutionary biology and we will illustrate this here using his famous hawk–dove game example (Maynard Smith & Price 1973; Maynard Smith 1982).

The game is played between two hypothetical morphs of an animal species: aggressive "hawks" and peaceful "doves." We assume that the two morphs produce offspring that are identical to themselves at a rate determined by their respective fitness. This means we implicitly assume that the hawks and doves are haploid and reproduce asexually.

The animals engage in pairwise contests over an essential resource, such as a territory. Gaining the

resource increases the expected fitness of the winner by an amount V. Total fitness is not determined solely by whether the resource is won or not, so we can view V as a fitness component that adds to the baseline fitness W_0 of the average individual. The genetically determined phenotype of the aggressive hawks is that of a fighter. In a contest, it will fight to gain the resource if necessary. If a hawk meets another hawk in a contest they will both fight until one of them is injured. The winner gets the resource worth V whereas the loser pays a fitness cost C (e.g. reduced survival due to injury). The expected payoff (change in fitness) for a hawk (H) contesting another hawk is denoted by $E(H, H) = (V - C)/2$. There is a 50:50 chance of winning or losing. The genetically determined phenotype of the dove is to be peaceful and it will never fight. When a dove (D)

Table 4.3 Pay-off of the hawk and dove morph depending on the phenotype of the opponent.

	Hawk opponent	Dove opponent
Pay-off for hawk	$(V-C)/2$	V
Pay-off for dove	0	$V/2$

meets another dove in contest they will share the resource and get an expected pay-off $E(D, D) = V/2$. A peaceful dove facing an aggressive hawk will retreat and will experience no change in fitness $E(D, H) = 0$, leaving the resource to the winning hawk $E(H, D) = V$. The pay-offs of the different contests are summarized in Table 4.3.

We will use invasion fitness analysis to find possible stable equilibriums. In evolutionary game theory, a stable equilibrium is referred to as an evolutionarily stable strategy (ESS). A population of (nearly) only doves would have resident fitness equal to $W_0 + V/2$; rare mutant hawks would (nearly) only face doves in contests and would thus have invasion fitness $W_0 + V$. Obviously $V > V/2$ so the dove strategy can never be an ESS—a population of doves can always be invaded by mutant hawks. A population of (almost) only hawks would have resident fitness equal to $W_0 + (V-C)/2$; rare mutant doves would (virtually) only face hawks and have invasion fitness $W_0 + 0$. For the hawk strategy to be an ESS, the fitness gain from winning the resource would have to be on average larger than the fitness loss from injury, that is, $(V-C)/2 > 0$ or $V > C$ (Figure 4.15). If $V < C$ then neither hawk nor dove is an ESS. A polymorphic population of hawks and doves could be evolutionarily stable, however, at the point where both strategies do equally well.[1]

Let p be the frequency of hawks and q that of doves ($q = 1 - p$). If a stable polymorphic equilibrium exists then $W_H = W_D$ at that equilibrium; otherwise, the strategy with the higher fitness would increase in frequency at the expense of the other. So, a stable equilibrium requires that:

$$pE(H,H) + qE(H,D) = pE(D,H) + qE(D,D)$$

Setting in for the expected pay-offs yields:

$$p(V-C)/2 + qV = p(0) + qV/2$$

and solving with respect to p yields:

$$\hat{p} = V/C \tag{4.13}$$

Equation 4.13 gives the frequency of hawks at the polymorphic, stable equilibrium (or ESS) for when fighting is very costly ($V < C$).

Although an abstract example, the hawk–dove game illustrates an important point about frequency-dependent selection; namely, that mean population fitness is not necessarily maximized. In this scenario, mean population fitness would have been highest if everyone in the population played the dove strategy ($V/2$). However, because hawks would do better than doves when rare (invasion fitness = V), the hawks would increase in frequency and steadily reduce mean population fitness (Figure 4.15). When $V > C$, the "hawk allele" would become fixed and mean population fitness would be reduced to $(V - C)/2$. Hence, Wright's adaptive landscape metaphor is not universally true. Natural selection is not necessarily a climbing process to the nearest peak in the landscape when selection is frequency dependent.

4.4 Mutation–selection balance

4.4.1 Approximating the mutation–selection equilibrium

We shall once more return to the one-locus viability selection model with constant fitness and analyze what happens when we introduce mutation to the equation. Specifically, we shall examine the expected frequency of an allele that is selected against but which repeatedly reappears due to mutation, a so-called recurrent mutation. This mutation–selection equilibrium is thought to be an important factor explaining the maintenance of genetic variation within populations. It also has applied values. In medicine, many genetic diseases are thought to be maintained in a mutation–selection balance (e.g. Repping et al. 2003), and in conservation biology harmful mutations are a concern particularly

---

[1] In a game with only two pure strategies, when neither of the pure strategies are evolutionarily stable, the polymorphic equilibrium always will be (Maynard Smith 1982).

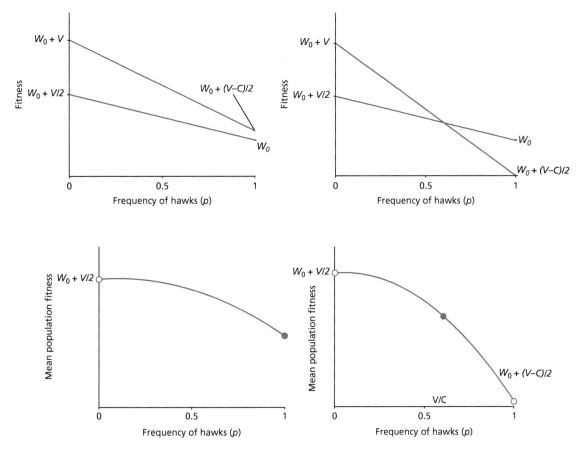

Figure 4.15 Graphical analysis of the hawk–dove game. The left panels show a case when $V > C$, in which the hawk strategy (blue graph) is a pure ESS whereas a population of doves (red graph) can be invaded by hawks (top left panel). Mean population fitness is not maximized at equilibrium (bottom left panel). The right panels show a case when $V < C$, in which neither pure strategy is an ESS but where the polymorphic equilibrium where $W_H = W_D$ is evolutionarily stable (top right panel). Again, mean population fitness is not maximized at equilibrium (bottom right panel). ESSs are marked with blue dots, unstable equilibriums are marked with red open dots.

in small and inbred populations (Kohn et al. 2006; Hedrick & Garcia-Dorado 2016).

In this exercise, we shall introduce a selection coefficient s and a constant h that gives the pattern of dominance in the heterozygote. We will consider three genotypes from two alleles, where one, A_1, is harmful; the fitnesses of the genotypes are summarized in Table 4.4.

When $h = 0$ the harmful allele A_1 is fully recessive and when $h = 1$ it is fully dominant. With partial dominance, $0 < h < 1$. We then introduce mutation: A_1 mutates to A_2 at a rate v and A_2 mutates to A_1 at a rate μ. The frequency of the

Table 4.4 Genotypic fitness with selection coefficient s and dominance converter h.

Genotype	Fitness
A_1A_1	$w_{11} = 1 - s$
A_1A_2	$w_{12} = 1 - hs$
A_2A_2	$w_{22} = 1$

harmful A_1 allele after one generation of selection is given as:

$$p_{t+1} = \frac{pw_1^*}{\bar{w}} + \mu q - vp \qquad (4.14)$$

Substituting the fitness terms with expressions for genotypic and mean population fitness using the selection coefficient and the dominance constant yields:

$$p_{t+1} = \frac{p[p(1-s)+q(1-hs)]}{p^2(1-s)+2pq(1-hs)+q^2 \cdot 1} + \mu q - vp$$

Multiplying into the parentheses and using that $p + q = 1$ and that $p^2+2pq+q^2 =1$ yields:

$$p_{t+1} = \frac{p(1-sp-hsq)}{1-sp^2-2hspq} + \mu q - vp \qquad (4.15)$$

We are seeking an equilibrium frequency in which $\Delta p = 0$ or $p_{t+1} = p$. However, equation 4.15 is not particularly easy to work with. We shall therefore do some approximations similar to those we did when we determined invasion and resident fitness. Because the mutant allele A_1 is selected against, its equilibrium frequency will be very low. We can therefore ignore some of the expressions above. First, at equilibrium $\bar{w} \approx 1$ so we can ignore the entire denominator (see equation 4.14 to remind yourself where the denominator came from). Second, because mutation rates are typically low there will be vanishingly few of the A_1 alleles that will mutate to A_2 at equilibrium, so $vp \approx 0$. Without introducing much error, we can therefore approximate the equilibrium $\hat{p}$ by setting $p_{t+1} = p$ and substituting p_{t+1} with the approximation of 4.15:

$$\hat{p} - s\hat{p}^2 - hs\hat{p}\hat{q} + \mu\hat{q} \approx \hat{p}$$

Replacing q with $1-p$ and multiplying into the parentheses yields:

$$\mu - \mu\hat{p} \approx s\hat{p}^2 + hs\hat{p} - hs\hat{p}^2 \qquad (4.16)$$

Let us first consider the case when A_1 is recessive. Here, $h = 0$ so hsp and hsp^2 in equation 4.16 are also zero. Moreover, because both the equilibrium frequency of A_1 and the mutation rate are small numbers, we have that $\mu p \ll \mu$ so we can ignore the former without introducing much error. At the mutation–selection equilibrium we therefore have the following approximation:

$$\mu \approx s\hat{p}^2$$

And solving for p to get the equilibrium frequency yields:

$$\hat{p} \approx \sqrt{\frac{\mu}{s}} \qquad (4.17)$$

We now turn to the case when h is positive and investigate how we can approximate equation 4.16 in this case. As above we are dealing with multiplication of very small numbers so $\mu p \ll \mu$ and $p^2 \ll p$. We ignore the super-small components in equation 4.16 and get:

$$\mu \approx hs\hat{p}$$

We find the approximate equilibrium frequency of A_1 at mutation selection balance by resolving for p:

$$\hat{p} \approx \frac{\mu}{hs} \qquad (4.18)$$

In Figure 4.17, we have plotted all mutation–selection equilibriums from $h = 0$ to $h = 1$ using a moderate selection coefficient ($s = 0.1$) and an allelic mutation rate of 10^{-5} based on equations 4.17 and 4.18. We can see that increasing the amount of dominance of A_1 (moving h away from zero) very rapidly reduces the genetic variation maintained at mutation–selection equilibrium. For this reason, we can predict that genetic variation maintained by mutation–selection balance largely involves recessive alleles. However, the equilibrium frequency of a recessive, deleterious recurrent mutation can be non-trivial (Figure 4.16).

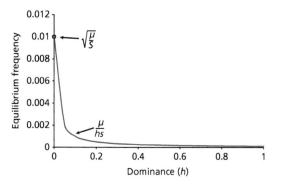

Figure 4.16 Mutation–selection equilibrium for different values of h when $s = 10^{-5}$.

4.4.2 Genetic load

We can define **genetic load** as the degree of reduction in mean population fitness due to the presence of deleterious alleles, that is:

$$\text{Genetic load} = 1 - \bar{w} \qquad (4.19)$$

Using the selection coefficient s and the dominance constant h as above we have from equation 4.15 that:

$$\begin{aligned}\text{Genetic load} &= 1 - \bar{w} = 1 - (1 - sp^2 - 2hspq)\\ &= sp^2 + 2hspq\end{aligned}$$

When A_1 is recessive, $h = 0$ and thus $2hspq = 0$, so:

$$\text{Equilibrium load} = s\hat{p}^2 = s\left(\sqrt{\frac{\mu}{s}}\right)^2 = \mu \quad (4.20)$$

The equilibrium load equals the mutation rate of the deleterious allele and is consequently not dependent on the strength of selection. At equilibrium, a lethal allele does not reduce mean population fitness more than a mildly deleterious allele. Why? The equilibrium frequency of the deleterious allele is very different in the two cases. When s is large the equilibrium frequency is very low, but each A_1A_1 individual present would contribute substantially to the reduction in mean population fitness. When s is small the equilibrium frequency is much higher, but each A_1A_1 individual only reduces the mean population fitness by a small amount. The two effects cancel each other out so at equilibrium mean population fitness is reduced by the same amount.

4.5 Selection at other stages of the life cycle

4.5.1 Sexual selection

The analyses above illustrate that it is relatively straightforward to analyze viability selection. Unfortunately, other forms of selection can be much harder to analyze; in many cases, analytical tools simply break down and we have to rely on computer simulations to learn how these forms of selection work. The benefit of viability selection is that it is reasonable to assume Hardy–Weinberg equilibrium

at the start of each generation, which allows us to track allelic changes quite easily.

We may classify selection according to when during the life cycle there is differential fitness. Viability selection takes place between birth and adulthood. Next, selection may take place as males and females form mating pairs. Such **sexual selection** is defined as differential mating success between members of one sex (Andersson 1994; West-Eberhard 2014). Members of one sex (usually males) may compete among themselves for access to mating, a process referred to as intrasexual selection. For instance, in herds of red deer (*Cervus elaphus*), the dominant male has almost exclusive access to mate with the females in the herd (Clutton-Brock et al. 1982). Sexually selected traits such as rutting pitch, body size, and antler size determine the dominance hierarchy among the males and hence mating access. Males that have closely matched indirect signals of strength may engage in fierce physical fights to win dominance.

By contrast, intersexual selection involves sexual preferences exerted by one sex (usually females) for characteristics of the opposite sex (Andersson 1994). Secondary sexual traits, such as the plumage and courtship display of the male superb fairywren (*Malurus cyaneus*), are the result of this form of selection (Figure 4.17). Intersexual selection involves coevolution between genes for mate preferences and secondary sexual traits, so models of intersexual selection would involve at least two loci (Kirkpatrick 1982).

Figure 4.17 The colorful male (left) of the superb fairy wren (*Malurus cyaneus*) looks dramatically different from the drab female (right). Exaggerated secondary sexual traits like these are the hallmark of traits favored by intersexual selection. Photo courtesy of Peter Nydegger.

Sexually selected traits are often bizarre and do not seem optimal for survival. Bright colors, conspicuous displays, and elaborate ornaments such as the peacock's tail would rather seem to hinder survival as they reduce agility and are likely to attract the attention of potential predators. One may therefore wonder why females should prefer males with such extreme traits, especially in species where the male provides nothing but sperm and does not contribute to parental care. Darwin (1871) was indeed puzzled by these traits as they seemed to contradict predictions from the theory of natural selection. His idea was that exaggerated secondary sexual traits do not evolve because they increase survival but simply *because* they are fancied by the opposite sex. The advantage of the traits has to do with increased mating success and not survival. Fisher (1930) elaborated on this idea and suggested that female preferences for an exaggerated male trait and the male trait itself would co-evolve, leading to a runaway process in which both the trait and the preference become more and more extreme. Choosy females would tend to get sons that inherit their father's extreme trait and daughters who inherit their mother's preference for that trait. Hence, both the trait and the preference would increase in frequency. Only when the cost of the trait in terms of reduced survival becomes sufficiently high would the runaway process be brought to a halt. Theoretical models show that such *Fisherian runaway sexual selection* indeed can work (Lande 1981; Kirkpatrick 1982; Sharma et al. 2016).

Another theory is that exaggerated secondary sexual traits somehow signal good genes. Females would benefit from choosing males with elaborate traits because her offspring would inherit these good genes. According to the *handicap hypothesis* (Zahavi 1975; Zahavi & Zahavi 1997) exaggerated male traits are honest indicators of the male's genetic qualities *because* they are costly to produce. Only the best males would be able to produce the most extreme traits and yet survive. Theoretical models suggest that the handicap principle could work under certain circumstances (e.g. Grafen 1990). Yet, despite intensive efforts it has proven difficult to obtain definite empirical evidence for the theory (Számadó & Penn 2015).

4.5.2 Fertility selection

If we continue working through the life cycle, we find that differential fitness can occur at fertilization. **Fertility selection** (also called fecundity selection) occurs when the number of offspring produced by a mating pair depends on their combination of genotypes. The entity we assign fitness to in fertility selection is therefore not individuals but mated pairs. The rhesus factor in humans illustrates this point (Avent & Reid 2000). The rhesus factor codes for the D antigen and human individuals either have it, they are rhesus positive (*Rh+*), or they don't and are rhesus negative (*Rh−*). Fertility selection occurs because a *Rh−* mother can develop antibodies against her *Rh+* fetus that attack the fetus's blood. Such an attack breaks down red blood cells in the fetus, which may cause serious illness, brain damage, and even death of the fetus or neonate. These *Rh+* and *Rh−* phenotypes are determined by one locus with two alleles: *rr* are *Rh−* whereas *Rr* and *RR* are *Rh+*; hence *R* is dominant. The hemolytic disease and, hence, low fitness only occur when *rr* females cross with *Rr* or *RR* males. The opposite cross, *Rr* or *RR* female with *rr* male, results in offspring with high fitness. Table 4.5 summarizes the fitness and frequencies of the various crosses and the expected genotypes of their offspring.

A full analysis of this (or any) case of fertility selection is quite complicated, but we can gain some insight from an invasion fitness analysis. First, assume that *r* is the common allele in the population except for a low frequency of the mutation *R*. Nearly all mating pairs would thus be *rr × rr* so resident fitness is 1. Most carriers of *R* would be heterozygous and, assuming equal sex ratio, half the crosses between a carrier of *R* would be between a female *Rr* and a male *rr*, with fitness 1, and half between female *rr* and male *Rr*, with fitness $1 − 0.5s$ (only half of their offspring would be *Rh+* and be likely to develop hemolytic disease), so the invasion fitness is $1 − 0.25s$. *R* should not be able to invade, so $R = 0$ ($r = 1$) is a stable equilibrium. Conversely, assume that *R* is the common allele in the population except for a low frequency of the mutation *r*. Nearly all mating pairs will be *RR × RR* so resident fitness is again 1. Most carriers of *r* would be *Rr* but a small

Table 4.5 Fitness and frequency of different mating pairs of the *Rh* blood factor system.

(F × M)	Fitness	Frequency	Offspring genotype		
			rr	*Rr*	*RR*
rr × *rr*	1	g_{rr}^2	1		
rr × *Rr*	$1 - 0.5s$	$g_{rr}g_{Rr}$	0.5	0.5	
rr × *RR*	$1 - s$	$g_{rr}g_{RR}$		1	
Rr × *rr*	1	$g_{Rr}g_{rr}$	0.5	0.5	
Rr × *Rr*	1	g_{Rr}^2	0.25	0.5	0.25
Rr × *RR*	1	$g_{Rr}g_{RR}$		0.5	0.5
RR × *rr*	1	$g_{RR}g_{rr}$		1	
RR × *Rr*	1	$g_{RR}g_{Rr}$		0.5	0.5
RR × *RR*	1	g_{RR}^2			1

proportion would be both female and *rr*. They would most likely mate with the common *RR* males and have fitness 1 – 2s. Hence, *R* = 1 (*r* = 0) is also a stable equilibrium and the mutant *r* would not be able to invade.

The *Rh* polymorphism is therefore puzzling. Selection should act to remove the rarer allele because of the maternal–fetal incompatibility. It is likely that some other factor is responsible for maintaining the polymorphism. For instance, the rhesus factor *R* may have a beneficial effect on a different phenotypic trait. When a gene affects more than one phenotypic trait we call it **pleiotropy**—i.e. the gene has a pleiotropic effect on multiple traits. When a trait is affected by opposing selective forces we have **antagonistic selection**. In the case of the human rhesus factor the alleles *r* and *R* are likely maintained by a peculiar case of balancing selection in which positive and negative (antagonistic) selective forces balance at a polymorphic equilibrium.

4.5.3 Selection at the gametic level

Finally, selection can occur among haploid gametes prior to fertilization and formation of a zygote. Characteristics of the sperm may affect its probability of being the one that fertilizes the egg and eggs may be selective as to which sperm is accepted. If we focus on a single allele A_1 at a heterozygous locus, a haploid sperm cell either carries this allele or it carries the other one. According to Mendel's

first law there is a 50 percent chance that A_1 will be present in the sperm that fertilizes the eggs. Clearly, an allele that could improve these odds would be favored by selection. Such alleles do exist and are known as **segregation distorters** or meiotic drive elements (Feldman & Otto 1991; Larracuente & Presgraves 2012). A classic segregation distortion element consists of two tightly linked key loci with two alleles segregating at each locus; a killer locus which either produces a sperm killing toxin K_1, or not K_2, and a target locus which is either resistant to the toxin T_1, or susceptible T_2 (Figure 4.18). The segregation distorter is the K_1T_1 gene complex that produces the toxin and is resistant to it. A double heterozygous male will largely produce sperm that are either K_1T_1 or K_2T_2. This is because recombination rarely will take place between closely linked loci. Thus, recombinant allele combinations or *haplotypes* (K_1T_2 and K_2T_1) will be strongly under-represented (see chapter 6). Following gametogenesis K_1 bearing sperm will kill T_2 bearing ones so that more than 50 percent of the sperm will be K_1T_1. These would therefore have a correspondingly higher probability of being present in the sperm that eventually fertilizes the egg. In a well-studied segregation distorter system in *Drosophila melanogaster* nearly all sperm are K_1T_1 following selective sperm killing (Larracuente & Presgraves 2012).

In many animals, sperm from different males may compete for fertilization. This would be the case when females regularly copulate with more

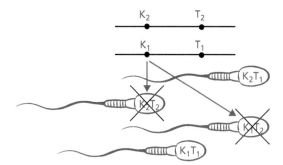

Figure 4.18 Segregation distortion by selective sperm killing. Sperm with the K_1 allele at the killer locus kill sperm with the susceptible T_1 allele at the target locus with a toxin whereas T_1 sperm are resistant to the toxin. The tightly linked K_1T_1 gene complex thus gets overrepresented in the ejaculate relative to K_2T_2. Recombinant K_1T_2 and K_2T_1 sperm would be rare due to tight linkage, but the former would also be targeted by the sperm killing toxin.

than one male and in aquatic animals with external fertilization. In such scenarios interesting forms of sexual selection may take place: **sperm competition** among male gametes for fertilization, and **cryptic choice** among females or their eggs as to which sperm is accepted for fertilization (Parker 1970; Eberhard 1996; Birkhead & Møller 1998; Arnqvist 2014).

Sperm competition may affect a number of traits related to fertilization success, including sperm morphology, sperm motility, and sperm count/testis size. Hosken & Ward (2001) showed experimentally that testis size responds to variation in intensity of sperm competition in the yellow dung fly (*Scathophaga stercoraria*). In a high sperm competition treatment, females were mated with three males in succession before laying a clutch, whereas in the low sperm competition treatment females copulated only once. After 10 generations of each differential treatment male flies in the high sperm competition line had evolved significantly larger testes than those in the low competition line.

Mechanisms of cryptic choice by females or her eggs are not very well understood. In several species, it has been found that sperm from conspecific males have a higher probability of fertilizing the egg than have heterospecific sperm. In Atlantic salmon (*Salmo salar*) and trout (*Salmo trutta*) such conspecific sperm precedence is apparently controlled by the female ovarian fluid, a semi-viscous protein rich fluid that baths the eggs, in which conspecific sperm are favored (Yeates et al. 2013).

Cryptic choice is apparently not restricted to discrimination between con- and heterospecific sperm, however (Eberhard 1996). In several species females appear to selectively choose sperm from genetically unrelated males, although the specific mechanism for cryptic choice usually remains obscure.

A number of proteins expressed in eggs and sperm are involved in the fertilization process. The genes coding these proteins are evolving very fast and show signs of being affected by positive or diversifying selection (Vacquier & Swanson 2011). This suggests that selection at the gametic level is strong and widespread. Yet, fertilization is as yet a surprisingly poorly understood biological process. How sperm competition, cryptic choice, and other selection processes drives the rapid evolution of fertilization proteins is likely to be an active and rewarding field of research in the years to come.

Study questions

1. Explain the difference between:
   a. Stabilizing and balancing selection.
   b. Disruptive and divergent selection.
   c. Directional selection and purifying selection.
   d. Fertility selection, viability selection, and sexual selection.
   e. Cryptic choice and female choice.
2. Discuss how F_{IS} and/or F_{ST} may be affected by the forms of selection mentioned in question 1a–c.
3. Consider the one-locus model of viability selection $p_{t+1} = \frac{pw_1^*}{\bar{w}}$. Find the equilibrium frequencies of A_1 and determine their stability when:
   a. $w_{11} = 0.9$, $w_{12} = 1$, and $w_{22} = 0.2$
   b. $w_{11} = 1$, $w_{12} = 0.9$, and $w_{22} = 0.8$
   c. $w_{11} = 1$, $w_{22} = 0.6$, and $w_{22} = 0.9$.
   d. What type of selection operates in a, b, and c?
   e. What will the frequency of A_1 be after one generation of selection when $p = 0.2$, $w_{11} = 1$, and $w_{12} = w_{22} = 0.6$?
4. Explain how different types of dominance affect the evolutionary dynamics and outcome in the one-locus model of viability selection.
5. Show that $\Delta p = p_{t+1} - p = \frac{pq}{2} \frac{1}{\bar{w}} \frac{d\bar{w}}{dp}$ and explain what the different components can reveal about the evolutionary dynamics and equilibriums of the model.

6. Multiple alleles at a locus. Assume a single locus with three alleles, A_1, A_2, and A_3. Let p_1, p_2, and p_3 denote the frequencies of A_1, A_2, and A_3, where $p_1 + p_2 + p_3 = 1$.
   a. Write down the Hardy–Weinberg frequencies of the different genotypes.
   b. Write down expressions for the marginal fitnesses of the three alleles.

7. Find the polymorphic equilibrium and calculate M and S to check if the stability criterion $M + S < 0$ holds when the frequency-dependent fitness functions are:
   a. $w_{11}(p) = 1 + p^2$, $w_{12}(p) = 3 + 2pq$, $w_{22}(p) = 1 + q^2$
   b. $w_{11}(p) = 3 - p^2$, $w_{12}(p) = 3 - 2pq$, $w_{22}(p) = 3 - q^2$
   c. $w_{11}(p) = 3 - 2p^2$, $w_{12}(p) = 2 - 4pq$, $w_{22}(p) = 3 - 2q^2$.
   d. What kind of selection operates at the polymorphic equilibriums?

8. Mutation selection balance.
   a. Consider a disease caused by a recessive allele that leads to early death ($s = 1$). The frequency of the allele is $1/200$. If the disease is at mutation–selection balance, what is the mutation rate? (Random mating is assumed.)
   b. Cystic fibrosis is caused by a recessive allele, and the disease affects roughly 1 out of 2500 Caucasian people (estimates vary). Until modern times, cystic fibrosis usually caused death in childhood ($s \approx 1$). Assuming random mating, calculate the mutation rate at mutation–selection balance. A reasonable mutation rate for a locus is 10^{-5}. Could other forms of selection than directional selection against cystic fibrosis be at work? (You may search Google Scholar for "advantage cystic fibrosis allele.")

9. Assume a single locus with two alleles, A_1 and A_2, where the frequency of allele A_1 is p. Assume Hardy–Weinberg frequencies of zygotes.
   a. Let $w_{11} = 0.5$, $w_{12} = 1$, and $w_{22} = 0.75$. Calculate the genetic load at $p = 0.5$.
   b. Calculate the genetic load at the equilibrium allele frequency.
   c. Let $w_{11} = w_{12} = 1$, and $w_{22} = 0.75$. Calculate the genetic load at $p = 0.5$.
   d. Calculate the genetic load at the equilibrium allele frequency.

10. Right or wrong? Determine whether the following general claims about single-locus, biallelic traits are correct or not.
   a. It is possible to have a stable equilibrium at which no heterozygotes exist.
   b. At a polymorphic equilibrium, all genotypes have equal fitness.
   c. Under complete dominance, all genotypes will have equal fitness at a polymorphic equilibrium.
   d. Under incomplete dominance, at least some of the genotypic fitnesses must differ at a polymorphic equilibrium.
   e. The heterozygote can be at a disadvantage at a frequency-independent polymorphic equilibrium.
   f. The heterozygote can be at a disadvantage at a stable frequency-dependent polymorphic equilibrium.
   g. At a polymorphic equilibrium maintained by frequency-dependence, all morphs must have equal fitness.
   h. The marginal fitnesses of the two alleles may differ at a polymorphic equilibrium.

CHAPTER 5

The power of natural selection

Adaptations are the products of natural selection. They are traits that evolved because they proved useful at some level. Often adaptations evolved because they enhanced the survival or reproductive output of individuals, but selection can also operate at other levels—from genes, via individuals, to populations and species. Sometimes a genetic change has positive effects on all levels. A mutation that increases the survival of its carrier would increase in frequency because of those beneficial effects; populations that become fixed for that allele may be less prone to extinction, which in turn may increase the longevity of that species. Other times there can be conflicting fitness effects at the different levels. In this chapter, we shall explore the power of natural selection in shaping the living world: We shall investigate the complexities of multilevel selection, study biological solutions to heterogeneity and unpredictability in the environment, and learn how interactions between species can shape evolution. However, we start this chapter by investigating factors that may constrain adaptive evolution.

5.1 The limits of natural selection

5.1.1 Genetic constraints

In the preceding chapters, we have seen many examples of imperfect adaptations. In many cases, it is the nature of heredity that acts as a limiting factor. It is reasonable to assume that trichromatic vision is a beneficial trait in diurnal primates. For instance, discriminating red from green makes it much easier to spot ripe fruits, which are important sources of food for many primate species. Yet, in most New World monkeys only one X-linked photopigment gene is present. Only females that are heterozygous for photopigments of different wavelengths have the optimal phenotype of trichromatic vision. Males, which only have one X-chromosome, and homozygous females are red–green color blind. This heterozygote advantage (complicated here by sex-linked inheritance) is one example of a **genetic constraint**. The optimal genotype cannot go to fixation because, due to Mendelian segregation, heterozygotes cannot reproduce heterozygotes only. The problem can only be solved by gene duplication and fixation of either allele in each of the resulting gene copies, which is actually what has occurred in primates: in the ancestor of the Old World monkeys and apes, and in the howler monkey (see chapter 2).

Genetic constraints may also prevent a population from climbing to the highest fitness peak in an adaptive landscape. In chapter 4 we saw that underdominance (heterozygote disadvantage) can have this effect. A novel mutation which is favorable in homozygotes (A_2A_2), but has lower fitness than the ancestral genotype (A_1A_1) in heterozygotes A_1A_2 cannot go to fixation by natural selection. This is true even if the A_2A_2 genotype has higher fitness than A_1A_1 (Figure 5.1). Framed in the metaphor of the adaptive landscape, the population is unable to traverse the valley of lower fitness in order to reach the optimum adaptive peak by natural selection alone. Constraints on natural selection will also

Evolutionary Genetics: Concepts, Analysis, and Practice. Glenn-Peter Sætre & Mark Ravinet, Oxford University Press (2019).
© Glenn-Peter Sætre & Mark Ravinet 2019. DOI: 10.1093/oso/9780198830917.001.0001

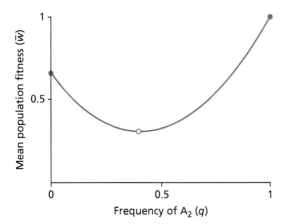

Figure 5.1 Underdominance is one example of a genetic constraint. A novel, mutant allele A_2 which is beneficial when homozygous A_2A_2 cannot spread in a population by selection when the heterozygote has low fitness.

arise when many genes affect a phenotype. Selection on one of these genes will lead to a correlated response in the others and all the genes involved are unable to evolve as independent units. Interactions between alleles at different loci can cause complex, multidimensional adaptive landscapes that further constrain adaptive evolution. We will explore more complex examples of multilocus evolution further in chapter 6.

Pleiotropy, when one gene affects more than one trait is another example of a genetic constraint. In chapter 2 we saw that ε-crystalin is a structural protein in the eye lens of birds and crocodiles and that the same protein functions as a digestive enzyme in their intestine. A mutation that improves the protein's properties as a digestive enzyme may impair its properties as a structural protein in the eye lens and vice versa. There is likely to be a **trade-off** between the two different functions of the gene, constraining optimal functionality in both of the phenotypic traits it is involved in.

5.1.2 Historic and ontogenetic constraints

A population may be poorly adapted to its environment because it lacks the necessary genetic variation that natural selection is able to act upon. Conspicuous examples of such a mismatch between the need for adaptation and actual response to this need can

sometimes be seen following rapid environmental change. There can be a **time-lag** before an organism manages to respond to the new conditions—failing to adapt quickly enough, or at all, may mean it will go extinct. Highly inefficient seed dispersal in many North and South American fruit species, such as the avocado (*Persea americana*), provides a good example of a poor adaptive response to environmental change (e.g. Guimaraes Jr. et al. 2008). Typically, fruits are effective structures for seed dispersal. Mammals and birds eat the fruits and spread the seeds when they later defecate. The seeds are swallowed by the consumer and are *scarified* as they pass through the animal's digestive tract. That is, the seed coat is weakened, and this helps the embryo to germinate. This process is crucial for dispersal; seeds in rotting fruit will often fail to germinate because they have not been scarified.[1] However, no living animal in the Americas is big enough to swallow the huge seed of the wild avocado and so it spreads very inefficiently. The majority of fruits fall to the ground and rot. Even if a seed germinates, it would usually have to compete with its parent for light and nutrients because its dispersal distance is almost zero. Why then has the avocado evolved in such a way? The most likely explanation is that the large size of the fruit and seed is an adaptation for dispersal via large animals that are now extinct. The megafauna of the Americas included giant sloths and elephants that would have been perfect partners for seed dispersal. These species went extinct some 13 000 years ago and the avocado and several other plant species that relied on having their seeds spread by the largest of the megafauna have not been able to respond to their absence by evolving fruits with smaller seeds that extant animals are able to swallow and spread. This constraint on adaptation therefore arises as a result of historical contingency.

Evolution by natural selection occurs without foresight. Since it is a blind process, one can liken it to tinkering with the available genetic variation. This often leads to odd (and sometimes less efficient) solutions to the problems organisms face. Many strange adaptations can only be understood in light of the evolutionary history of the organism

---

[1] Avocado farmers can increase germination success by artificial scarification of the avocado seeds.

in question. Ancestral shapes and body plans often set the stage for the course evolution is able to take and often this appears to have acted to limit optimal design. Flatfish are benthic (bottom dwelling) fishes that gain protection from predators by having a flat body and camouflage colors on their dorsal (upper) side. The obvious body plan for any benthic dwelling fish species is that observed in skates and rays—a downward facing flat belly (ventral side) and a camouflage-colored dorsal—i.e. a dorsoventrally compressed body shape. Skates and rays evolved from sharks which are also slightly dorsoventrally compressed. Hence, evolution likely made this compression more extreme during the evolution of these bottom dwellers. Bony fishes, however, are typically laterally compressed. Hence, a bony fish that tries to hide flat on the ocean floor would be much better off laying on its left or right side than on its belly. This explains the origin of the bizarre, asymmetric body shape of flatfish such as flounders and halibut as they adapted to a benthic life style (Figure 5.2).

Figure 5.2 Skates and rays (top) have a dorsoventrally compressed body plan because they evolved from slightly dorsoventrally compressed ancestors. In contrast, the asymmetric head and body of benthic flounders (bottom) evolved because they are the descendants of ventrally compressed ancestors.

Laying on the side would mean that one of the eyes would be facing the ocean floor where it is of no use. Natural selection has solved this problem by moving one of the eyes to the opposite side of the skull. A flatfish larva has normal symmetry with one eye on either side of the skull. During ontogenesis one of the eyes wanders to the opposite side, along with other asymmetric adjustments of the skull, gills, and mouth until the adult shape is reached (e.g. Fuiman 1997). Such "best of a bad situation" solutions are due to **ontogenetic constraints**. In short, since flatfishes are descended from a laterally compressed ancestor, the evolution of skull asymmetry was a more feasible evolutionary pathway than evolving dorsoventral compression.

Despite popular misconceptions of evolution as progressive and resulting only in "the survival of the fittest," our own bodies exhibit numerous seemingly suboptimal designs that are best understood as the result of ontogenetic constraints. An obvious example is that natural human birth requires the squeezing of a large skull through a narrow pelvis—a painful and often dangerous process. This is most likely because we are the descendants of quadrupedal ancestors with much smaller skulls.

5.1.3 The problem of stasis in evolution

Sometimes natural selection causes rapid evolutionary change. As we saw in the previous chapter, the peppered moth evolved camouflage colors to dark surfaces within a few decades when industrial pollution removed lichens from trees and made many surfaces dark and sooty. In contrast, other organisms have hardly changed morphologically in millions of years. Fossils of horseshoe crabs from the Ordovician period, about 445 million years ago, are remarkably similar to modern representatives (Rudkin et al. 2008) (Figure 5.3).

Evolutionary stasis, in which a species remains similar over long evolutionary time spans, appears to be common in the fossil record. In 1972 paleontologists Niles Eldredge and Stephen Jay Gould published an influential but controversial paper where they presented the theory of **punctuated equilibria** (Eldredge & Gould 1972). The theory was inspired by observations from fossil bearing strata. In these strata, a common observation is that

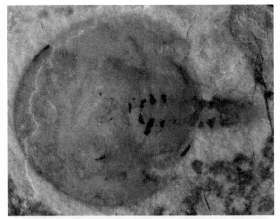

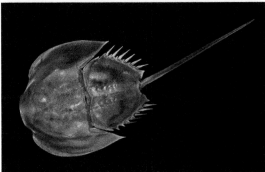

Figure 5.3 The horseshoe crab is often regarded a living fossil because its body shape has remained remarkably similar over hundreds of millions of years. Top panel: the fossil horseshoe crab *Lunataspis aurora* from the late Ordovician, about 445 million years ago. Image ℗ The Manitoba Museum, Winnipeg, MB. Bottom panel: a modern horseshoe crab, photo courtesy of Didier Descouens.

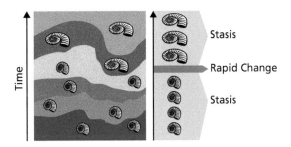

Figure 5.4 It is commonly observed that one fossil phenotype is abruptly replaced by a different phenotype in a fossil bearing stratum (left). According to the theory of punctuated equilibria, this pattern emerges because the fossil species has experienced phases of evolutionary stasis interrupted by a phase of rapid evolutionary change (right). See main text for alternative hypotheses for the pattern.

species remain relatively unchanged over long periods of time before they suddenly are replaced by another, notably different species (Figure 5.4).

Many factors could explain such a pattern. The fossil record is incomplete and, except in rare cases, has low temporal resolution. Therefore, a change that appears suddenly and abruptly in the fossil record could have taken thousands of years to occur. Moreover, the sudden appearance of a different species in a stratum may reflect range shifts and competitive exclusion rather than evolutionary change. Furthermore, the fossil record typically captures only morphological characteristics from taxa.[2]

Accordingly, we can only determine morphological stasis from fossil data—information on behavioral, ecological, and physiological adaptations over such a long period of time is largely missing. However, Eldredge and Gould argued that the fossil record may accurately reflect the paths of evolution: Normally, time is able to pass without a great deal of evolutionary change (stasis). Then bursts of rapid evolution take place in connection with speciation events (Figure 5.5).

Assuming that the fossil record does accurately reflect stasis and accepting that some species such as the horseshoe crab do indeed show morphological stasis, what could cause this? Evolutionary stasis could be the result of genetic constraints. The influential evolutionary biologist, Ernst Mayr, held that the genomes of living organisms are tightly co-adapted entities maintained through purifying selection and gene exchange in large, interbreeding populations (Mayr 1942, 1954). According to this view, a change in one gene would most likely disturb the fine-tuned and complex network of gene interactions that natural selection has sculpted over millions of years—so selection would act against it. Alternatively, intraspecific gene flow would swamp such a change, meaning it is unable to establish. Mayr held that speciation and evolutionary change for the most part occur under very special circumstances in which adaptive gene interaction networks break down, causing a **genetic revolution**. This would occur during founder events, when a small number of individuals colonize a new area,

---

[2] There are several examples of trace fossils, such as footprints, which can shed light on behavior and ecology.

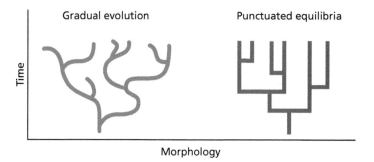

Figure 5.5 Two models of phenotypic change during evolutionary history. a) Under gradual evolution change occurs at any time, although the speed of change may vary. b) According to the punctuated equilibrium model evolutionary change occurs in bursts in connection with speciation events followed by long periods of little or no change (stasis).

such as an island. Following a founder event, the population size would be small, causing a **founder effect**, an increased rate of genetic drift. According to Mayr, random genetic drift would tend to break up gene interaction networks. Natural selection would then finally be able to sculpt novel, fine-tuned gene interaction networks and a new species would emerge relatively rapidly. Eldredge and Gould's punctuated equilibria theory was certainly inspired by Mayr's thoughts on speciation.

There are mixed opinions on Mayr's theory of genetic revolutions in small founder populations (e.g. Barton & Charlesworth 1984; Templeton 2008). It is true that genes do interact and that random changes by genetic drift can alter these interaction networks, setting the stage for natural selection to take evolution in a new direction. On the other hand, there are also theoretical reasons to expect adaptive evolution to be faster in large, outcrossing populations than in small founder populations. Assuming a constant mutation rate μ per locus per generation, the number of new mutations in a population would be proportional to its size $2Ne\mu$. Advantageous mutations would thus arise at a proportionally higher frequency in a large population than in a small one. Moreover, smaller populations would be more prone to lose adaptive genetic variation through the increased rate of genetic drift they would experience (see chapter 3). Finally, there are now countless examples of recent or contemporary adaptive evolution taking place from a range of different species. This is in stark contrast to the apparent stasis observed in the fossil record: Organisms can evolve rapidly and readily do so. Hence, although genetic constraints can limit the evolutionary potential of a species in certain directions of phenotypic space it is probably not the main cause of stasis in evolution.

Stabilizing selection would tend to keep a species from changing. If a species lives in an environment that does not change a great deal over long periods of time there would be little or no selection pressure for evolutionary change. Our planet has seen a lot of environmental change in the 445 million years since the Ordovician period when the horseshoe crabs first appeared. However, the oceans have provided muddy benthic floors where prey items such as worms and molluscs have been plentiful during such a long period. It is reasonable to argue that a body plan that functioned well on muddy sea floors during the Ordovician period may function equally well today. Analyzing 450 fossil time series, Voje (2016) demonstrated that taxa that can be characterized as having experienced evolutionary stasis evolve equally much in terms of amount of morphological change per unit of time as those experiencing directional change. The difference is that the former species fluctuate around mean trait values whereas the latter experience change in a particular direction (Figure 5.6). Stasis thus appears to represent the persistence of adaptive zones and ecological niches over millions of years. Apparent stasis is the result of net selection in effect being stabilizing.

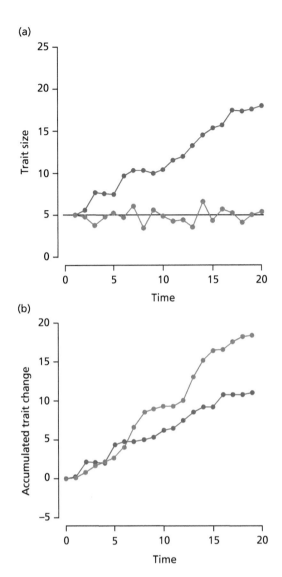

Figure 5.6 Fossil lineages experiencing evolutionary stasis (blue lines) experience a similar amount of morphological change as those experiencing directional change (red lines). The former experience fluctuations around mean trait values, whereas the latter experience change in a particular direction (top panel). Accumulated change per unit time was found to be similar in the two groups (bottom panel). Modified from Voje (2016).

5.2 Level of selection

5.2.1 Level of sorting, adaptation, and immortal genes

At what level is natural selection operating? This has been a recurrent discussion in evolutionary biology. However, such debates have rarely been fruitful because biologists may mean different things when they talk about selection.

First, natural selection involves sorting. Some individuals survive to reproduce, others die at a young age; some taxa have high speciation rates, others rarely speciate or have high extinction rates. It is legitimate to say that selection operates at the level where sorting takes place. Individual selection occurs when there is sorting in terms of differential survival and reproduction between individuals. When we first introduced selection, it was at the individual level. In a similar manner, **species selection** occurs when one taxon has a higher speciation rate and/or lower extinction rate compared to a different taxon. For instance, among ungulates the proportion of even-toed species (antelopes, cattle, deer, etc.) has increased during the Tertiary and Quaternary periods (the last 65 million years) relative to odd-toed ones (horses, rhinos, tapirs, etc.) due to differences in the ratio of speciation and extinction events in the two groups. Importantly, the number of toes in ungulates is a convenient way of grouping species in the clade based on their descent from common ancestors with the odd- or even-toed characteristic. The number of toes itself is unlikely to be responsible for the different speciation/extinction ratios among the groups—as we shall see in a moment, this has probably arisen due to divergent adaptations at the individual level.

Second, natural selection results in adaptation. In this sense, it is legitimate to say that selection operates at any level where the process results in an adaptation. For example, segregation distortion is an adaptation at the gametic level. It is favored because sperm carrying the distorter will have higher fertilization success than those that do not. However, most adaptations appear to be individual adaptations—that is, traits that enhance the survival probability or reproductive success of individuals. Nonetheless, such individual adaptations can have correlated effects at higher levels of organization. Returning to ungulates, those with even toes have a more efficient digestive system than odd-toed ungulates and this may be part of the reason why the former has diversified into a larger number of species. However, it is very unlikely that the digestive system of even-toed ungulates evolved *because* it enhanced the speciation/extinction ratio of these species. Taxon sorting is simply a consequence of ancient adaptations at the individual level. This means that when we focus on the level of adaptation, species selection is an epiphenomenon. It can explain trends in the fossil record but it does not bring us any closer to understanding how and why an adaptive trait evolved. The concept of species selection is not without use, however; it helps to explain why traits are maintained during evolutionary history (see section 5.3.4).

Finally, one could argue that all kinds of selection can be reduced to the level of the gene. Individuals die, but their genes may linger for long evolutionary time through their descendants. Richard Dawkins popularized a gene-centered view of evolution in his best-selling books *The Selfish Gene* and *The Blind Watchmaker* (Dawkins 1976, 1986). In his language genes, or *replicators*, are potentially immortal. They compete among themselves for survival and transmission. Building well-functioning but transient individual organisms or *vehicles* is an effective way for replicators to linger. Individuals can be likened to temporary *survival machines*, built and used by mindless replicators for their own selfish means.

Dawkins' extreme reductionism has provoked many biologists. Organisms and ecosystems are arguably much more than individual genes competing for transmission. Moreover, a gene-centered view may seem to suck all meaning out of life. However, although it may be unpleasant to think of ourselves as nothing but dispensable survival machines for our genes it is a view of evolution that has its merits. A gene-centered view can help us think correctly when we strive to understand complex phenomena such as genetic conflict and kin selection.

5.2.2 Genetic conflict and gene level selection

A segregation distorter is an adaptation at the gametic level. It did not evolve because it does anything good for an individual's survival or fecundity but because sperm with the distorter gain a transmission advantage by killing other competing sperm. On the contrary, a segregation distorter will often be harmful at the individual level. There is often a substantial reduction in fertility for males possessing the segregation distortion alleles K_1T_1 (Palopoli 2000). Accordingly, there is a genetic conflict between the segregation distorter and other genes in the genome. A mutation that suppresses or limits sperm killing and restores male fecundity would be favored by selection at the individual level.

Selective sperm killing does not just harm the gene complex (K_2T_2) targeted by the segregation distorter. Alleles that are closely linked to the complex are also at risk. This is because they are typically inherited together with the target. Hence, mutations that silence selective sperm killing should tend to accumulate close to K_2T_2. For an allele that is closely linked to the segregation distorter (K_1T_1) the situation is the opposite. It will be more likely to be inherited alongside the killer allele (K_1) and therefore would benefit from an increased rate of selective sperm killing. In other words, it too experiences beneficial segregation distortion. As predicted, it is commonly observed that the two key loci of segregation distorters (K and T) are associated both with silencers and enhancers of sperm killing (Palopoli 2000).

The outcome of this genetic conflict ultimately depends on the relative fitness consequences at the gametic and at the individual level. If the fitness costs at the individual level are low the distorter may go to fixation. At this point, selective sperm killing would cease because there would be no target alleles left to attack. Mendelian inheritance would be restored. Alternatively, if the fitness consequences at the individual level are severe the segregation distorter may become completely suppressed by silencers. Again, Mendelian segregation would be restored. At intermediate levels of individual fitness costs the antagonistic selective forces at the gametic and individual levels may balance sufficiently for the distorter to be maintained at a low to moderate frequency in the population for a long time. This may even result in repeated events of evolutionary moves and countermoves due to the appearance of enhancer and silencer alleles linked to K_1T_1 and K_2T_2 respectively.

Mendel's first law is not a universal law of nature. Often, it is the result of genomic conflicts in which selfish alleles are kept at bay by selection at the individual level. One way to view Mendelian inheritance is as a delicate truce between competing alleles that sometimes breaks.

Segregation distorters are not the only selfish genetic elements that occur. Transposable elements (TEs) parasitize the replication and transcription machinery of the cell in order to proliferate in our genomes. At best, they do only accidental good to the survival or reproduction of the individual, but more often they cause great harm (see chapter 2). Like with segregation distorters, defense mechanisms against TEs at the individual level have evolved. These include epigenetic silencing of active TEs and post-transcriptional degradation or inhibition by specialized RNA molecules. The structure and size of the eukaryote genome is to a large extent a consequence of repeated cycles of genomic conflict between the host genome and TEs throughout evolutionary history.

Genomic conflicts can also arise between genomic elements that owe their transmission success to the survival and reproductive success of the individual. Such conflict occurs whenever genetic elements only have partly overlapping interests with respect to transmission. Mitochondria are important organelles supplying energy to the cell. They carry their own DNA, called mitochondrial or mtDNA, and are inherited through the maternal line. Sons inherit their mitochondria from their mother but there they stop. Males do not contribute mitochondria to their offspring.[3] The difference in pattern of inheritance of mtDNA and nuclear DNA can set the stage for a genomic conflict. A mutation in mtDNA that is harmful to males in terms of survival or fecundity, but not to females, would be favored by selection at the mitochondrial level. Females with such mutant mitochondria would produce an excess of daughters that are able to propagate the mitochondria and fewer sons, which are dead ends with respect to mtDNA transmission. This phenomenon is called **mother's curse** (Gemmell et al. 2004). It is clearly not an adaptation at the individual level. This is most obvious for males as they experience reduced survival or fecundity, but it is not usually beneficial for females to possess mother's curse mitochondria either. The mother's curse mutation would eventually lead to a skewed sex ratio in the population with an excess of females. Competition for mates

would thus be stronger among females than among males. Hence, at this stage a female would be better off producing sons to secure grandchildren. At the individual level, there is balancing selection for an even sex ratio (Fisher 1930). An important point here is that this genetic conflict is not primarily between the sexes but between the mtDNA and the nuclear genome. Nuclear modifiers that restore male fitness (and an even sex ratio) would be favored by selection at the individual level. Mother's curse mutations are common in nature. One human example is the condition known as *Leber hereditary optic neuropathy*, a mitochondrial genetic disease that causes blindness in young adult males (Man et al. 2002).

Genetic conflict is a corollary to Mayr (section 5.1.3). Genomes are not fine-tuned, co-adapted entities in which their constituent genes collaborate to produce well-functioning individual organisms. They are battlegrounds. Sometimes, however, the better option is to team up and collaborate in the fight for survival and transmission.

5.2.3 Kin selection

JBS Haldane allegedly once uttered the following words after having done some calculations on the back of an envelope:

"I'm prepared to lay down my life for eight cousins or two brothers."

He had calculated and deduced the logic of kin selection. **Kin selection** is a theory that seeks to explain altruistic behavior (Hamilton 1964a; 1964b; Maynard Smith 1964). In other words, why would an individual act in a way that seems to favor others but not obviously themselves? A famous example of biological altruism is the reproductive behavior of bees, ants, and other **eusocial** insect species, in which a large number of individual females forfeit their own reproduction and instead help their sister, the queen, to raise her offspring. As the (probably apocryphal) quotation from Haldane alludes to, kin selection can be understood as selection favoring traits through their positive fitness effects on genetic relatives (receivers) despite negative fitness consequences for the ones performing the altruistic acts (actors). The mathematics underlying this principle

[3] Occasionally some male mitochondria can be transferred at fertilization (e.g. McCauley 2013).

is actually very simple and can be presented in the form of Hamilton's rule (Hamilton 1964a; 1964b; West et al. 2007). An altruistic allele will increase in frequency by kin selection in a population when the following equation is satisfied:

$$rB > C \qquad (5.1)$$

That is, it will spread when the benefit in terms of average increased fitness to the receiver (B), weighted with degree of relatedness between the actor and the receiver (r), exceeds the average cost in terms of reduced fitness for the actor (C).[4] The **coefficient of relatedness** r is a measure of how closely related two individuals are and can be defined as the probability that two genetic relatives share an allele that is **identical by descent**. That is, not only do the two relatives share an allele, say A_1, but they have inherited it from a common relative. As an illustrative example, we will focus on a particular allele at a diploid locus in a human female. Let us call her Mary. Half her eggs would carry a copy of that allele. Hence, there is a 50% probability that it will end up in any of Mary's offspring. Thus r = 0.5 between parent and offspring. If one of Mary's offspring inherited this allele the probability that a brother or sister also inherited it is likewise 50%. Thus r = 0.5 also between siblings. What would r be between first cousins? To calculate this, we need to count the conditional probabilities that it is shared between the line of closest relatives connecting the two cousins. Let us say that Mary's son Bobby has a cousin called Becky, who is the daughter of Mary's brother Mike (Figure 5.7). What is the coefficient of relatedness between Bobby and Becky? First, there is a 50% chance that Bobby inherited an allele from his mother Mary (r = 0.5). Second, there is a 50% chance that Mary shares that allele with her brother Mike if she has it, so the coefficient of relatedness between Bobby and uncle Mike would be r = 0.5·0.5 = 0.25. Third, and finally, there is a 50% chance that Becky would have inherited that allele from her father if he possesses

it. Hence, the coefficient of relatedness between Bobby and Becky becomes r = 0.5·0.5·0.5 = 0.125. There is a 12.5% probability that these two cousins will share a particular allele that is identical by descent, ultimately inherited by both from a common grandparent.

Kin selection was formulated by Hamilton as a special case of individual selection, in which individuals have **inclusive fitness** rather than individual fitness. Mary's (expected) inclusive fitness would equal the number of surviving offspring she produces, plus the number of extra offspring genetic relatives of Mary's genetic relatives produce as a consequence of her altruistic acts. This is then weighted by degree of relatedness between Mary and her relatives, minus any extra offspring Mary is only able to produce as a consequence of help received from relatives. Although mathematically correct, inclusive fitness is quite a complex concept and unsurprisingly, it is hard to measure in the wild. Kin selection can, however, be formulated and understood using only the direct fitness concept that we introduced in chapter 4 (Taylor et al. 2007) and many authors now prefer this direct fitness approach over inclusive fitness when modeling kin selection.

Despite the simplicity of Hamilton's rule, kin selection is often misunderstood. According to the theory an actor would on average benefit indirectly in terms of fitness from acting altruistically through the positive fitness effects it has on genetic relatives. But is this always the case?

We can test this with a thought experiment. Say that a mutation A occurred in an individual male that caused him to forfeit his own reproduction and rather stay home and help his parents raise younger siblings. Say that he could have expected to raise two offspring if he had started a family on his own (C = 1, see footnote 4), which is the mean fitness in the population, and by helping his parents they succeed in raising three more offspring than they would have managed on their own (B = 3). The coefficient of relatedness between siblings is 0.5, so Hamilton's rule appears to be satisfied: 0.5·3 > 1. Would the A-allele increase in frequency? Obviously not. No one would inherit the altruistic allele and it would disappear with the altruistic male because he does not reproduce. Clearly, carriers of an altruistic

[4] Note that when we measure the cost to the actor C as number of offspring he/she forsakes by acting altruistically it takes into count the relatedness between the actor and those offspring. The coefficient of relatedness between parent and offspring is 0.5. Hence, if an altruist forfeits two offspring from helping relatives raise theirs, C equals 1.

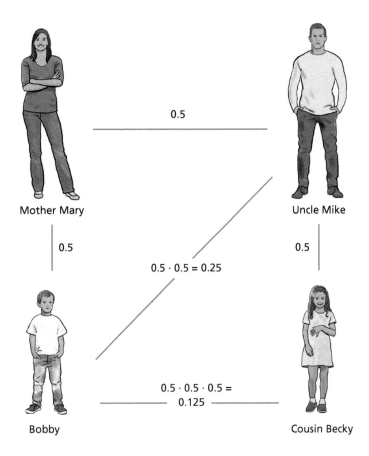

Figure 5.7 Calculating the coefficient of relatedness r between genetic relatives. Due to independent Mendelian segregation, the likelihood that two individuals share an allele identical by descent equals the conditional probability that they have inherited it from a common recent ancestor.

allele must reproduce and on average produce more offspring than those that do not carry it if this allele is to increase in frequency in the population.

Switching to a real-world example, how then can we explain the origin of sterile castes in eusocial insects? Consider bees with workers that remain sterile to nurture the offspring-producing queens—these are essentially the same as the male in our thought experiment. In order for such a caste to evolve, the queen must carry the altruistic allele too. The allele has to be differentially expressed between the queen and the workers: help when in a worker, exploit help when in a queen. Genes are indeed expressed differently in workers and queens and this arises when larvae that are nursed to become queens are fed large amounts of royal jelly, which epigenetically triggers the development of queen

morphology and behavior (Kucharski et al. 2008). Perhaps we shouldn't focus solely on why some individuals behave altruistically towards relatives. Instead we might be best placed to ask why cooperation and division of labor evolves among genetically related individuals.

The problem with our previous thought experiment above is that we misrepresented the significance of the coefficient of relatedness. Relatives do not help each other because they are closely related *per se* or because they have some mystical interest in the fate of each other's genes. Rather, the point is that in a situation where an altruistic allele is rare in a population it would be much less rare among close kin, because kin inherit alleles from each other. The coefficient of relatedness can be understood as the probability that a pair of relatives would share a copy of not just any allele, but *the altruistic allele itself*, when it is still rare in the population. Hamilton's rule represents the threshold condition for when the invasion fitness of a rare altruistic allele would be higher than the resident fitness of the ancestral, "selfish" allele.

We are getting closer to understanding kin selection, but we are not quite there yet. Let us make a second thought experiment where we correct the mistakes from the first one. We let carriers of a rare altruistic allele A reproduce, so that when they do so there is a 50% chance that parents and offspring, as well as siblings, would share the allele. We consider a simple population with non-overlapping generations, which at one point in time consists of only one group of siblings in which some carry the A-allele and others the non-altruistic, or selfish, alternative allele S. For simplicity, we assume that the organism is haploid albeit sexually reproducing. Hence, a mating between an A and an S would result in offspring in which half are A and half are S. Altruists help all their siblings and this increases the fitness of the recipients at some cost to the actor. Say that without receiving any help, an individual would on average produce one offspring ($W_0 = 1$). We set the costs and benefits so that Hamilton's rule would be satisfied: $B = 1$, $C = 0.25$ and $r = 0.5$. From Figure 5.8, however, we see that it is S that increases in frequency, even if Hamilton's rule is (apparently) satisfied. Carriers of the selfish allele receive the benefit of altruism without paying the cost (Table 5.1). Does this mean that Hamilton's rule is false? No, it does not. But it shows that $rB > C$ alone is not a

Table 5.1 Fitness and frequency of altruist (A) and selfish (S) genotypes in a population of five haploid full siblings in which $W_0 = 1$, $B = 1$, and $C = 0.25$.

	A	S
Frequency before selection	$3/5 = 0.6$	$2/5 = 0.4$
Fitness (number of offspring) per individual	$W_0 + B - C =$ $1 + 1 \cdot 2 - 0.25 \cdot 4 = 2$	$W_0 + B$ $1 + 1 \cdot 3 = 4$
Frequency after selection	$6/(6+8) = 0.429$	$8/(6+8) =$ 0.571

sufficient condition for a rare altruistic allele to increase in frequency in a population.

The problem with this second thought experiment is that Hamilton's rule only applies to populations that are structured into temporal family groups. In Figure 5.9 we use the same fitness parameters as in Figure 5.8 and Table 5.1 and compare groups of selfish individuals with a group containing altruists. Individuals in groups of selfish individuals would not receive any help from their siblings and would thus have fitness $W_0 = 1$. When Hamilton's rule is satisfied, the altruistic allele A would increase in frequency because family groups with altruists are more productive than are family groups of only selfish individuals. Kin selection involves selection at two levels; the individual level and the level of the family group. There is selection against altruists at the individual level (Figure 5.8) but selection for family groups with altruists (Figure 5.9). There is a clear parallel here between genetic conflicts and kin selection. In both cases, there is antagonistic selection

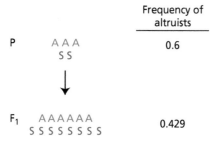

Figure 5.8 Change in number and frequency of altruists (A) in a population consisting of one group of interacting siblings in which some are selfish (S). Although the number of altruists increases, their frequency in the population decreases. Modified from Rice (2004).

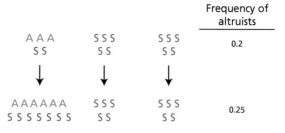

Figure 5.9 Change in number and frequency of altruists (A) in a population consisting of several groups of interacting siblings. The frequency of altruists increases in the population because groups with altruists are more productive that groups of only selfish (S) individuals. Modified from Rice (2004).

between two different levels of organization: gene versus individual, and individual versus family group, respectively. In both cases, it is possible for the higher level to win the conflict: Genomes in which the constituent genes collaborate to make successful individuals, and individuals that collaborate to make successful family groups. An altruistic allele can increase in frequency in a population if its negative fitness effect on individuals is more than offset by its positive effect on the productivity of family groups those individuals reside within. With kin selection, it is the beehive that becomes better adapted, not the individual bees. Cooperation and division of labor makes more productive hives. The beehive or the ant society can be thought of as a superorganism and it is this superorganism that is the target of kin selection with respect to level of adaptation. The individual bees and ants can be likened to the individual cells in a multicellular organism.

We may ask similar questions as to why the cells in our body collaborate. A large number of organisms are single celled, whereas we can be likened to huge colonies of collaborating cells. Our somatic cells are genetic dead ends. Yet they work together for the greater good, each tissue specialized to perform different tasks, thereby contributing to making a well-functioning body. During development, cells may even commit suicide through programmed cell death—the ultimate altruistic act. This is essential for the function of the "colony". There is a notable difference to eusocial insects with this example though—the cells in our body are more closely related to each other than are the bees in a hive, they are genetically identical clones. Kin selection has therefore been able to shape collaboration and division of labor to a much higher level of sophistication in multicellular organisms than in beehives. Indeed, the cells that constitute a multicellular organism are so intimately co-adapted and integrated that it is rather meaningless to think of them as individuals. They are integral parts of a higher level of organization, the multicellular individual (Michod 2007).

Multicellularity almost certainly evolved through kin selection. Present day slime molds may serve as a model system to understand the early stages of evolution of multicellularity. For most of their lives these creatures live as single-celled amoebas. However, when food becomes scarce the cells aggregate into multicellular fruiting bodies. These fruiting bodies consist of stalk cells that eventually die and spores that become spread and found a new generation of single-celled amoebas (Hudson et al. 2002). The stalk cells are thus altruistic. When the fruiting body consists of clonal cells from the same genetic lineage there is no conflict of interest between the level of the individual cell and the level of the fruiting body. Sometimes, however, amoeba cells from different genetic lineages aggregate in a fruiting body. When this occurs, there will be competition between the lines to become spore cells (Figure 5.10). Kinship between cells and benefits of cooperation and division of labor therefore explains the evolution of multicellularity.

Multicellularity, and secondly eusociality, can be viewed as the masterpieces of evolution by kin selection. However, kin selection is also relevant for understanding a range of behaviors and traits in social organisms, including in our own species. In chapter 4 we saw that conspicuous warning colors in poisonous or unpalatable prey species would be subject to positive frequency-dependent selection. When the trait is common in the population most predators will have learned that the signal is associated with something unpleasant and avoid attacking such prey. However, when the signal is rare most predators will be naïve; instead the conspicuous colors will make the prey easily visible and hence prime targets for being attacked. How could warning colors have evolved when they appear to be selected against when rare? The answer, in many cases, is likely to be kin selection. In insect species, such as butterflies, females typically lay their eggs in clusters. Hence, when the caterpillar larvae hatch they will be surrounded by a large number of siblings that start to forage together (Figure 5.11). When a warning signal is rare in a population it would nevertheless be *locally* common in such a group of foraging siblings because the siblings would have inherited the warning signal from a common recent ancestor. Hence, although naïve predators may attack a few of the strikingly colored siblings they would soon learn. Caterpillars in warning colored sibling groups may therefore have higher survival on average than those in groups without such colors, even when the warning signal is rare in the population.

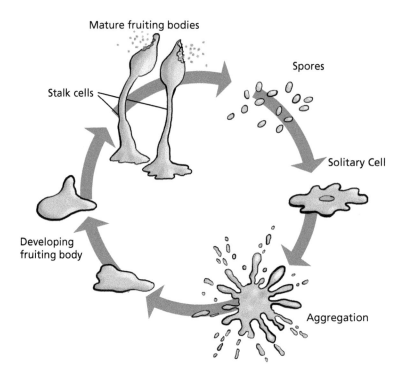

Mature fruiting bodies

Spores

Stalk cells

Solitary Cell

Developing
fruiting body

Aggregation

Figure 5.10 Simplified life cycle of slime molds. Amoeboid, solitary cells aggregate at food shortage to form multicellular fruiting bodies. Stalk cells die, whereas spore cells disperse and form a new generation of single-celled amoebas. Kin selection can explain the altruistic behavior of stalk cells.

Figure 5.11 Warning colors may increase in frequency by kin selection when rare. These caterpillars are siblings and enjoy protection by sharing a striking warning color. Photo courtesy of David Howarth.

5.2.4 Group selection

Sorting—one of the prerequisites for selection—can take place at the group, population, or species level. For example, groups can go extinct and be formed through colonization. However, can this sorting lead to adaptations at the group level? Most evolutionary biologists hold that such **group selection** would usually be a weak force in adaptive evolution. First, why would groups that are not made up of family members tend to share genetic properties that could be favored by a mechanism similar to kin selection (West et al. 2007)? Second, the turn-over rate of groups (founding/extinction) would usually be much slower than that of individuals (birth and death). Hence, adaptations that benefit individuals at the expense of the group would easily invade (Maynard Smith 1964; Williams 1966).

Before the 1960s biologists hardly questioned whether selection usually would act to the benefit of the species, group, or the individual. The fact that there could be conflicts of interest between levels of organization had not yet been discussed. However, with the publication of the book *Animal Dispersion in Relation to Social Behavior* (Wynne-Edwards 1962) this changed. Wynne-Edwards explicitly advocated the idea that adaptations are to the benefit of groups and not individuals. According to his view, a group of birds in an area of forest would restrain from laying too large clutches in order not to overexploit their resources, because that would threaten the survival of the group. It is, however, easy to see that such group level adaptations would be unstable and easily invaded by selfish individuals. A mutant individual that is able to lay a larger clutch than average would have more offspring than the other birds. The mutation would spread and go to fixation, even if this ultimately lowered the productivity of the group of birds. In short, individual selection would triumph over group selection.

Despite the inherent problems with the idea of group selection some evolutionary biologists still argue that it is likely to be an important force in evolution (e.g. Sober & Wilson 1998; Wilson 2005; Nowak et al. 2010). Indeed, much more sophisticated mathematical models of group selection than Wynne-Edwards' rather naïve verbal considerations have been put forward, and these appear to demonstrate that group level adaptations indeed can evolve (e.g. Nowak et al. 2010). However, careful re-examination of such group selection models shows that they are mathematically equivalent with kin selection models (West et al. 2007; Marshall 2011). It appears to be theoretically unlikely that group level adaptations can evolve except among close kin. Kin selection and not group selection explains altruism and adaptation at higher levels of organization.

5.2.5 Individual adaptations to living in groups

Social animals often possess traits and behaviors that appear to make them well suited for living in groups. The level of aggression is often low among group members: they may share food, collaborate in hunting, and cheaters or free-riders may be punished. We are social animals and this is apparent from our own evolutionary history. This does not necessarily imply that these traits have evolved through kin selection (or group selection). The group is an important part of the environment for the individuals living in it. Hence, individual selection would tend to favor traits that make them successful at living within a group (West et al. 2007). **Reciprocal altruism** is one example of such a trait.

It can explain certain types of altruistic behavior that do not require relatedness among the actor and the receiver in order to be favored by selection (Trivers 1971). An individual may benefit from helping another individual in the group at some immediate cost, if the favor is later repaid. Reciprocal altruism is widespread in human societies. We exchange favors all the time. However, such reciprocity puts high demands on cognitive skills. One needs to remember past actions and who did what in groups that can be quite large. This is likely beyond the cognitive abilities of many non-human animals. However, some interesting mammal examples have been reported (e.g. Wilkinson 1988).

The common vampire bat (*Desmodus rotundus*) of Central and South America roosts in small colonies in hollow trees or caves during the day and fly out to feed during the night. They feed primarily on blood from large mammals such as cattle, horses, and tapirs. As they return to the roost in the morning some of the bats may have been unsuccessful in finding food. Those that have been successful will regurgitate some of the blood they have gathered in order to share it with the unsuccessful ones. A bat that has received food from an individual in the group one night appears to remember the favor and will later be inclined to repay the favor when the roles are reversed and he or she is the one that has been successful in feeding (Wilkinson 1984). The cost of sharing food is likely low compared to the benefit of receiving. Vampire bats are small animals and cannot survive long without eating. Hence, receiving food after an unsuccessful night could mean the difference between life and death. Successful bats, on the other hand, may collect more blood than they actually need and so are unlikely to lose much from sharing their surplus.

One problem with the evolution of cooperation, such as food sharing in the vampire bat, is that it is vulnerable to cheating. What mechanism could stop a bat from parasitizing on the good will of the others? As we saw in chapter 4, game theory has been applied to evolutionary biology in order to understand the evolution of evolutionarily stable strategies—and the evolution of cooperation is an example of this. The problem of cooperation is often illustrated with the **prisoner's dilemma** game (Luce & Raiffa 1957). Imagine two prisoners that are the suspects of two crimes, of which one is minor and the other serious. They have agreed to keep quiet about the serious crime. The police have strong evidence against the two for the minor crime, but not for the serious one. Therefore, both are taken aside, one by one, and offered a deal: if they implicate the other prisoner for the serious crime, they will be freed of all charges. If one prisoner implicates the other but not vice versa, he will go free. The implicated prisoner would then be sentenced for the serious crime alone and serve ten years. If both implicate each other, they will get five years each. Finally, if both collaborate and keep to their initial agreement they would only serve one year each for the minor crime.

The prisoners must choose between two strategies. They may defect on their initial agreement and implicate the other (strategy D) or they may collaborate (strategy C) by sticking to their initial agreement and keep quiet (Table 5.2).

What would a rational prisoner choose? Assuming that the other prisoner has kept his promise and collaborated, the best response would be to defect. You would go free rather than spending one year in prison. Assuming that the other prisoner has defected, it would be better for you to defect than to stick to the agreement. You would spend five years in prison rather than ten. Regardless of what the other prisoner has chosen the best response would be to defect. Thus, if they act rationally, both would defect and spend 5 years in prison, even if they would have been better off by collaborating (one year in prison).

Interestingly, when human subjects participate in psychological experiments that are structurally similar to the prisoner's dilemma game they tend to behave nicer and more collaborative than the rational prediction. We are inclined to cooperate, but we also tend to punish people that refuse to cooperate. The prisoner's dilemma can be extended and modified to represent situations more frequently encountered in real life, in which humans (or animals) would have to choose between cooperating or defecting. If there are multiple encounters rather than just one, cooperative strategies tend to fare better. One simple, and now famous, collaborative strategy is called **tit-for-tat**. In a repeated prisoner's dilemma game the tit-for-tat strategist would always start by cooperating, but retaliate and defect at next encounter if the opponent defects. Tit-for-tat came out as the winning strategy in a computer simulation tournament in which game theorists anonymously submitted strategies that competed in a repeated prisoner's dilemma game (Axelrod 1984).

It is not hard to draw parallels to human societies, where reputation is important. People are inclined to trust you and be willing to collaborate with you if you have a good social reputation based on your previous actions in life. Free-riding and cheating tend not to be winning strategies in the long run (Santos et al. 2018).

5.3 Adaptation to a heterogeneous and unpredictable world

5.3.1 Phenotypic plasticity and learning

The world is far from static and accordingly, the optimal phenotype may differ under fluctuating and varying conditions. Natural selection may therefore favor plastic traits that are able to change depending on environmental cues. **Phenotypic plasticity** is the ability of a single genotype to produce more than one alternative phenotype in response to environmental conditions (West-Eberhard 1989). Changes in conditions can encompass a wide range of phenomena; the freshwater physid snail *Physella virgata* will adjust the shape and size of its protective shell depending on whether sunfish, its predators, are present in the water or not (Langerhans & DeWitt 2002). The sunfishes leave chemical cues in the water that induce reduced growth and the development of a rotund shell shape in the snail (Figure 5.12). Both traits are beneficial in environments containing these predatory fish. Reduced growth is associated with anti-predator avoidance behavior. The snails crawl out of the water in response to the chemical cues of fish predation. Out of the water, snails are safe from predatory fish, but cannot feed and so

Table 5.2 Payoff matrix for the prisoner's dilemma game.

	D	C
D	−5	0
C	−10	−1

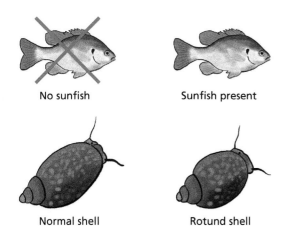

No sunfish   Sunfish present

Normal shell   Rotund shell

Figure 5.12 Phenotypic plasticity in physid snails. Chemical cues from predatory sunfish induces growth of a rotund shell shape and smaller size.

their growth rates are reduced. A rotund, robust shell shape is also much more difficult for fish predators to mechanically crush via biting than the more elongated normal shell shape. Hence, snails with the rotund shell phenotype are more likely to survive attacks from predatory fish.

These altered phenotypes increase survival when predatory fish are encountered but they clearly come with a cost. Reduced growth reduces fecundity and a rotund shell appears to make the snails more vulnerable to other predators such as crayfish. Thus, the alternative phenotype is only induced when sunfishes are present. Ironically, however, not all sunfishes are snail eaters; unfortunately, physid snails are unable to distinguish the chemical cues from predatory and non-predatory sunfishes. Hence, maladaptive responses occur when only non-predatory sunfishes are present in the environment (Langerhans & DeWitt 2002). Adaptations are not always foolproof.

Phenotypic plasticity is particularly widespread among sessile organisms such as plants (Schlichting 2002; Donohue 2013; Palacio-López et al. 2015; Turcotte & Levine 2016). Mobile organisms such as beetles or birds can actively move away from an area where conditions are unfavorable and seek out areas that fit them better. A plant does not have that luxury and must cope with the conditions where it grows. Indeed, plants often show an array of plastic responses to environmental variation. They can shape their leaves adaptively according to local

humidity and level of sun radiation, adjust the investment in roots according to the content of nutrients and water in the soil, or switch between outcrossing or self-fertilization depending on the abundance of insect pollinators (see chapter 3). As an example, Nakayama et al. (2014) investigated the molecular mechanisms of phenotypic plasticity in North American lake cress (*Rorippa aquatica*). This semi-aquatic plant develops thread-like, branched leaves when submerged in water but simple leaves (not unlike those of a basil) in terrestrial conditions. Nakayama and co-workers showed that genes involved in regulating the gibberellin phytohormone pathway were differentially expressed under different environmental conditions, thus affecting leaf development.

The study of the genetic and molecular basis of phenotypic plasticity is still in its infancy, but is receiving growing interest (e.g. Nicotra et al. 2010; Sun et al. 2018). Plasticity involves sensing external cues that induce developmental process. These responses involve epigenetic modifications, likely of specific "plasticity genes." We currently lack general answers to a range of questions regarding the evolution of phenotypic plasticity. Which traits are likely to show adaptive plasticity? To what extent does plasticity contribute to adaptive diversification (Price et al. 2003)? Which genes are involved? How are the epigenetic modifications orchestrated? In-depth studies of particular model systems and comparative work that takes advantage of the latest developments in molecular genetic technologies is likely to give us valuable new insights in the years to come. We will return to epigenetics and phenotypic plasticity in more detail in chapter 10.

An important route for plastic responses to environmental heterogeneity and unpredictability is that of learning. Animals imprint, learn from trial and error, from imitating others, or by association in order to cope with their environment. Finding and handling food, avoiding predators and other dangers in the environment, and finding and attracting a partner are examples of activities that often have strong learning components in animals (e.g. Griffin 2004; Slagsvold & Wiebe 2007; Verzijden et al. 2012). Learning adds flexibility to animal life but it is not limitless. Natural selection shapes the learning abilities of animals in adaptive ways. Positive and negative feedback loops ensure that learned responses (usually) increase fitness rather than the opposite. For instance, an animal may learn to utilize a particular food resource because its sensory and nervous system "tells" it that it tastes good and satisfies hunger and it may learn to avoid items that contain toxins because they taste unpleasant or make the animal sick.

Learning abilities are not only products of natural selection. Learning may also affect the course of evolution (Baldwin 1896). A learned behavior may affect an animal's survival or reproductive success and may therefore affect the genetic makeup of a species through natural selection. This is the so-called **Baldwin effect**. Early humans discovered fire and started to cook food maybe as early as 1.9 million years ago (Wrangham et al. 1999). This was likely a learned behavior but it affected our subsequent evolution in significant ways. Cooking makes nutrients more readily accessible and cooked food is much easier to digest than raw food. Hence, humans spend much less time feeding and less energy on digesting food compared to our closest relatives among the great apes. Correlated evolutionary changes that followed the transition from raw to cooked food include smaller teeth and jaws, and a shorter intestine (Wrangham & Conklin-Brittain 2003). Compared to its size the brain is an extremely expensive organ and humans expend a much larger proportion of their energy budget on brain metabolism than other primates. The increased nutritional value of cooked food likely relaxed constraints on brain size evolution and thus set the stage for the evolution of the extraordinary human brain (Leonard et al. 2007).

5.3.2 The paradox of sex

The prevalence of sexual reproduction in nature is somewhat of an evolutionary puzzle; asexual reproduction is much more effective (Maynard Smith 1978). Consider a sexually reproducing population in which females on average produce two offspring and where males do not contribute with any parental care. Assuming a balanced sex ratio, each female on average produces one son and one daughter. Now, consider a mutant female that is similar in all ways except that she reproduces asexually through

parthenogenesis. If her reproductive output is similar to that of the sexual females she would produce two daughters which also would reproduce clonally. Therefore, after only two generations the mutant female would have produced four times as many granddaughters as the average sexual female (Figure 5.13). There is a twofold cost of producing males in sexually reproducing organisms because males do not bear their own offspring. Clearly, all else being equal the asexual mutant would soon replace the sexually reproducing ones in the population because of this huge selective advantage.

At the gene level, sexual reproduction is associated with another cost. In offspring of sexually reproducing organisms, half the genes come from the father and half from the mother. The coefficient of relatedness between a sexually reproducing mother and her offspring is only $r = 0.5$. In contrast, the clonal daughters of an asexually reproducing female would be identical to their mother, $r = 1$. Under sexual reproduction a given allele only has a 50% chance of being passed on to the next generation each time an offspring is produced, whereas an allele is guaranteed transmission under asexual reproduction. An allele that caused a female to switch from sexual to asexual reproduction would thus have a huge transmission advantage.

Given these obvious costs of sexual reproduction, why do so many organisms reproduce sexually? Why haven't they been replaced by asexual clones? Clearly, the advantages of sex on shorter or longer time scales must be very large to offset these costs. Theories that seek to explain the evolutionary

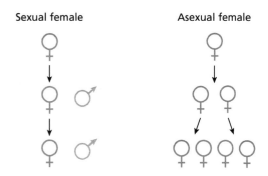

Figure 5.13 The twofold cost of sex. All else being equal, after two generations, an asexual female would have given rise to four times as many granddaughters as an equally productive sexual female.

maintenance of sexual reproduction fall into two main categories. Those that hypothesize that sexual reproduction is an individual adaptation that outperforms asexual reproduction during the lifetime of an individual and those that hypothesize that it is maintained due to long-term advantages that extend beyond the lifespan of any single individual.

5.3.3 Sexual reproduction as an individual adaptation

One obvious consequence of sexual reproduction and recombination is that it causes variation. The offspring of a sexually reproducing female will be genetically variable whereas the clonal offspring of an asexually reproducing female will be identical. **The lottery hypothesis** likens offspring with tickets in a lottery (Williams 1975). A sexual female buys different tickets (different genotypes) whereas an asexual female buys several copies of the same ticket (identical clones). In a heterogeneous and unpredictable environment buying different tickets may be the safest bet. Some of the variable offspring of a sexual female are likely to have genotypes that enable them to survive under prevailing conditions, whereas the asexual female risks losing the entire clutch if that genotype is a poor fit. Asexual reproduction is therefore a case of essentially putting all your eggs in one basket. A related hypothesis states that the genetically variable offspring of a sexual female would compete less intensively among one other than would identical clones—allowing them to extract more food and resources from the environment and thus have higher survival on average (Bell 1982). Theoretical models suggest that the advantage of variable offspring could offset the disadvantage of sex if environmental conditions are highly unpredictable. However, it seems unlikely that environmental conditions in general are so variable and unpredictable that the lottery hypothesis is the only explanation for the maintenance of sex. The hypothesis may nonetheless explain a reduction in the cost of sexual reproduction. However, some aspects of the environment are predictably unpredictable.

Pathogens such as viruses and bacteria have short generation times and thus often evolve fast compared to their host. According to the **Red Queen**

hypothesis[5] sexual reproduction is maintained because it helps sexually reproducing individuals in resisting parasites through continuously changing gene combinations (Hamilton 1980; Hamilton et al. 1990). In an **arms race** between hosts and parasites, the fitness associated with a particular immunity allele is expected to constantly change. A parasite would never be able to specialize on a particular host genotype because sex and recombination constantly reshuffle alleles and generate genotypes that are novel to the parasite. Morgan et al. (2011) tested the Red Queen hypothesis in the nematode *Caenorhabditis elegans*. Populations of these animals are composed of males and hermaphrodites. The hermaphrodites reproduce either through self-fertilization or through outcrossing with males. Morgan and co-workers manipulated strains of *C. elegans* genetically to be (a) obligate self-fertilizing or (b) obligate outcrossing. In addition, they kept (c) un-manipulated wild strains containing self-fertilizers and out-crossers at a ratio of about 80:20. Replicas of the three kinds of strain were then subjected to three different parasite treatments (1) control (no parasites), (2) evolution (repeated exposure to a fixed, non-evolving strain of the pathogenic bacterium *Serratia marcescens*) or (3) coevolution (30 generations of exposure to a coevolving strain of *S. marcescens*). The results strongly supported the Red Queen hypothesis. In the coevolution treatment (but not the other two treatments) all obligate self-fertilizers strains went extinct within less than 20 generations. They were not able to respond evolutionarily to the evolving bacteria. In the wild type strains in the coevolution treatment, the proportion of out-crossers quickly rose from about 20% to more than 70%. An initial rise in the proportion of out-crossers also occurred in the evolution treatment where the bacteria were not allowed to coevolve, but this rise was followed by a decline after about eight generations (Figure 5.14). Sexual reproduction and outcrossing were thus apparently favored by selection in the face of coevolving bacteria. The

[5] vanValen (1973) developed a general evolutionary theory which is also known as the Red Queen hypothesis. This theory proposes that organisms must constantly evolve in order to keep up with coevolving organisms. The Red Queen hypothesis for the maintenance of sex can be regarded as a special case of the more general theory.

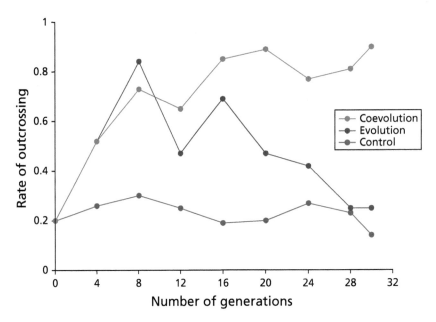

Figure 5.14 Change in rate of outcrossing in wild-type *C. elegans* over 30 generations. Populations were exposed to three different treatments: control (no bacteria, red line), evolution (fixed strain of the bacterium *S. marcescens*, purple line), and coevolution (coevolving *S. marcescens*, blue line). Modified from Morgan et al. (2011).

experiment demonstrates that selection against coevolving parasites *can* be sufficient to offset the cost of sex.

Sexual reproduction includes two processes: recombination at meiosis and outcrossing in the form of union of haploid gametes from two different individuals to form a diploid zygote. Bernstein et al. (1985) suggested that recombination and outcrossing evolved and is maintained by selection because, respectively (1) these processes allow more efficient DNA repair and (2) they enable masking of deleterious mutations. This is the **DNA repair and complementation hypothesis**. The DNA of a chromosome can be physically damaged in the form of breaks, fork formation, and modified bases. Such damage is common in nature and can have serious fitness consequences as it interferes with replication and transcription. During recombination, damage is effectively repaired enzymatically when the undamaged homologous chromosome acts as a template for the chromosome that has become damaged (Figure 5.15). There is a large and immediate benefit of such recombinational repair during meiosis because progeny with undamaged DNA will have higher survival. Indeed, recombination is

also prevalent in many organisms that do not outcross, such as in self-fertilizing hermaphrodites, suggesting that DNA repair may be the ultimate function of recombination.

The second component of sexual reproduction is outcrossing. According to the DNA repair and complementation hypothesis outcrossing is maintained because it effectively masks deleterious, recessive mutations. This advantage of outcrossing is easily seen during episodes of inbreeding, in which rare recessive deleterious alleles become expressed in homozygotes, leading to inbreeding depression (see chapter 3).

In hermaphroditic plants, some species are self-incompatible and thus obligate out-crossers, others are self-compatible and may switch adaptively to self-fertilization when cross-pollination is prohibited, and yet others are obligate self-fertilizers. Goldberg et al. (2010) demonstrated that the proportion of outcrossing, self-incompatible species remains high despite frequent transitions from self-incompatibility to self-compatibility and despite short-term advantages of being able to self-fertilize when cross-pollination is constrained. Obligate outcrossing plants remain common because they have

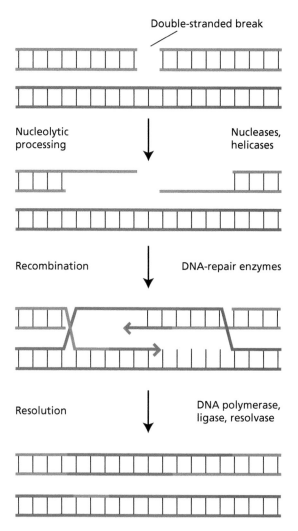

Double-stranded break

Nucleolytic
processing

Nucleases,
helicases

Recombination

DNA-repair enzymes

Resolution

DNA polymerase,
ligase, resolvase

Figure 5.15 Enzymatic repair of a DNA break during homologous recombination.

lower extinction rates compared to self-compatible ones. Hence, outcrossing also has long-term advantages and is being maintained by species selection (see section 5.3.4 below).

The various hypotheses for the maintenance of sex are not mutually exclusive. It is likely that variability of offspring in heterogeneous and unpredictable environments, benefits against coevolving parasites, DNA repair, and masking of deleterious mutations all contribute to tip the balance in favor of sexual over asexual reproduction. Moreover, the relative importance of each factor may vary between organisms and environmental conditions.

Additionally, the prevalence of sexual reproduction in nature may in part be due to a decreased risk of extinction and an increased rate of speciation in sexual organisms compared to asexual ones.

5.3.4 Long-term advantages of sex and recombination

Genetic consequences of sexual reproduction are likely to reduce the likelihood that sexual organisms go extinct and increase their rate of diversification (increased rate of speciation). Accordingly, due to sorting at higher levels (populations and species) the prevalence of sexual reproduction in nature may increase even if it is less effective than asexual reproduction in the short run.

Because sex combines genes from two individuals, advantageous gene combinations can be obtained more easily and quickly in a sexually reproducing population than in an asexual one (Williams 1975). Suppose that the ancestral genotype at two loci is ab and that two random mutations at either locus are advantageous so that AB would have the highest fitness. In a sexually reproducing organism the advantageous AB combination could be obtained through mating between individuals that have either of the advantageous mutations aB × Ab. In contrast, there is no way in which aB and Ab can be combined in an asexual population; both mutations would have to occur within the same lineage (Figure 5.16). Adaptive evolution may therefore be faster in sexually reproducing organisms; accordingly, these may be better able to cope with environmental change, which in turn reduces their risk of extinction. We will see in chapter 8 that adaptations are often linked to speciation, so faster adaptive evolution may also increase speciation rate and diversification rate of sexual organisms.

Sexually reproducing organisms can also more easily escape the accumulation of harmful mutations than asexual ones (Muller 1932, 1964; Felsenstein 1974). Suppose that a deleterious mutation occurs in a population of an asexual organism. If the fitness is reduced only slightly, the mutation may go to fixation through genetic drift in a finite population. At this point a second slightly disadvantageous mutation may occur that also goes to fixation by drift. Over time, more and more harmful mutations would accumulate in the population. Once fixed, such mutations cannot be removed, except in the unlikely case of a back-mutation. Eventually, the population would therefore be so loaded with harmful mutations that it would go extinct. This phenomenon is referred to as **Muller's ratchet** (Figure 5.17).

Sexual reproduction

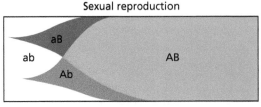

Asexual reproduction

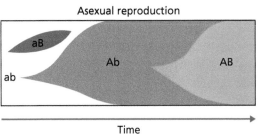

Time

Figure 5.16 Schematic model illustrating that novel beneficial genotypes may be created more rapidly with sexual than asexual reproduction. The ancestral genotype is ab and beneficial mutations can occur at both loci (A and B respectively). The beneficial AB genotype is recombined rapidly in a sexual population (top), whereas the two alleles must arise independently in an asexual population (bottom).

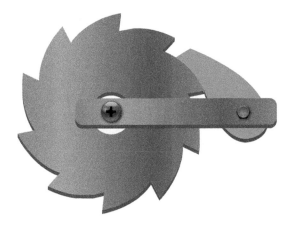

Figure 5.17 Muller's ratchet. A ratchet is a device that can only be moved in one direction. Likewise, in the absence of recombination deleterious mutations would accumulate in an asexual population. Once, fixed they cannot be removed.

A ratchet can only move in one direction. If you move the ratchet one click forward there is no turning back. Likewise, fixed deleterious mutations are irreversible in an asexual population and so they represent one step closer to extinction. In a sexually reproducing population, however, offspring with fewer deleterious mutations than their parents can easily form following recombination, whenever mutation free regions from either of the parental genomes are combined during meiosis (Figure 5.18). Sexually reproducing organisms are therefore much less likely to be driven to extinction through the accumulation of deleterious mutations.

Finally, sexual reproduction may facilitate purging of deleterious mutations (Kondrashov 1988). Such purging may be particularly effective when strong sexual selection is operating (Agrawal 2001; Siller 2001). Sexual selection usually involves competition among males for mating access and/or female choice. Under such sexual selection, male mating success often depends on individual condition, which at least in part may be due to variation in mutation load. Accordingly, deleterious mutations would effectively be removed from a strongly sexually selected population as males with the highest mutation load would be the least likely to mate. Lumley et al. (2015) tested this hypothesis using laboratory populations of the flour beetle (*Tribolium castaneum*). Strains of beetles were reared for up to seven years under conditions that differed solely in the strength of sexual selection. The strains were then subjected to a regime of strong inbreeding for 20 generations with enforced mating between siblings. Strains that had experienced strong sexual selection prior to the inbreeding regime were resilient to extinction and maintained high fitness. In contrasts, strains that had experienced weak or no sexual selection showed rapid fitness declines under the inbreeding regime and went extinct within less than 10 generations.

Sexual reproduction is an ancient trait, but it most likely originated as an individual adaptation that outperformed asexual reproduction under the prevailing conditions. It is difficult to guess how and why it originally evolved but once present, a number of short- and long-term advantages are likely contributing to maintaining it and increasing its relative prevalence among living organisms.

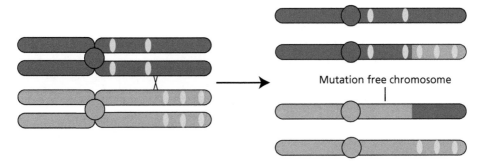

Figure 5.18 Removal of deleterious mutations (yellow) by recombination.

5.4 Evolving communities

5.4.1 Coevolution among mutualists

Other organisms arguably constitute the most important components of an organism's environment. Organisms affect each other through predation, parasitism, competition, and mutualism. Moreover, since organisms are not static components of the environment, but themselves evolving entities, they can affect each other's evolution. Organisms may coevolve (Ehrlich & Raven 1964).

Mutualistic coevolution provides some fascinating examples of natural history. The Attini ant tribe, which includes the well-known leafcutter ants, depend on the cultivation of fungi for food. In return, the ants nourish the fungi, protect them from pathogens and competitors, and aid their dispersal (Mueller et al. 2005; Mehdiabadi & Schultz 2010). The prime group of cultivated fungi belong to the family Lepiotaceae, although some ant species have switched to cultivate other fungi. The mutualistic relationship varies in sophistication between species pairs. "Lower attines" cultivate Lepiotaceae species that are very similar to wild fungi, and are not totally dependent on fungus cultivation for survival. In contrast, "higher attines," including leafcutter ants, cultivate highly derived Lepiotaceae species in areas where wild fungi would not be able to survive. The ants themselves are totally dependent on the fungus for survival. Optimal conditions for the fungi are constructed by the ants, in nests deep below the surface, and the fungi are fed freshly cut grasses or leaves. The fungi, in turn, have evolved special nutrient-rich structures called *gongylidia*, designed for being harvested by the ants and fed to their larvae.

When ant queens disperse to form new colonies, they bring a cluster of fungal mycelium with them and start building a new fungal garden. Attini ants are masters at **niche construction** (Day et al. 2003; Kylafis & Loreau 2008). They construct their own environment in their underground nests where their commensal fungi thrive. Such niche construction has evolutionary consequences because it affects the selection pressure on the organisms involved.

There are other coevolving players too in the attine ant/Lepiotaceae fungus system (e.g. Mehdiabadi & Schultz 2010). Fungi in the genus *Escovopsis* parasitize the Lepiotaceae, thus threatening the survival of both the mutualistic fungus and the ants. The ants fight the parasitic fungi, in part by actively removing *Escovopsis* spores. In addition, the ants harbor mutualistic bacteria, *Pseudonocardia*, on their cuticle, which produce an antibiotic substance that specifically targets and inhibits the growth of the *Escovopsis* fungi (Currie et al. 1999).

As seen above, coevolution can result in reciprocal adaptations among interacting species, although few can match the fungus gardening ants and their associates in level of sophistication. Another famous example is the coevolution between flowering plants and their insect pollinators which are rewarded with nectar (and pollen) for their pollinating service (Ehrlich & Raven 1964). The high diversity of flowering plants and successful groups of pollinating insects, such as butterflies and bees, is a testament to the evolutionary significance of this mutualistic coevolution.

However, as with cooperating individuals within a species, mutualistic relationships between species are vulnerable to cheating whenever evolutionary interests are only partly overlapping. Indeed, there are flower species that do not reward insects with nectar (see chapter 4), but still receive pollinators because they resemble flowers that do (Schaefer & Ruxton 2009). Flowers and pollinators may also have subtler conflicting interests. A flower would benefit if it could increase the likelihood that a pollinator would visit a flower of the same species as itself rather than a different one to ensure successful pollination, whereas a pollinator may be equally well off visiting any nectar-rewarding flower species. Plants in the genera *Citrus* and *Coffea* spike their nectar with caffeine (Wright et al. 2013). Wright and co-workers showed that honeybees rewarded with caffeine were three times as likely to remember a learned floral scent as those only rewarded with sucrose. In this way, the plant secures pollinator fidelity and thus increases its reproductive success.

5.4.2 Coevolution among natural enemies

Coevolution between natural enemies: predators and prey, herbivores and plants, and hosts and parasites, has caused a range of adaptations and counter-adaptations in each party. Nature is indeed, red in tooth and claw. Predator adaptations include acute senses to detect prey, extreme speed like in the cheetah (*Acinonyx jubatus*) or the peregrine falcon (*Falco peregrinus*) to hunt down their prey, killing weapons such as sharp teeth, claws, or venom as in snakes and spiders, and lures such as the bola of a bolas spider (Mastophoreae) (Figure 5.19): The bolas spider swings its bola which reeks of pheromones resembling those of a female moth. Male moths are attracted to the female-resembling smell and movements but instead get stuck to the sticky bola (e.g. Haynes et al. 2002).

Prey counter-adaptations include camouflage color as in the peppered moth, chemical defenses and warning colors as in the *Heliconius* butterflies, and protective armor or spikes as in armadillos (Dasypodidae) or hedgehogs (Erinaceinae). Prey may also possess speed and senses that match those of their predators for detecting and escaping them. Plants may invest in chemical defense against grazers, which may evolve resistance against the toxins as a counter-move, or they may evolve thorns which are countered by thick, horny skin.

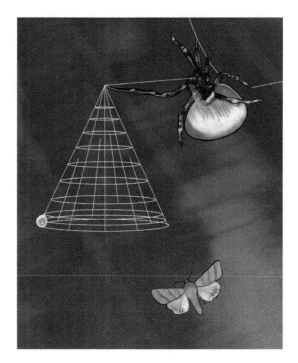

Figure 5.19 The bolas spider catches male moths by swinging its sticky bola that smells like a female moth.

Parasites likewise possess a range of adaptations to enter their host. The anthrax bacterium (*Bacillus anthracis*) eventually kills its bleeding herbivore host (e.g. a zebra). Bacterial spores get caught in the extra lush vegetation that grows in the soil fertilized by the rotting carcass—an irresistible meal for the next herbivore that passes the spot, which in turn gets infected (Turner et al. 2014). Vertebrates have an impressive defense against parasites with their immune system, which can generate antibodies against thousands of pathogens.

Coevolution between natural enemies readily leads to an arms race, in which an adaptation in one of the parties triggers a counter-adaptation in the other party, which induces adaptations to overcome that counter-adaptation and so forth. The outcome of such arms races varies. If one of the parties is unable to counter an adaptation in the other party it may go extinct. An equilibrium may be reached where further adaptation in any party becomes too costly. Finally, the arms race may lead to an escalation with more and more elaborate adaptations and counter-adaptations appearing over evolutionary time. Coevolutionary arms races are important sources for evolutionary progress. Impressive adaptations such as the vertebrate eye and immune system owe their complexity to past and ongoing episodes of coevolutionary arms races with their antagonists.

5.4.3 Coevolution between competitors

Coevolution between competing species may lead to arms races similar to what we observe between natural enemies. For instance, plant species compete for light. The extreme height of many tree species is the result of arms races to outgrow competitors for a spot in the sun (Falster & Westoby 2003). However, a more common evolutionary outcome is that competition is reduced over time. When two species have overlapping niches, leading to competition for resources, natural selection may favor traits that cause them to specialize on different resources to reduce competition. This can lead to **character displacement** (Brown & Wilson 1956; Grant 1972; Schluter 2000a; Pfennig & Pfennig 2012) in which species diverge in phenotypic traits in geographic regions where they co-occur. The Darwin finch species *Geospiza fortis* diverged in beak size from a closely related competitor *G. magnirostris* over the course of only 22 years on one of the Galápagos islands following the arrival of the latter species (Grant & Grant 2006). The two species initially competed intensively for seeds, but apparently because *G. magnirostris* was more effective at processing the large and tough seeds of the common plant *Tribulus cistoides*, *G. fortis* specialized on smaller seeds and evolved a smaller beak as a response to the competition (Figure 5.20).

Conversely, when competition is relaxed, for instance due to local extinction of a competitor or colonization of a new habitat where the competitor is absent, we may observe **competitive release**, in which a species' ecological niche is expanded (Grant 1972). On Gotland in the Baltic Sea the coal tit (*Periparus ater*) is the only titmice in coniferous forests, whereas on the Swedish mainland they share the habitat with the larger crested tit (*Lophophanes cristatus*) and willow tit (*Poecile montanus*). On the Swedish mainland, the coal tits forage on the outer twigs of the trees, leaving the rest of the trees to the

G. magnirostris

Figure 5.20 Character displacement in Darwin's finches. The species *G. fortis* (top panel) evolved a smaller beak (lower, left) after experiencing secondary contact with the large beaked *G. magnirostris* (lower, right) on a Galapagos island.

larger titmice. On Gotland, in the absence of its competitors, the coal tit has expanded its use of foraging sites to the inner parts of the trees and reaches much higher population densities (Alerstam et al. 1974). Associated with the change in behavior, coal tits on Gotland have evolved a larger body size and longer tarsi (Alatalo & Gustafsson 1988).

Study questions

1. Evolutionary adaptations can be impressive, but are rarely the most perfect solutions one could imagine to the challenges organisms face. Review factors that can constrain adaptive evolution.
2. Explain the rationale behind the theory of punctuated equilibria and suggest alternative hypotheses that can explain the pattern of stasis interrupted by a rapid morphological transition in a fossil stratum.
3. Explain the difference between group selection and species selection. Discuss whether the maintenance of sexual reproduction can be understood as a form of group selection.
4. Discuss similarities and differences between genetic conflicts and kin selection.
5. Calculate the coefficient of relatedness r between half-siblings and second cousins.
6. Imagine a large random-mating diploid population of birds, with a clutch size of two. The

birds are monogamous, so members of a clutch are full sibs. In typical members of the population (genotype aa), the probabilities of survival of the older and younger sib are 1 and k, respectively, where $k < 1$. A rare dominant allele A, for "altruistic" behavior has the following effects. If present in the older sib, it causes the sib to be less greedy for food, so that its survival probability becomes $1 - c$ and that of its sib becomes $k + b$, where $k + b < 1$. The A-allele has no effect when present in the younger sib. (Derived from Mock & Parker 1997.)

a. Show that the conditions under which the A-allele will increase in frequency is $b/2 - c > 0$. (Hint: when A is rare and mating is random, most mating in the population would be $aa \times aa$, but almost all matings involving altruists would be $aa \times Aa$.)

b. Suppose the A-allele goes to fixation. How many chicks on average survive in each clutch? Suppose that a new, dominant, "selfish" mutation S arises which makes the carrier not limit its greed (i.e. survival when in an older sib is 1). Will the S-allele increase in frequency in the population?

c. Is there any frequency-dependent selection in this kin selection model? (Remember that selection is acting at two levels.)

7. Assume that the psychological feeling of guilt or bad conscience has a genetic basis. What evolutionary mechanism(s) could potentially explain the evolution of such a trait? Explain why.

8. Environmental factors affect the phenotype. Discuss how natural selection may sculpt phenotypic responses to environmental variation in an adaptive way. How can we determine if an environmentally induced plastic response is adaptive or not?

9. Discuss differences and similarities between learning and phenotypic plasticity.

10. Explain why asexual reproduction is more efficient than sexual reproduction.

11. What are the potential short- and long-term benefits of sexual reproduction?

12. Define coevolution. Discuss how coevolution may play out differentially depending on the nature of the species interaction; that is, whether it is taking place between mutualists, natural enemies, or competitors.

CHAPTER 6

Multilocus evolution

Most phenotypic traits are affected by a multitude of genes, which may interact in complex ways. This means that the single locus model we explored in chapters 3 and 4 is not always able to capture the full complexity of genetic evolution. In many cases, multiple genes are involved and so in this chapter, we formalize the analysis of multilocus evolution. We shall introduce concepts such as linkage disequilibrium and epistasis, both of which are necessary to properly understand multilocus evolution. We shall further introduce you to the currently highly active field emerging as a result of a crossover between quantitative genetics and genomics. We will explore methods such as quantitative trait locus (QTL) analysis and genome wide association study (GWAS) that allow us to associate phenotypic variation with likely causative genes and that have made important advances in our understanding of the genetic underpinnings of disease. We start gently, however, by just adding one biallelic locus to the single locus model of chapters 3 and 4.

6.1 The two-locus model

6.1.1 Linkage equilibrium and disequilibrium

In chapter 3 we introduced the Hardy–Weinberg null model as a tool to investigate the effects that various demographic and evolutionary processes have on allele and genotype frequencies. A comparable null model for the two-locus case needs to take into account one complicating factor, namely that recombination can occur between the pair of

loci. This means the ideal conditions for our model require the following assumptions:

1) Individuals in the population mate randomly.
2) The population is infinitely large.
3) No selection occurs.
4) No mutation occurs.
5) No gene flow occurs; and, specific to the multilocus case:
6) Recombination breaks down associations between the alleles at the two loci.

When these conditions are met, we expect two loci to be in **linkage equilibrium (LE)**. Two loci are in LE when the genotype at one locus is independent of the genotype at the other locus; that is, when the alleles at the two loci segregate independently. Mendel's second law assumes that loci are in LE (see chapter 1). Because we are interested in the effects of evolutionary and demographic processes on the genetic makeup of populations, deviations from the equilibrium are of interest to us. For this reason, the concept of **linkage disequilibrium (LD)** is more frequently discussed in the literature than the equilibrium case. Linkage disequilibrium occurs when genotypes at the two loci are *not* independent of each other—in other words, when allele frequencies are correlated. Violations of one or more of the assumptions above can cause a pair of loci to be in linkage disequilibrium. The concept of LD can be a little confusing the first time you encounter it, but don't let that dissuade you; we will demonstrate it clearly in the next section.

Evolutionary Genetics: Concepts, Analysis, and Practice. Glenn-Peter Sætre & Mark Ravinet, Oxford University Press (2019).
© Glenn-Peter Sætre & Mark Ravinet 2019. DOI: 10.1093/oso/9780198830917.001.0001

Defining linkage equilibrium and disequilibrium

In the one-locus model we tracked alleles and genotypes. With two loci, genotypes include a minimum of four alleles from two different loci, such as $A_1A_2B_1B_2$. Even with only two alleles per locus the number of different genotypes is quite high (nine or ten depending on how we define genotypes—i.e. whether we distinguish different double heterozygotes). From a multilocus genotype like this, we can derive a **haplotype**—i.e. the set of alleles at each locus inherited from a parent, or in other words, the allelic combination of the parental gametes. Since alleles at two loci may or may not segregate independently, it is more convenient to track haplotypes than alleles. With two loci, A and B, each with two alleles there are four different haplotypes: A_1B_1, A_1B_2, A_2B_1 and A_2B_2. We denote the frequency of a haplotype as x_{ij} where the suffix indicates the alleles it contains. Thus, for instance, the frequency of A_1B_1 is x_{11}. At linkage equilibrium (i.e. under the six assumptions above) we expect the alleles at the two loci to segregate independently. The expected frequency of the different haplotypes at LE thus equals the probability of combining the alleles each of them contains. In other words, the expected frequency of a haplotype equals the product of the allele frequencies included in that haplotype. Let's demonstrate this more clearly: For two loci with two alleles each, we can denote the frequency of A_1 and A_2 as p_A and q_A respectively, and the frequency of B_1 and B_2 as p_B and q_B. The expected frequency of the four haplotypes at LE is thus simply:

$$E(x_{11}) = p_A p_B$$
$$E(x_{12}) = p_A q_B$$
$$E(x_{21}) = q_A p_B \qquad (6.1)$$
$$E(x_{22}) = q_A q_B$$

An individual that is double heterozygote at our two loci ($A_1A_2B_1B_2$) may have inherited the A_1B_1 haplotype from one parent and A_2B_2 from the other. Alternatively, it might have inherited the haplotypes A_1B_2 and A_2B_1 from its parents. We distinguish the two possibilities by denoting the two genotypes as A_1B_1/A_2B_2 and A_1B_2/A_2B_1 respectively. Since there are two haplotype combinations that yield the double heterozygote genotype there are ten different

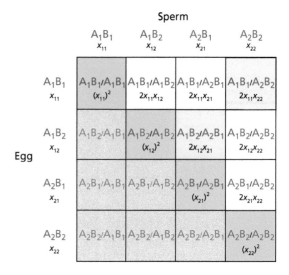

Figure 6.1 Genotypes and expected genotype frequencies at linkage equilibrium in the two-locus model. The diagonal (yellow) shows genotypes in which similar haplotypes are combined at fertilization. Below the diagonal (grey) are genotypes that are also produced by the opposite cross, shown above the diagonal. There are, therefore, 10 different ways that haplotypes can be combined into ordered genotypes (yellow, white, and blue). Marked in blue are the two genotypes in which recombination between locus A and B will yield haplotypes that differ from the parental haplotypes in the resulting gametes.

haplotype combinations, or *ordered genotypes*, in the case of two loci with two alleles each (Figure 6.1).

Just like we derived the expected Hardy–Weinberg genotype frequencies by randomly combining alleles of sperm and egg at fertilization, we find the expected two-locus genotype frequencies under ideal conditions (the six assumptions above) by randomly combining the four different possible haplotypes from either parent. There is only one way of combining haplotypes that are similar in the two gametes (yellow diagonal in Figure 6.1), but two ways of combining different haplotypes (half of them are shown in grey in Figure 6.1). This is because genotypes such as A_1B_1/A_2B_1 can result both from fertilization of an A_1B_1 sperm with an A_2B_1 egg, and from the opposite cross (A_2B_1 sperm with A_1B_1 egg). The genotype frequencies shown in black in Figure 6.1 are the genotype frequencies expected at linkage equilibrium, again, when there is no association among the alleles at the two loci. These are the two-locus equivalents of the expected genotype frequencies at Hardy–Weinberg equilibrium in the one-locus model (i.e. $E(g_{11}) = p^2$, $E(g_{12}) = 2pq$, and $E(g_{22}) = q^2$), given that the haplotype frequencies also are in linkage equilibrium.

Recombination between locus A and B during meiosis (see chapter 2 for an in-depth treatment of the mechanics of recombination) can reshuffle the parental alleles and yield different haplotypes in the resulting gametes. Recombination is antagonistic to associations among alleles—if LD does occur between loci, recombination will work to reinstate LE. Note, however, recombination will only result in different haplotypes in the gametes (Figure 6.2) for the two double heterozygous genotypes (marked in light blue in Figure 6.1). Recombination still occurs with the other genotypes, but it does not result in different haplotypes being produced.

In the one-locus model, one generation of random mating is sufficient to bring an imbalanced population to Hardy–Weinberg equilibrium. Linkage equilibrium is approached more slowly and the speed depends on the rate of recombination between the two loci—i.e., the further apart two loci are, the more likely recombination will occur between them. This is moderated by a number of different factors, including physical proximity in the genome, the presence of recombination rate modifying genes, and also genome structure. As a general rule of thumb though, increasing physical distance in the genome is a good predictor of recombination rate. Nonetheless, even if the two loci are segregating totally independently, for instance because they are completely physically unlinked and occur on different chromosomes, the probability is still only 50 percent that recombination will occur.

Suppose we have made two large, inbred lines of a hypothetical butterfly species to remove heterozygosity and that we mix the two lines to investigate how an imbalanced population would approach LE. We focus on two loci with phenotypic effects that are easy to identify. Let us assume that locus A determines the color of the forewing and locus B the color of the hindwing so that A_1B_1/A_1B_1 are all blue and A_2B_2/A_2B_2 are all red, and that heterozygotes at either locus have intermediate purple wing parts. At the start of our hypothetical experiment we have one entirely blue and one entirely red inbred line (Figure 6.3). We assume that the mixed population is large enough that we can ignore genetic drift and that the other LE assumptions are met (random mating, no selection, no mutation, no gene flow, and the loci are free to recombine).

Clearly, individuals from the blue line will only produce gametes with the A_1B_1 haplotype and those from the red line only the A_2B_2 haplotype. If there are equally many blue and red butterflies in the joint population and mating is random, it is equally likely that a blue butterfly will mate with another blue as with a red butterfly and vice versa for a red butterfly. Half of the offspring would therefore be double heterozygotes with A_1B_1 inherited from one parent and A_2B_2 from the other and have purple hind- and forewings (A_1B_1/A_2B_2), 25 percent would

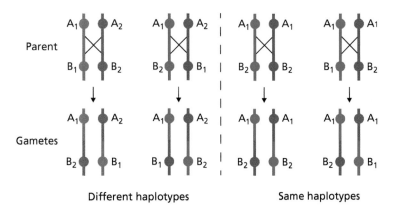

Figure 6.2 Recombination between the biallelic loci A and B yields gametic haplotypes that differ from the parental ones only in the two double heterozygous genotypes (left panel). In the other eight genotypes, recombination results in gametic haplotypes similar to those of the parent. Two examples are shown in the right panel.

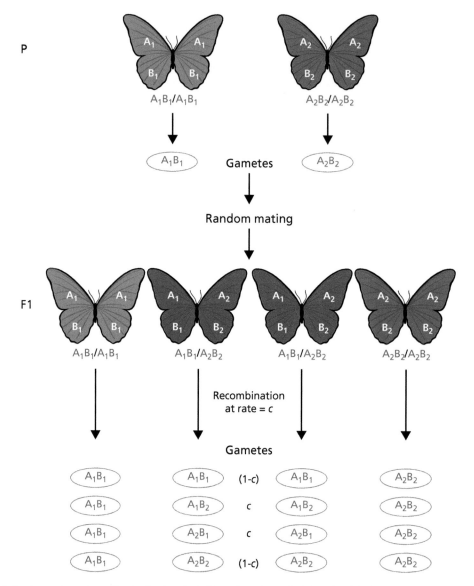

Figure 6.3 Random mating and recombination in a hypothetical butterfly species in which the color of the forewing and hindwing are determined by two different loci A and B. At the start of the experiment we mix an all-blue inbred line with an all-red line and allow them to mate randomly. Although Hardy–Weinberg equilibrium is achieved at each of the loci after one generation of random mating (F1), the population is still in linkage disequilibrium. Recombinant haplotypes accumulate gradually and LD decays over successive generations at a speed determined by the recombination rate c.

be blue A_1B_1/A_1B_1 and 25 percent red A_2B_2/A_2B_2. If we focus on only one locus, each of them is now in Hardy–Weinberg equilibrium. For instance, the frequency of A_1 is $p = 0.5$ in our example, so at HWE:

$$E(g_{11}) = p^2 = 0.25$$

$$E(g_{12}) = 2pq = 0.5$$

$$E(g_{22}) = q^2 = 0.25$$

which is what we would find in a large population as reasoned above. In contrast, the haplotypes and two-locus genotypes are by no means in LE. The only haplotypes present in the F1 generation are A_1B_1 and A_2B_2. The recombinant A_1B_2 and A_2B_1 haplotypes are completely absent and only three of the ten possible genotypes from Figure 6.1 are present.

However, if we allow the butterflies to mate randomly in successive generations, recombination will start to break up the associations between A_1 and B_1, and between A_2 and B_2 respectively. The double heterozygotes in the F1 generation will produce gametes in which some are recombinant A_1B_2 and A_2B_1 (Figure 6.3). The proportion of recombinant haplotypes will depend on the recombination rate c. If A and B are located on the same chromosome, recombination will occur whenever an odd number of crossing-over events occur between the two loci. The recombination rate can maximally be $c = 0.5$ and the closer the two loci are in physical proximity on a chromosome, the lower the probability that recombination will occur.

As we track the change in the frequency of each of the haplotypes per generation we need to take recombination into account. Let us start with the haplotype A_1B_1. From Figure 6.1 we see that there are four genotypes that contain the haplotype A_1B_1 that consequently would produce A_1B_1 gametes, namely those in the top row. All the gametes from the first genotype (A_1B_1/A_1B_1) would be A_1B_1, whereas half the gametes would be A_1B_1 in the next two genotypes ($A_1B_1/A_1/B_2$ and A_1B_1/A_2B_1). Recombination would complicate things in the last genotype in the top row of Figure 6.1 (A_1A_1/A_2B_2). Half the gametes would be A_1A_1 when the loci do not recombine (at rate $1 - c$). When recombination occurs, recombinant haplotypes (A_1B_2 and A_2B_1) would be produced instead at the rate c. Finally, however, A_1B_1 is gained (half the gametes) when recombination occurs (at rate c) in the other double heterozygote genotype (A_1B_2/A_2B_1). Adding it all together, we have that the frequency of A_1B_1 in the next generation would be:

$$x'_{11} = (x_{11})^2 + 0.5 \cdot 2x_{11}x_{12} + 0.5 \cdot 2x_{11}x_{21} \\ + 0.5 \cdot 2x_{11}x_{22} \cdot (1-c) + 0.5 \cdot 2x_{12}x_{21} \cdot c \quad (6.2)$$

We can rearrange this to get:

$$x'_{11} = x_{11}(x_{11} + x_{12} + x_{21} + x_{22}) \\ - c(x_{11}x_{22} - x_{12}x_{21}) \quad (6.3)$$

Note that the first parenthesis in equation 6.3, $(x_{11} + x_{12} + x_{21} + x_{22})$, equals 1, the sum of the frequency of all the haplotypes present. Further, we define the second parenthesis as the parameter D.

$$D = x_{11}x_{22} - x_{12}x_{21} \quad (6.4)$$

Therefore, we have that:

$$x'_{11} = x_{11} - cD \quad (6.5)$$

Using similar calculations for the other haplotypes, we have:

$$x'_{11} = x_{11} - cD$$
$$x'_{12} = x_{12} + cD$$
$$x'_{21} = x_{21} + cD$$
$$x'_{22} = x_{22} - cD \quad (6.6)$$

The parameter $D = x_{11}x_{22} - x_{12}x_{21}$ (equation 6.4) is a measure of degree of linkage disequilibrium (Lewontin & Kojima 1960). Clearly, from the formula there is linkage disequilibrium ($D \neq 0$) whenever there is an imbalance in the frequency of the haplotypes A_1B_1 and A_2B_2 relative to A_1B_2 and A_2B_1, because any imbalance is indicative of non-independent segregation.

In the F1 generation in the butterfly example in Figure 6.3 there is an excess of the haplotypes A_1B_1 and A_2B_2 ($x_{11} = x_{22} = 0.5$) relative to A_1B_2 and A_2B_1 ($x_{12} = x_{21} = 0$). Therefore, in this example:

$$D = x_{11}x_{22} - x_{12}x_{21} = 0.5 \cdot 0.5 - 0 \cdot 0 = 0.25$$

We can also define linkage disequilibrium D as the difference between the observed frequency of a haplotype x_{ij} and its expected frequency $E(x_{ij})$ at linkage equilibrium (see equation 6.1). By definition we have that:

$$D = x_{11} - E(x_{11}) = x_{11} - p_A p_B$$
$$-D = x_{12} - E(x_{12}) = x_{12} - p_A q_B$$
$$-D = x_{21} - E(x_{21}) = x_{21} - q_A p_B$$
$$D = x_{22} - E(x_{22}) = x_{22} - q_A q_B \quad (6.7)$$

where p_A and q_A are the respective frequencies of A_1 and A_2 and p_B and q_B those of B_1 and B_2.

In the example from Figure 6.3 we have that the frequency of each allele at the two loci is $p_A = q_A = p_B = q_B = 0.5$ and only two haplotypes (A_1B_1 and A_2B_2) are present in the F1 generation at equal frequencies (0.5). Thus:

$$D = x_{11} - E(x_{11}) = 0.5 - 0.5 \cdot 0.5 = 0.25.$$

Note that $D = 0.25$ in this example no matter which of the haplotypes in equation 6.7 we use. The two ways of defining linkage disequilibrium (equations 6.4 and 6.7) are mathematically equivalent.

Decay of linkage disequilibrium by recombination

The linkage disequilibrium parameter D ranges from 0.25 (maximum excess of the haplotypes A_1B_1 and A_2B_2) to −0.25 (maximum excess of the haplotypes A_1B_2 and A_2B_1). At linkage equilibrium $D = 0$. The degree of linkage disequilibrium will change over time by recombination. If we assume that our six assumptions are met, we have from equation 6.3 that D_1, the degree of linkage disequilibrium after one generation of random mating and recombination, is:

$$D_1 = x'_{11}x'_{22} - x'_{12}x'_{21}$$

where primes indicate haplotype frequencies the next generation. Using equation 6.6 and that $x_{11} + x_{12} + x_{21} + x_{22} = 1$, we have that:

$$D_1 = (x_{11} - cD)(x_{22} - cD) - (x_{12} + cD)(x_{21} + cD)$$
$$D_1 = x_{11}x_{22} - x_{12}x_{21} - cD(x_{11} + x_{12} + x_{21} + x_{22})$$

$$D_1 = D - cD$$
$$D_1 = D(1-c) \tag{6.8}$$

This leads to the general relationship:

$$D_t = D(1-c)^t \tag{6.9}$$

where t is the number of generations. In Figure 6.4 we have plotted the decay of linkage equilibrium [$D_t = D(1 - c)^t$] for different recombination rates c. Note that even loci that are segregating independently ($c = 0.5$), such as loci that are on different chromosomes, require several generations to reach linkage equilibrium. Tightly linked loci will remain in linkage disequilibrium a long time.

One inconvenience with the parameter D is that its maximum value depends on the allele frequencies at the two loci. Lewontin (1964) suggested to standardize the parameter to the maximum value it can take:

$$D' = D / D_{max} \tag{6.10}$$

Note that D_{max} would be equal to the lesser of $E(x_{12}) = p_Aq_B$ or $E(x_{21}) = q_Ap_B$ when D is positive and to the lesser of $E(x_{11}) = p_Ap_B$ or $E(x_{22}) = q_Aq_B$ when D is negative. The standardized parameter D' ranges from 0 at linkage equilibrium to 1 at **perfect linkage disequilibrium**. In the F1 generation in Figure 6.3 $D' = 0.25/0.25 = 1$. Standardizing the parameter D' in this way is convenient for cases where we want to compare the degree of linkage disequilibrium between different pairs of loci.

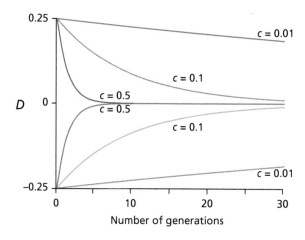

Figure 6.4 Decay of linkage disequilibrium $D_t = D(1 - c)^t$ at different rates of recombination c.

6.1.2 Estimating haplotype frequencies from genotype data

We can relatively easily estimate LD between pairs of loci from genotypic data. This is a common requirement in evolutionary genetics and genomics studies, as increased LD might indicate selection, population size changes, or recent gene flow (see next section). However, estimating LD is not without pitfalls. The definition(s) of linkage disequilibrium outlined above are easily applicable to pairs of single nucleotide polymorphisms (SNPs) because these are typically biallelic markers. SNP markers with three or even four alleles do occur but these are much rarer than the biallelic case and are easy to exclude from a dataset. In some cases, we can gather information about LD from the way markers are genotyped; with RAD-loci for example, we know that two SNPs on a single RAD-locus are likely to be in LD as they are in close physical proximity. However, normally when we genotype genome wide SNPs from real populations we do not know the haplotypes of the individuals. For instance, an individual may be heterozygous at two SNPs on a given chromosome but we have no direct way of knowing whether the ordered genotype should be A_1B_1/A_2B_2 or A_1B_2/A_2B_1. When we discuss haplotypes in this manner, we can say we have no phase information on our SNP genotypes—they are unphased. This complicates estimates of LD that, as we have seen, is defined as an imbalance in haplotype frequencies. If we know the genotype of the parents it is sometimes possible to reconstruct the haplotypes of the offspring. This has been done using parent–offspring trios and is particularly successful in humans (Harris & Nielsen 2013). However, often it is difficult to obtain such data from wild populations—it is either impossible to sample parent–offspring trios or far too expensive to sequence multiple individuals for this purpose. Instead, what we usually do is to use a statistical method based on the assumption that haplotypes are randomly joined into genotypes to obtain the likely haplotype combinations of our genotyped individuals—this is known as statistical phasing. Software such as *Beagle* (Browning & Browning 2007) and *Shapeit* (O'Conell et al. 2014) yield good estimates of the actual haplotypes according to simulation studies (Marchini et al. 2006, 2007). Furthermore, programs such as *Shapeit* can use information from sequence reads to inform phasing—for example if SNPs occur within the same read or on different strands of a mate-pair library (Delaneau et al. 2013).

6.1.3 Factors that cause linkage disequilibrium

We have seen that physical linkage (assumption 6 in the LE model) slows down the decay of linkage disequilibrium. Let us have a closer look on how violations of the other of our six assumptions affect LD.

Non-random mating

Inbreeding readily causes LD (Weir & Cockerham 1969; Golding & Strobeck 1980). In chapter 3 we saw that the effect of inbreeding is to reduce the frequency of heterozygotes relative to the Hardy–Weinberg expectation. Therefore, in the two-locus case, inbreeding will similarly reduce the frequency of the double heterozygote genotypes A_1B_1/A_2B_2 and A_1B_2/A_2B_1. As we have seen, these are the only genotypes in which recombination leads to different haplotypes in the gametes than in the parent. Thus, inbreeding slows down the decay of linkage disequilibrium because it reduces the frequency of the genotypes through which disequilibrium is broken down. Moreover, because genetic relatives share alleles of recent common ancestry, haplotypes specific to interbreeding relatives will tend to accumulate, causing linkage disequilibrium.

Assortative mating can cause a pair of loci to remain in linkage disequilibrium. Assortative mating may result from preference alleles at one locus that correspond to variation in a phenotypic trait determined by a different locus. In chapter 3 we saw that the Gouldian finch exhibits assortative mating based on head color. The preference for black or red color appears to be genetically determined in these birds (Pryke 2010). Crossing experiments suggest that red or black head color and the preference for these colors are determined by a pair of sex chromosome-linked loci that are in strong LD. Physical linkage between the two loci and strong assortative mating likely contribute to maintain the association between the preference and color alleles in these birds.

Genetic drift

Genetic drift changes allele frequencies in a random manner and this can cause an *imbalance* in haplotype frequencies and thus LD. Simulation studies show that the degree of LD can become quite substantial just by random sampling of alleles in finite (small) populations (e.g. Slatkin 1994). However, over time alleles would become fixed by genetic drift (see chapter 3). By definition, two loci can only be in LD if they are both polymorphic. For instance, if B_2 is lost by genetic drift in a population in which A_1 and A_2 are still segregating, individuals would either be A_1B_1 or A_2B_1 so there can be no imbalance in haplotype frequencies.

Isolated populations would diverge across loci by genetic drift. For instance, population 1 may become fixed for A_1 and B_1, whereas population 2 becomes fixed for A_2 and B_2. If the two populations later come together there would be strong LD in the joint population. Most individuals would either have the haplotype A_1B_1 or A_2B_2. When we sample a population that recently was subdivided there will be an excess of genotypes and haplotypes specific to the two ancestral populations (heterozygote deficiency relative to HWE and LD). They are in effect two different populations living in the same location. If mating is random, Hardy–Weinberg equilibrium would be restored after only one generation of random mating, whereas LD will be maintained longer. Loci that are physically linked (i.e. located close to each other on the same chromosome) will remain in linkage disequilibrium for a longer time than loci further away from each other following such secondary contact. Studies of variation in LD across the genome in natural populations can therefore give us information about their demographic history including the timing of admixture events (Slatkin 2008). For instance, if secondary contact between two previously isolated populations was established a long time ago we would expect that only pairs of loci that are in close proximity to each other are still in LD. This is because recombination would have restored linkage equilibrium at loci further apart. If secondary contact was established more recently, however, even loci far apart would still be in LD. The size of such blocks of loci in LD can therefore be used to estimate when secondary contact was

established. This is one of the methods which has been used to test the hypothesis that our ancestors interbred with Neanderthals tens of thousands of years ago (e.g. Noonan et al. 2006). Indeed, pairs of SNPs at loci suspected to be of Neanderthal origin are in strong LD, consistent with ancient admixture between the two human lines (Wall et al. 2009). Similar approaches using linkage blocks—i.e. regions of the genome in high LD—can also be informative for estimating the timing of admixture between human populations in more recent history, tracing events such as the spread of the Mongol Empire into Europe or later European colonialism (Hellenthal et al. 2014). We will explore the application of such methods further in chapter 9.

The decay of LD following secondary contact can be further retarded if there is selection for population-specific gene combinations, selection against gene combinations of mixed ancestry, and/or if there is assortative mating among individuals derived from either of the ancestral populations. When such selection and assortative mating is sufficiently strong, LD will not decay—the ancestral populations have become different species. We shall explore the evolution of barriers to gene exchange and speciation further in chapter 8.

Natural selection

Multilocus selection can be complex and yield surprising results. We will therefore discuss it more thoroughly in a separate section (section 6.2). In general, however, multilocus selection can cause LD whenever certain allele combinations (haplotypes) at two or more loci are associated with higher fitness than others. When balancing selection maintains alleles at multiple interacting loci, LD can even be maintained infinitely at an evolutionary equilibrium (Rice 2004; see section 6.2.1).

Natural selection and genetic drift can interact in interesting ways. Typically, when selection is affecting two or more loci that are physically linked, genetic drift will create small amounts of LD that has the potential to weaken selection. This is the **Hill–Robertson effect** (Hill & Robertson 1966). The Hill–Robertson effect significantly reduces the efficiency of selection in removing deleterious mutations from linked loci and the magnitude of the effect accumulates with the number of linked loci.

Therefore, natural selection may favor mutations that increase the recombination rate (Slatkin 2008) and indeed there is a relationship between polymorphism and recombination rate in many species (Comeron et al. 2007). Felsenstein (1974) suggested that recombination and sexual reproduction might actually have evolved in the first place to overcome the Hill–Robertson effect (see section 5.3 for a discussion of the evolution of sex and recombination).

Mutations

When a novel mutation such as a point mutation first appears, it is in perfect LD ($D' = 1$) with other polymorphic loci on the chromosome it occurred. Consider for instance a monomorphic locus B where the ancestral B_1 allele mutates to B_2 during gametogenesis and a nearby polymorphic locus where the alleles A_1 and A_2 segregate. The new mutant B_2 allele would be perfectly associated with the A-allele that happened to be present at the chromosome the mutation occurred (say A_2). If the B_2 allele is not lost by genetic drift (or selection), recombination would eventually break up the association between A_2 and B_2 and LE would be approached gradually. However, for closely linked loci, LD would be maintained a long time because recombination would be rare. The persistence of strong LD between a mutant allele and closely linked loci can be used to map interesting mutations, such as those causing a genetic disorder or disease (see section 6.3.5).

An inversion, the 180° flip of a segment of a chromosome (see chapter 2), is a particularly interesting type of structural mutation with respect to LD and recombination (Kirkpatrick 2010). Recombination between the inverted and non-inverted segment is greatly reduced in heterozygote offspring. This is because recombination in heterozygotes results in aneuploid gametes that are inviable (Figure 6.5). As a consequence, the inverted and ancestral segments are effectively isolated from each other.

The inverted segment may become lost by genetic drift or selection, or it may go to fixation. Sometimes, however, both the inverted and non-inverted segment are maintained in the population by balancing selection. In such cases, because the genes within the region hardly recombine, the inverted and non-inverted version will diverge genetically through selection and genetic drift.

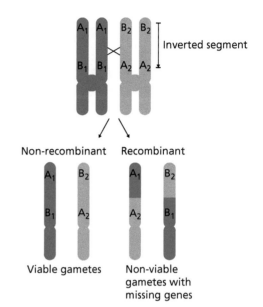

Figure 6.5 Recombination in individuals that are heterozygous for an inverted and a non-inverted segment of a chromosome results in aneuploid, inviable gametes with missing genes. This strongly reduces the (effective) rate of recombination.

Such structural rearrangements can have quite profound effects on phenotypes and evolution. For example, an inversion event explains the genetics of a most peculiar polymorphism in male plumage ornamentation, reproductive behavior, and body size in the ruff (*Calidris pugnax*) (Küpper et al. 2016; Lamichhaney et al. 2016).

These shorebirds aggregate on leks—i.e. an area where courtship and mating take place. The aggressive "independent" male morph has elaborated plumage ornamentation (ruffs and head tufts) of variable but predominantly dark color (Figure 6.6). They defend small territories on the lek against competing independent males and attract females with a spectacular courtship display. The non-territorial "satellite" morph is often accepted within the territory of an independent male. The satellites have light-colored ruffs and head tufts. They co-display with the independent male to attract females. However, they behave submissively and allow the independent male to act dominantly. Females apparently find such demonstrations of dominance attractive, so an independent male benefits in terms of mating success from accepting a satellite on his territory. The satellite benefits from the arrangement

Figure 6.6 Three male morphs of the ruff (*Calidris pugnax*). The smaller female mimic "faeder" (left), the submissive "satellite" (center), and the dominant "independent" (right). The polymorphism in plumage ornamentation, size, and behavior is caused by genes located within a segment of chromosome 11 that exhibits inversion polymorphisms that protect the genes from recombining. Photo courtesy of Simon Fraser University.

too because he gets access to females without having to defend a territory of his own. Finally, the rare and smaller "faeder" morph lacks male ornamentation altogether and instead resembles a female. Through female mimicry they gain access to the territories of independent males and pursue a strategy of sneak copulations when real females solicit matings from ornamented, displaying males. The three mating tactics and associated variation in plumage ornamentation and body size are apparently maintained by negative frequency-dependent selection (Widemo 1998; Jukema & Piersma 2006).

The polymorphism in plumage, mating behavior, and size is perfectly associated with a large inversion on chromosome 11 that contains about 100 genes (Küpper et al. 2016; Lamichhaney et al. 2016). The independent morph is homozygous for the ancestral non-inverted chromosome, whereas the faeders and satellites are heterozygous for two different versions of the inversion (Figure 6.7). The breakpoint of the inversion disrupts the reading frame of an important gene. Therefore, homozygotes for the inversion die at an early stage of development. This explains why satellites and faeders are always heterozygous for the inversion.

The inverted and non-inverted copies are strongly differentiated. Estimates based on F_{ST} suggest that the inversion that distinguishes faeders from independents occurred about 3.8 million years ago. Then, about 520 000 years ago, a rare recombination event occurred in a heterozygote that resulted in the inversion variant possessed by the satellite morph (Figure 6.7). Although such recombination would usually be detrimental (Figure 6.5), a viable gamete could be produced in some extremely rare occasions if vital genes are not deleted in the process. In the case of the satellite chromosome this appears to have happened because of "lucky" multiple crossing-overs that allowed essential genes to be maintained in the resulting gamete (Lamichhaney et al. 2016).

Genetic differences between the two inversion variants and the non-inverted copy explain the morphological and behavioral differences between the three morphs. Among the genes that are located within the inverted segment is a gene for melanin color-pigment synthesis as well as genes that are essential in the metabolism of sex hormones (Küpper et al. 2016; Lamichhaney et al. 2016). These are **candidate genes** for the plumage and behavioral

Ancestral chromosome

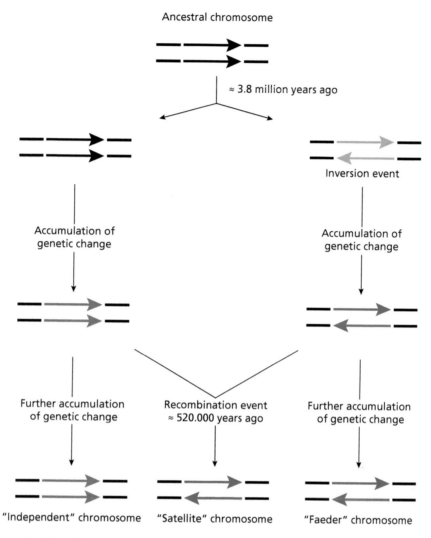

≈ 3.8 million years ago

Inversion event

Accumulation of
genetic change

Accumulation of
genetic change

Further accumulation
of genetic change

Recombination event
≈ 520.000 years ago

Further accumulation
of genetic change

"Independent" chromosome   "Satellite" chromosome   "Faeder" chromosome

Figure 6.7 Reconstruction of the evolutionary history of the three chromosome segments underlying the polymorphism in mating behavior, size and plumage in the ruff (*Calidris pugnax*). Modified from Lamichhaney et al. 2016.

differences that distinguish the morphs, although the full details of the genetics of the phenotypic polymorphisms remain to be explored.

Because the three versions of the chromosome segment essentially do not recombine, alleles that distinguish the three morphs, i.e. fixed differences at pairs of loci, are in perfect LD ($D' = 1$). Such segments of non-recombining genes, like those in the ruff, are referred to as **supergenes** because they protect beneficial gene combination from being dissociated. Female mimicking sneaker behavior is only

beneficial if you also possess genes that make you resemble a female in size and color, and submissive satellite behavior is only beneficial if your plumage honestly signals your non-aggressive and cooperative intentions. Hence, recombination between the relevant genes would likely have resulted in phenotypes with low fitness.

Gene flow

When a population receives migrants from a genetically differentiated population there will typically be linkage disequilibrium between pairs of loci at which the two populations differ in allele frequency. To take an extreme example, if a population that is fixed for A_1 and B_1 starts to receive immigrants from a population fixed for A_2 and B_2, it will possess an excess of the A_1B_1 and A_2B_2 haplotypes for the same reason as discussed above for the case of recent admixture. Random mating between residents and immigrants and recombination will eventually counteract LD, but gene flow between genetically differentiated populations is a common source of LD in natural populations. Indeed, increased LD in populations thought to be exchanging genes is often used as means of determining whether gene flow is occurring (Nosil 2012). Furthermore, LD among alleles derived from migration from a different population can be used to determine the timing of gene flow, as recombination should break down ancestry blocks from the migrant population at a predictable rate (Baird et al. 2003).

6.2 Multilocus selection

6.2.1 Additive and non-additive fitness

Genes that contribute to a phenotypic trait may interact in different ways with respect to how they affect fitness. We have **additive fitness** when fitness is the sum of contributions from alleles at different loci. As a hypothetical example, let us say that the butterfly species in Figure 6.3 lives in an environment that is dominated by a blue background color. An all-blue butterfly would be better camouflaged against the background than a purple one, and an all-red butterfly would be the poorest fit to such an environment (Figure 6.8). We assume now that the alleles at the locus for the color of the forewing and the hindwing would contribute to fitness in an additive manner. The more blue alleles (A_1 and B_1) the individual possesses, the higher fitness. Because the forewing is larger than the hindwing in our hypothetical butterfly, we may assume that adding

Figure 6.8 Illustration of additive fitness. Against a blue background, a hypothetical butterfly species has higher fitness the bluer its wing color and lower the redder, due to differences in camouflage. Forewing color is determined by locus A and hindwing color by locus B. We have assumed that adding a blue A_1-allele to the forewing-locus increases fitness by 0.2 whereas adding a blue B_2-allele to the hindwing-locus increases fitness by only 0.1, due to the different size of the two wing parts.

a blue A_1 allele increases fitness more than adding a blue B_1 allele (Figure 6.8). When fitness is additive, two locus selection (with two alleles at each locus) behaves very much like the one-locus case but with four "alleles," since we have four haplotypes. In our hypothetical butterfly example with the blue background, the population would eventually become fixed for the blue A_1 and B_1 alleles.

However, very often loci interact in a non-additive way. Let us say that our hypothetical butterfly species does not rely on camouflage for protection, but on mimicry. Say there are two poisonous model species, one all blue and the other all red—our butterfly therefore benefits from resembling one of these (Figure 6.9). In this case the fitness contribution of say a blue hindwing depends on the color of the

forewing. A blue hindwing is associated with high fitness if the forewing is also blue (good mimicry), but low fitness if the forewing is red (poor mimicry). The fitness contribution of an allele at one of the loci depends on the genotype at the other locus. This is an example of epistatic fitness, a non-additive interaction between alleles at different loci on fitness. **Epistasis** is also used to describe interactions between multilocus alleles on phenotype. Note that in the hypothetical butterfly example with mimicry the phenotypes are determined additively whereas fitness is epistatic.

Intuitively, there should exist an internal equilibrium (likely unstable unless frequency-dependent selection stabilizes the system, see chapter 4) in this second hypothetical example, in which both all-blue and all-red mimics are favored and intermediates

Figure 6.9 Illustration of epistatic fitness. A hypothetical butterfly species gains protection from mimicking either a blue or a red poisonous model species. The fitness contribution of an allele at one of the loci depends on the genetic background, namely the genotype at the other locus. See main text for further explanation.

(poor mimics) are selected against. This would result in disruptive selection for either all-blue or all-red mimics. It is also easy to see that the population would be in linkage disequilibrium at this evolutionary equilibrium. There would be an excess of A_1B_1 and A_2B_2 haplotypes relative to A_1B_2 and A_2B_1. So clearly, when there is epistasis for fitness, LD can be maintained even at an evolutionary equilibrium.

6.2.2 Epistasis and the evolutionary process

Epistasis between loci in which the fitness of a genotype at one locus depends on the genotypes at other loci, is thought to be common in nature. Most traits are probably determined by two or more genes interacting in complex ways to produce phenotypes. Indeed, it is possible that most genes interact with one another (Boyle et al. 2017). This means that, in most cases, epistasis is an extremely complex phenomenon that extends well beyond the simple two locus models we provide here. Nonetheless, simple models still help direct our thinking and give us

a basis from which to begin to understand such a complex network of interactions.

Epistasis may constrain adaptive evolution (e.g. Whitlock et al. 1995). In the butterfly example with mimicry (Figure 6.9) there are two fitness peaks (populations with all-blue and all-red butterflies) with a fitness valley in between (individuals with mixed wing colors are poor mimics and have low fitness). Suppose that our hypothetical population has become fixed for the blue alleles at the two loci but that it would be even better off mimicking the red butterfly. Say, the red model species is more poisonous and/or more numerous than the blue one, thus offering the mimic better protection. However, because of the fitness valley between the two peaks the population would be unable to evolve all-red color by natural selection alone. Epistasis is thought to often cause such multiple peaks in the adaptive landscape and thus constrain adaptive evolution to a particular outcome.

In some cases, however, epistasis may actually facilitate adaptive evolution (e.g. Hansen & Wagner

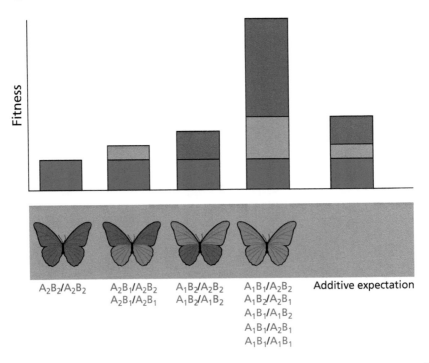

Figure 6.10 Illustration of positive epistasis between beneficial alleles. The hypothetical butterfly lives in a blue environment and benefits from having blue camouflage color. Adding alleles that yield blue color to only one wing part improves fitness only slightly because the butterfly still contrasts against the background with the other wing part. In this example of positive epistasis, alleles that yield blue color on both wing parts act synergistically, yielding higher fitness than the additive expectation, thus accelerating the fixation of the beneficial A_1 and B_1 alleles. Note that we have assumed that A_1 and B_1 are dominant.

2001; Carter et al. 2005; Phillips 2008). This is the case when there is **positive epistasis** among beneficial alleles at two or more loci. We can use our hypothetical butterfly species to illustrate this scenario too. Assume that the ancestral butterfly is red colored but that it would be better off with an all-blue color because it lives in a blue environment (camouflage). Assume further, just for simplicity, that this time the blue alleles at both loci are dominant. When we considered this example before, we first assumed that beneficial blue alleles acted additively—i.e. that any addition of a blue allele improved fitness on a blue background (see Figure 6.8). However, it may

be more likely that blue alleles interact epistatically. Why is this? Adding a blue allele to the hindwing locus may improve fitness only slightly because the butterfly still contrasts with the background with its other wing part. Likewise, adding a blue allele to the forewing locus only would also be insufficient to make it camouflaged against the background. However, a blue allele at both wing part-loci might improve fitness beyond the additive expectation, because contrast against the background is nearly completely eliminated (Figure 6.10). Positive epistasis between beneficial alleles is essentially cases when two alleles together at a pair of loci lead to a more fit

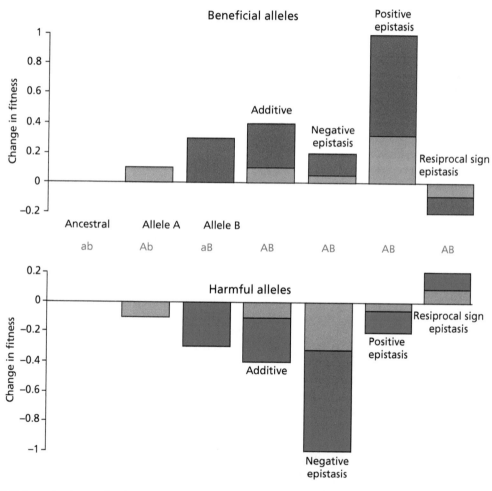

Figure 6.11 Interactions between beneficial (top panel) and harmful dominant alleles (bottom panel) at two loci. Genotypes containing both dominant alleles (AB for short) show additive fitness when their fitness is the sum of fitness contribution from either allele (Ab and aB), they show negative epistasis when fitness is lower than expected from the contribution of either allele alone, they show positive epistasis when fitness is higher than expected from the contribution of either allele alone, and reciprocal sign epistasis when the joint action of the two alleles are opposite in terms of change in fitness compared to when they occur alone.

phenotype than expected from either allele alone; that is, a case where $1 + 1 > 2$. Fixation of the beneficial blue alleles at the two loci would be faster when fitness is positive epistatic than when it is purely additive because selection would be more effective. Positive epistasis between beneficial alleles will therefore facilitate adaptive evolution.

Conversely, when two alleles together at two loci lead to a less fit phenotype than expected from either allele alone there is **negative epistasis** (Figure 6.11). When there is negative epistasis between beneficial alleles, adaptive evolution is retarded compared to the additive case. In contrast, negative epistasis between deleterious alleles (that is, both deleterious alleles present at the two loci yields a less fit phenotype than expected from either allele alone) would facilitate their removal from a population because purifying selection would become more effective. Positive and negative epistasis are examples of **directional epistasis**. Finally, alleles at two loci may show **reciprocal sign epistasis** when they occur together. In such cases a genotype containing, say, two beneficial alleles would be less fit than the genotype containing neither, or a genotype containing two alleles which are harmful when they occur alone is more fit than the genotype containing neither (Figure 6.11). Clearly, epistasis can have a range of effects on the course of evolution depending on its nature.

6.3 Quantitative genetics in the genomic era

6.3.1 The architecture of the genotype–phenotype map

An important challenge in evolutionary genetics is to map the genetic basis of phenotypic characters. In earlier chapters we have seen examples in which variation at a single locus corresponds to discrete variation in a phenotypic character. However, a large number of phenotypic traits, perhaps the majority, exhibit continuous variation and are affected by a large number of interacting genes, as well as by environmental factors. Such characters are called quantitative traits. **Genetic architecture** is a common way to refer to the genetic basis for phenotypic differences between individuals or populations. The

term is broad and includes knowledge about how many and which genes are involved in producing a given phenotype as well as how gene interactions, such as pleiotropy and epistasis, further affect that phenotype. Unraveling the genetic architecture of phenotypic traits is important because it informs us about the evolutionary potential of populations and how they may respond to selection or demographic changes. Further, it can help us understand the biology of important traits such as genetic disorders and relevant traits in animal and plant breeding. Both of these have real-world implications beyond basic science in medicine, agriculture, and food production. However, reconstructing the actual architecture of the genotype–phenotype map of quantitative traits, that includes all contributing genes and how they interact through biochemical pathways and with environmental variation, and how these interactions change during the course of evolution, is a formidable task and we are still in an early phase of such endeavors.

6.3.2 Sources of phenotypic variance and the concept of heritability

A natural starting point when seeking to unravel the genetic architecture of a quantitative trait is to quantify its variation and to determine to what extent the observed variation is due to genetic and/or environmental factors. For continuously varying traits such as body length, beak height, or IQ we can describe the variation found in a population by calculating the mean trait value and its variance. Let X_i be the measured value of a continuous trait such as body length in the ith individual and n the total number of measured individuals. The mean value is simply the sum of values divided by sample size:

$$\bar{x} = \frac{\sum X_i}{n} \tag{6.11}$$

and the variance the average sum of squared deviations from the mean value:

$$V = \frac{(X_1 - \bar{x})^2 + (X_2 - \bar{x})^2 + \ldots + (X_n - \bar{x})^2}{n-1}$$

$$= \frac{1}{n-1} \sum_{i=1}^{} n_i (X_i - \bar{x})^2 \tag{6.12}$$

Variance is a useful parameter to describe variation in a quantitative trait because the total variance can be partitioned into different sources that contribute to it. For any phenotypic trait that is affected by both genetic and environmental factors the total phenotypic variance in a population (V_P) is simply the sum of variance caused by these genetic (V_G) and environmental factors (V_E), That is:

$$V_P = V_G + V_E \qquad (6.13)$$

Genetic sources of variance can be further divided into several sub-categories, including additive variance (V_A), dominance variance (V_D), and epistatic, or interaction variance (V_I). Together the values of these subcategories yield the total amount of genetic variance for a given trait:

$$V_G = V_A + V_D + V_I \qquad (6.14)$$

Additive variance refers to the deviation from the mean phenotype due to additive genetic effects. **Dominance variance** involves deviation due to interactions between alternative alleles within a locus, and **epistatic variance** involves deviations due to interactions between alleles at different loci. Of these, the additive variance plays the most important role in evolutionary theory because the additive effects of all the alleles that affect a trait are responsible for the resemblance between parents and offspring. This is therefore the part of the genetic variance that will respond to selection within a population.

Let us dwell a little on the sources of variance. Consider first a simple, hypothetical case of a trait that is affected by only one locus with two alleles, for example body length. Assume that inheritance is additive and that the mean length of A_1A_1 is 170 cm, those of A_1A_2 175 cm, and A_2A_2 180 cm. Variation within each genotype would be environmental and contribute to environmental variance. The sources of this environmental variation may be differences in nutrition value or amount of food received during development as well as variation in exposure to diseases and other health issues that can retard growth. Even if two individuals have the same genotype and experience the same environment they may end up with a slightly different length due to developmental instabilities. Deviations due to developmental instability are incorporated in the environmental variance.

As for the genetic variation in this example, it is entirely additive and would contribute to the additive variance in the population. If we control for environmental variation, knowledge of genotype allows us to perfectly predict the phenotype. Moreover, the average offspring would be exactly intermediate between its parents when controlling for the environment. For instance, crosses between A_1A_1 (170 cm) and A_1A_2 parents (175 cm) yields offspring in which half are A_1A_1 (170 cm) and half A_1A_2 (175 cm), so the average offspring would be 172.5 cm.

Now, consider a case when A_2 is dominant so that A_1A_1 are 170 cm and A_1A_2 and A_2A_2 are 180 cm. There is still a positive relationship between the number of A_2 alleles and body length and offspring still tend to be more similar in phenotype to their parents than to the population average (parents with few A_2 alleles in their genotype on average produce shorter offspring than parents with many A_2 alleles). Therefore, some of the genetic variation is additive in nature and contributes to the additive variance of the population. However, because of dominance the relationship is not perfect. For instance, crosses between A_1A_1 and A_2A_2 parents yield offspring which are equally tall as the A_2A_2 parent rather than intermediate between the two parents. Some of the genetic variance is therefore due to dominance variance.

When more loci are affecting a trait, epistasis can similarly reduce the resemblance between parents and offspring and thus reduce the additive variance by introducing epistatic variance.

Finally, additive variance depends on allele frequencies. If most individuals in the population have the same genotype (say A_2A_2) there will obviously be less genetically determined variation in body length than if A_1 and A_2 are both common.

The proportion of the total variance in a trait that is due to additive variance in a population is an important parameter in evolutionary biology and is referred to as **heritability** in the narrow sense. We define this parameter as:

$$h^2 = V_A / V_P \qquad (6.15)$$

A second measure is heritability in the broad sense, which is defined as:

$$H^2 = V_G / V_P$$

or the total genetic variance divided by phenotypic variance. Of these, heritability in the narrow sense (from now on referred to as just heritability) is the more important parameter in evolutionary theory because it directly relates to how a trait would respond to (artificial or natural) selection.

In animal and plant breeding this property of heritability has long been appreciated in the famous **breeder's equation** (Lush 1937), stating that the response to selection (R), measured as the change in the mean value of the trait after one generation of selective breeding, equals the heritability of the trait (h^2) multiplied with the selection differential (S), which is a measure of the strength of (artificial) selection. That is:

$$R = h^2S \qquad (6.16)$$

With **truncation selection**, commonly used in animal and plant breeding, a fixed proportion of the population is chosen to reproduce and to found the next generation. Here, S is equal to the difference in mean trait value of these selected individuals and that of the population as a whole. By estimating the heritability of a trait of interest (e.g. seed count or milk yield) a breeder can use the breeder's equation to predict how much his or her crop or stock would improve from one generation of selective breeding.

The breeder's equation can also be used to predict evolutionary responses to selection in a natural population. However, since natural selection rarely operates in the form of truncation selection, the selection differential S has to be generalized. In evolutionary biology it is commonly defined as the covariance between the trait and relative fitness (Falconer & MacKay 1996).

If we knew all the alleles that contribute additively to the variation of a trait and how much each locus contributed to the phenotype we would know the heritability of that trait. However, even without any knowledge of the genetic details of a trait we can estimate heritability by comparing the similarity of genetic relatives while controlling for environmental variation. For livestock or crops, heritability can be estimated by comparing the similarity in trait values of genetic relatives relative to the population average, such as parents and offspring or among siblings, reared under identical conditions. For agricultural species one can also estimate the realized

heritability by using the breeder's equation. The strength of selection S is determined by the breeder and the response to selection R can be measured directly after one generation of selection. Hence, by rearranging the breeder's equation, the realized heritability can be estimated as:

$$h^2 = \frac{R}{S} \qquad (6.17)$$

Heritability can likewise be estimated in natural populations by comparing trait values of genetic relatives, but again we need to control for environmental variation. Boag (1983) estimated heritability in morphological traits in Darwin's finches by regressing offspring trait values on their respective mid-parent values. In such a regression analysis the slope of the regression line equals the heritability of the measured trait, provided environmental effects have been nullified (Figure 6.12).

Controlling for environmental effects is not always as straightforward in natural populations as in the livestock case. However, conducting **common garden experiments** is one means of overcoming the difficulty of estimating heritability from wild populations. In a common garden, genetically different individuals are transplanted from their native environment and

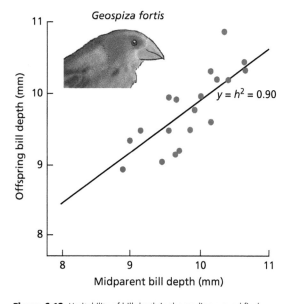

Figure 6.12 Heritability of bill depth in the median ground finch (*Geospiza fortis*) estimated by regressing bill depth of offspring on their mid-parent's trait values. The slope of the regression line equals the heritability of the trait. Redrawn from Boag (1983).

reared in a common environment in order to separate the effects of genetic and environmental influences on the phenotype. Obviously, this approach is limited to species which can easily be transplanted, reared, and maintained in controlled settings.

Heritability can also be estimated for traits in human populations, but it is clearly much more difficult in our own species because it is not possible to conduct experiments to isolate environmental and genetic factors. However, it is possible to control for these factors on a *post hoc* basis. Genetic relatives may have similar trait values both because of genetic inheritance and because they share a similar environment. Twin-studies are an important source for disentangling the relative importance of genes and environment in explaining variation in human traits. For instance, to estimate the heritability of a trait we may compare trait similarities between a large number of identical and non-identical pairs of twins. Identical twins are virtually genetically identical ($r = 1$), whereas non-identical twins are as genetically different as non-twin siblings ($r = 0.5$) but, like identical twins, experience the same environment in the womb. We can measure the similarity of a given trait, such as height, between twins and then quantify the difference in this similarity between identical and non-identical twin pairs. This difference should arise due to additive genetic variation, assuming the two twin sets experience similar environments on average (e.g. Grasby et al. 2017).

Naturally, heritability estimates from human populations attract considerable media and popular attention. For this reason, it is important to fully understand what heritability estimates of important human traits actually imply. If we take IQ as an example, the meaning of an estimated heritability of 0.8 can easily be misunderstood. A common misconception is that this would imply that IQ is determined 80 percent by genes and 20 percent by the environment. However, genetic and environmental factors affecting a trait are nothing like ingredients we mix when baking a cake. A heritability of 0.8 simply means that within the population investigated, at that specific point in time, 80 percent of the observed phenotypic variance can be ascribed to additive genetic variance. Stated even more plainly, 80 percent of the observed variation in IQ *among* the individuals tested is due to additive genetic variance.

A second common misconception is that heritability estimates can be extrapolated to say anything about the genetics of differences between groups or populations. This is plainly wrong. It is possible that the entire difference between two populations is due to genetic factors even if the heritability of the trait is zero in both populations. This would be the case if the two populations were fixed for different alleles so that all residual variation within each population was due to environmental factors. Presumably, skin color has a quite low heritability both among light skinned Sami of Northern Scandinavia and among dark skinned Maasai of Eastern Africa. Yet the difference between the groups in skin color is almost entirely due to fixation of different pigmentation alleles in the two populations and is genetic in nature. Likewise, even with a heritability of 100 percent within each of two populations, the difference in trait values between them could be entirely due to environmental factors. This would be the case if all variation within each population was due to additive effects but that the two populations experienced a systematic or discrete difference in environmental conditions. Body length generally has a high heritability within populations, often around 80 percent. Taller parents tend to produce taller children than do shorter parents. Yet, the difference in average stature (about 4–5 cm) between North and South Korean people is most likely entirely due to the much poorer average nutritional and health status of the northern population. Note also that the heritability of a trait can be quite different across populations. In more uniform environments a relatively larger proportion of the phenotypic variance would tend to be due to additive genetic factors compared to in more heterogeneous environments. It is entirely likely that heritability in body length would be higher in the more egalitarian South Korea compared to the northern population where wealth is more unevenly distributed.

6.3.3 Evolvability

In evolutionary biology, heritability has often been used as a predictor of the evolutionary potential of a population. Indeed, for a specific trait in a given population, such as milk yield in cattle, heritability has a high predictive value as seen above in the breeder's equation. The higher the heritability, the higher the response to the same strength of artificial selection. However, when we are interested in comparing the evolutionary potential of a number of different traits across populations or species, heritability is problematic (Houle 1992). One important issue is that the amount of additive genetic variance is scaled by a variance term (recall that $h^2 = V_A / V_P$). The parameter is therefore highly sensitive to differences in the overall phenotypic variance of the different traits and to correlations between additive variance and other variance components. If we compare two traits with a similarly high additive variance but in which one of the traits has a much higher phenotypic variance we may underestimate the evolutionary potential of the more variable trait if we only consider heritability. An alternative measure of the evolutionary potential of a trait, or its **evolvability**, is to rather scale additive genetic variance by the mean value of the trait. Hansen & Houle (2008) suggested the following definition of mean-scaled evolvability:

$$e_\mu = V_A / \bar{x}^2 \qquad (6.18)$$

Where $\bar{x}$ is the mean trait value before selection. Hansen et al. (2011) compared estimates of heritability and evolvability for a large number of traits in a large number of species by examining previously published empirical studies. They found that the correlation between the two measures were close to zero and argued that we should use evolvability and not heritability as a predictor of the evolutionary potential when comparing different traits and/ or populations or species. For instance, life history traits, such as age at maturation, longevity, and clutch size, generally have a much lower heritability than morphological traits. Yet, life history traits have a high evolvability, meaning that they readily respond to selection in terms of changes in the mean value of the traits. The low heritability of life history traits is likely caused by high levels of phenotypic variance that mask the high levels of additive genetic variance that actually are present (Hansen et al. 2011).

As we have seen, the amount of additive genetic variance in a trait affects its heritability and evolvability. One common misunderstanding is therefore that dominance and epistasis invariably constrain adaptive evolution of quantitative traits. This is, however, not the case. First, as we have seen, loci in which the alleles show partial or complete dominance and/or that interact epistatically do contribute to the additive variance of a trait. Second, as we saw in the two-locus case, certain types of epistasis actually facilitate adaptive evolution and can therefore increase the evolvability of a trait. According to Carter et al. (2005), the major determinant for the evolution of additive variance is directional epistasis. Positive epistasis in the direction of selection leads to an acceleration of evolvability, whereas negative epistasis slows it down. An important empirical question is therefore to what extent epistasis tends to be directional and, if so, whether positive or negative epistasis dominates.

Directional epistasis may help to explain a phenomenon which long has puzzled animal and plant breeders. When breeders select for a particular trait such as fruit size or milk yield, one would assume that genetic variation for that trait would quickly become exhausted. Yet, even after long-term selection agricultural species often continue to respond phenotypically to an artificial selection regime. Carlborg et al. (2006) investigated the genetic architecture of body size in two lines of chicken (*Gallus gallus*) that had been selected for high and low body weight for 42 generations. The authors identified one major locus for growth. One would think that genetic variation at this locus would soon become exhausted under the selection regime and thus that the chickens would cease to diverge in size after some generations of selection. Yet, even after 42 generations of selection they continued to diverge. Carlborg et al. (2006) ascribed this discrepancy to four other loci that interacted epistatically with the major locus. Specifically, allelic variants at these other loci were associated with higher weight in the high-weight line, whereas no such effect was found in the low-weight line. Thus, apparently positive epistasis between the major locus and the four other loci had released variation that selection could act upon in the high-weight line and thus generated a considerably larger selection response than predicted by a single-locus model.

6.3.4 Quantitative trait locus (QTL) analysis

Identifying the genes and genetic architecture that underlie a trait is a major focal goal of evolutionary biology. So, having verified that variation in a quantitative trait is heritable (and evolvable) the next logical step would be to home in on the genomic regions that harbor genes which contribute to controlling that phenotype. In quantitative trait locus (QTL) analysis (also known as QTL-mapping) we take advantage of the presence of strong linkage disequilibrium between alleles at a locus that affects a quantitative trait and genetic markers in the physical vicinity of that locus by using a controlled crossing scheme. A QTL is simply a specific genomic location that correlates with variation in a quantitative trait because it either harbors a gene that affects the trait or because it is physically linked to such a gene. This small detail is an important one—QTL analysis has been successful in identifying genes underlying traits (Colosimo et al. 2004—see later in this section) but sometimes it is only possible to get an indication of the possible candidates involved in adaptation (Fountain et al. 2016). So, QTL analysis should be seen as one of a series of methods used to begin unraveling genomic architecture underlying a trait of interest (Broman & Sen 2009).

The classical approach in a QTL analysis is to breed two strains of an organism that differ significantly in a heritable trait of interest. Let us take fruit size in strawberries as a hypothetical example (Figure 6.13). Here we could create two inbred lines of plants, one of which bears large fruits and another which bears small fruits, to produce a large number of large- and small-fruited plants. The rationale with this inbreeding approach is to obtain experimental lines which are highly homozygous across the genome and that would be homozygous for different alleles at genes that affect the focal trait. Then, the parental lines are crossed, resulting in heterozygous (F1) individuals. Finally, these F1 individuals are crossed using one of a few different crossing schemes to produce a second and genetically and phenotypically variable generation. For instance, F1s may be crossed with each other to produce F2 progeny (an F2 intercross design), or we may choose to backcross the F1s with either parental strain (F1-backcross design). How does this work? Assuming that parental strains are fixed for different alleles A_1 and A_2 at a given marker position (i.e. parental genotypes are A_1A_1 and A_2A_2 respectively), all F1 individuals will be heterozygous A_1A_2. In a backcross design, offspring from the cross to say the A_1A_1 parent will be either A_1A_1 or A_1A_2. In the intercross, we produce A_1A_1, A_1A_2, and A_2A_2

offspring (note that this pattern does not apply to markers with biased transmission—such as on the sex chromosomes). Different crossing schemes have different pros and cons with respect to efficiency in detecting as many QTLs as possible versus detecting the ones associated with the largest effect on the phenotype (Darvasi 1998). However, the main distinction is that an intercross makes it more straightforward to detect dominance in either parental allele, whereas in a backcross it is only possible to detect dominance from the parent the offspring is crossed to (Broman & Sen 2009).

Irrespective of our crossing scheme, our next step is to genotype all the crossed individuals for a number of genetic markers. If our earlier inbreeding approach worked well, we would hope that we can identify many informative markers in our final F2 sample. Loci that are homozygous in the grandparents we started our cross with will be heterozygous in the F1s and should, all things being equal, follow Mendelian segregation (i.e. 1:2:1) in the final generation (i.e. A_1A_1, A_1A_2, and A_2A_2, as explained above). We then measure the phenotypes of all our individuals and attempt to correlate variation in phenotype with variation in genotypes—i.e. an association between a given genotype and the phenotype, greater than expected by chance. Since our trait of interest here is fruit size, individuals with different fruit sizes should tend to possess different alleles at markers that are close to a gene that affects fruit size because there has only been one generation of recombination since the inbred lines were crossed. That is, there would be strong linkage disequilibrium between a causal gene and markers in its vicinity. In contrast, recombination between a causative gene and markers further away from it would have occurred in a large number of individuals resulting in much weaker LD. Here again, we see that LD is the key to identifying genomic architecture in this study design.

For a QTL analysis, we are able to test for phenotype–genotype associations on all our informative markers. In the era of high-throughput sequencing, this can be in the thousands (Glazer et al. 2015; Fountain et al. 2016). The most basic association test is an analysis of variance (ANOVA) test of phenotypes among genotypes at each locus—i.e. whether genotype identity can explain a significant proportion of the phenotypic variance. An ANOVA in this

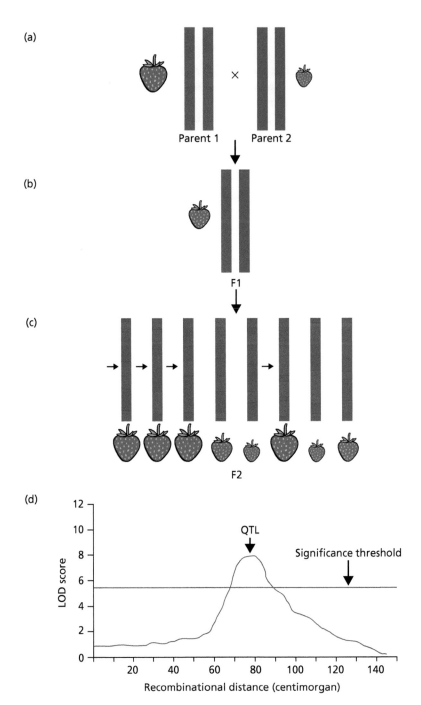

Figure 6.13 The principle of QTL mapping using an F2 intercross. (**a**) Parents from two inbred strains that differ markedly in a phenotypic trait of interest, here fruit size in strawberries, are crossed to produce heterozygote F1 offspring (**b**). Red and blue bars illustrate genetically different chromosomes in the two strains. Then the F1s are crossed according to one out of several crossing schemes to produce a generation of genetically and phenotypically variable offspring (**c**). In this example the F1s have been crossed to produce F2 progeny in an F2 intercross (for simplicity, only one chromosome is drawn per individual in the F2 generation). Genotypes at a large number of genetic markers along the chromosome are then correlated with phenotypic variation. In this example, individuals with alleles inherited from the red, large-berried strain near the position marked with an arrow all produce large berries. Hence, apparently a gene near this position affects fruit size. This position is a QTL. **d**) A corresponding QTL-plot shows an interval of degree of association (LOD-score) along the chromosome as a function of recombination rate (centimorgan). Note that it is not possible to be certain of the exact position, but rather we have a region of the genome where genotypes are associated with phenotypic variation in this trait. See text for additional explanation.

manner can be used outside the QTL analysis framework too and is often used by evolutionary biologists to test phenotype–genotype associations in wild populations (Lamichhaney et al. 2015; Ravinet et al. 2015). An ANOVA is a method that compares the variance among groups to the total variance in a dataset. Think back to our previous QTL example, with our strawberry plants; in our F2 generation we have measured the fruit size of each individual. Since we crossed parental lines with different fruit sizes, there is a range of fruit size in our F2 generation—in other words there is variance. Because we are trying to identify the genetic basis of this trait, we also know the genotypes of our F2s at our markers. For a given marker with A_1 and A_2 alleles, we can divide our F2 individuals into A_1A_1 and A_2A_2 homozygotes and also A_1A_2 heterozygotes; we can group them by their genotype. With an ANOVA, we compare the total variance with the variance explained by grouping individuals in this way. In short, we test whether genotype is able to explain a significant proportion of the variance in phenotype.

However, most QTL analyses make use of an interval mapping approach which performs association tests against markers anchored in a **genetic map**—i.e. a map of markers arranged in terms of their recombination frequency, rather than physical distance. Typically, this is shown as centimorgans (cM), where a single centimorgan between two markers represents a probability of 0.01 that a recombination event will occur within one generation (or alternatively, that in 100 generations, there is one recombination event). Conceptually similar to ANOVA, the fact that interval mapping takes marker location into account means that it is possible to identify an interval on the genetic map where a QTL might occur between two markers. Association tests with interval mapping use maximum likelihood (see chapter 9) to test against a null model of no association between a marker and a trait. This allows for the calculation of a LOD score for each marker—i.e. the logarithm of the odds ratio. The LOD score represents the ratio between the likelihood of two models; that a QTL exists at a given position (an association model) or that no QTL occurs anywhere at all in the genome (a null model). Highly positive LOD scores indicate the presence of a QTL. QTL studies therefore typically produce a plot, showing the position on the X-axis and a measure of the strength of correlation between the genotype of the markers and phenotype as LOD scores on the Y-axis (Figure 6.13). In addition, we typically calculate a threshold for the statistical significance of the correlations using a permutation approach. Strong peaks above a significance threshold therefore indicate the location of QTLs on the genetic map.

QTL crossing experiments have been an extremely useful means of identifying genome regions involved in adaptive traits. In a landmark series of studies, QTL analysis was used to identify the genetic basis for phenotypic variation in the threespine stickleback (*Gasterosteus aculeatus*) (Figure 6.14). Threespine stickleback are often mistaken as being a solely freshwater fish; however, they are in fact ancestrally marine. Following glacial retreat at the end of the Pleistocene (~15–20 000 years ago), marine stickleback have repeatedly and independently colonized and adapted to freshwater environments throughout the Northern Hemisphere (McKinnon & Rundle 2002). Marine stickleback are most easily identified by the thick armor plates they carry on the lateral flank of their body (Figure 6.14), largely thought to serve as protection from predators—this is known as the completely plated phenotype. However, a signature of freshwater adaptation is the loss of armor plates—resulting in a low-plated freshwater form (Figure 6.14). This loss, among other freshwater adaptations, has occurred multiple times, independently in most (although not all) freshwater populations; it is a classic example of parallel adaptation (Jones et al. 2012). In some cases, the loss is so extreme that fish exhibit no plates at all (Bell 1994). The exact reason for armor loss is not clear although it is likely a combination of reduced predation pressure and the lower availability of calcium ions in freshwater environments—in other words it is more costly to maintain fully plated armor in freshwater (Barrett et al. 2008). Using an F1 backcross design between two freshwater stickleback ecotypes, Peichel and colleagues genotyped and phenotyped all offspring to identify QTLs driving phenotypic variation across a number of traits, including armor plates (Peichel et al. 2001). They identified a suite of loci explaining 8–10 percent of the variance in plate number but did not find a locus of major effect. However, two follow-up studies combined a higher-density QTL analysis using a cross between a low-plated freshwater fish and completely plated marine fish and sequence data to identify a candidate gene on stickleback chromosome IV—*Ectodysplasin*; a QTL at this region explained 78 percent of the variance in plate morphology (Colosimo et al. 2004, 2005). This major effect locus is biallelic with complete (C) and low (L) alleles—therefore homozygote CC are completely plated, homozygote LL have a reduced armor phenotype and the heterozygote CL is typically partially plated. Subsequent genomics studies have repeatedly confirmed that this gene is under divergent selection between marine and freshwater populations and that the low-plated freshwater phenotype has evolved in parallel throughout the Northern hemisphere (Jones et al. 2012; see also chapter 10 for a more detailed summary of research into stickleback armor plates). In the majority of cases, such rapid parallel adaptation has occurred from standing genetic variation. The allele for low armor plates is apparently maintained at low frequency in the marine population due to ongoing gene flow between freshwater and marine populations (Schluter & Conte 2009).

Figure 6.14 Threespine stickleback (*Gasterosteus aculeatus*) exhibit variation in the number and size of armor plates along the lateral flank on their bodies. Fish residing in freshwater have evolved a reduced armor phenotype (upper) in parallel across the Northern Hemisphere whereas fish residing in the ocean have many and large plates (lower). QTL-analysis, combined with other methods, has revealed a candidate locus with a large effect on armor located on chromosome IV. The frequency of the allele associated with reduced armor is much higher in freshwater populations than in marine populations.

It is also possible to perform QTL analysis in natural populations without conducting such crossing experiments. To do so we need information on how the individuals in the population are related to each other—we need pedigree information for the families. The task is then to search for markers that co-segregate with phenotypic variation within families. Schielzeth et al. (2012) used such an approach to identify QTLs associated with beak color in the zebra finch (*Taeniopygia guttata*). These small birds have beaks that vary from pale orange to bright red (Figure 6.15). The authors first estimated the heritability of beak color, which was found to be 34 percent in their population. Second, they mapped variation in beak redness to 1404 SNP-markers genotyped in a large pedigree. This analysis revealed 4 QTLs, each located on different chromosomes, which together explained a large part of the additive genetic variance in beak color.

Provided it is possible to perform a crossing experiment and maintain organisms in a laboratory or establish a pedigree, it is fairly straightforward to identify QTLs. It is also relatively easy to determine the pattern of dominance for single QTLs, although as we noted, this depends somewhat on crossing design. For codominant QTLs heterozygotes would be intermediate between homozygotes in average trait value whereas they would be similar to one of the homozygotes in cases of complete dominance. It is also possible to identify epistasis between loci with

Figure 6.15 The zebra finch (*Taeniopygia guttata*) exhibits variation in beak color, ranging from bright red as in this male, to pale orange. According to a QTL analysis, variation at four loci on four different chromosomes contribute to the heritable variation in beak color (Schielzeth et al. 2012). Photo courtesy of Rainer Blankermann.

QTL analysis but this requires much larger sample sizes. This is because the number of genotype classes and statistical tests increases exponentially with the number of loci. A standard F2 QTL analysis models three genotypes (two homozygotes and the heterozygote) for, say, 10^3 genomic locations. A corresponding two-locus epistatic analysis models nine genotype classes (assuming that the phenotype of the two different double heterozygotes is similar) and 10^6 two-locus combinations. Ultimately population size limits our ability to use QTL analysis to reconstruct more complex genetic architectures such as higher order epistasis in which three or more loci interact non-additively.

6.3.5 Genome wide association study (GWAS)

QTL analysis is effective for identifying a genomic region that harbors a gene with a major effect on a trait of interest. However, it is only possible in species where raising them in captivity or in the lab is feasible or where it is straightforward to obtain a pedigree. One further and more fundamental disadvantage, however, is that it is a rather coarse-grained analysis. In a classical QTL experiment there is only one generation of recombination. Therefore, even markers relatively far away from a causative gene would co-segregate with the causative gene and thus associate with the trait. In other words, linkage disequilibrium is not sufficiently broken down. Consequently, we often fail to narrow down to specific candidate genes for trait variation in such studies because there can be many plausible genes in the QTL region. We home in on a relevant genomic region, but additional studies are often necessary for identifying candidate genes, like for the stickleback example described above (see chapter 10). Another problem is that the genetic variation that is actually important in causing phenotypic variation in a population may not be present in the parental lines used in a QTL experiment and thus go undetected. We can only detect the genetic differences that are present in the lines, strains, or individuals used to begin a cross. Finally, loci with small phenotypic effects can be hard to detect statistically with the QTL approach.

Genome wide association study (GWAS) is a more recent and potentially more powerful alternative to the QTL approach. GWAS takes advantage of the explosion in number of genetic polymorphisms that can be simultaneously genotyped and analyzed with high-throughput sequencing technology. In a typical GWAS analysis hundreds of thousands or even millions of genetic markers, typically SNPs scattered across the genome, are used. Like in a QTL analysis the goal is to identify associations between markers and a trait of interest, but since the density of markers is so high there is no need for crossing experiments or elaborate pedigree information to find associations. At a sufficiently fine scale, polymorphic sites will tend to be in linkage disequilibrium with each other, even in an outcrossing population, because recombination between closely linked sites would be extremely rare. Recall that a novel mutation would be in perfect LD with polymorphic loci at the chromosome it occurred and that closely linked polymorphisms will remain in LD a very long time. Hence, with a sufficiently high marker density and sufficiently large sample sizes we can associate genotypes and phenotypes directly from a population.

GWAS is a much-used tool in medical genetics where the goal is to identify SNPs that associate with heritable disorders and diseases in humans. A classical procedure is to genotype a large number of people that have a specific medical condition and compare with a large control group. Allelic variants that are significantly overrepresented in the focal group relative to the control group are candidates for causing the condition or being closely linked to a causative locus. As an example, several studies have set out to investigate the genetic architecture of Alzheimer's disease using GWAS (e.g. Harold et al. 2009; Lambert et al. 2009; Hollingworth et al. 2011) and several risk loci has been identified. Alzheimer's disease is the most common source of dementia. It is a chronic neurodegenerative disease that worsens over time. It is highly heritable with heritability estimates up to 76 percent in some populations (Harold et al. 2009). Pérez-Palma et al. (2014) conducted a meta-analysis where they combined previously published SNP data of 4569 individuals (2540 patients with Alzheimer's disease and 2029 controls) using hundreds of thousands of SNPs. When correlating genotype and phenotype for such large numbers of markers it is necessary to adjust the level of significance. The convenient $P < 0.05$ is of no use because thousands of SNPs would by chance show an apparent association with the trait at that level. Based on the number of correlations involved, Pérez-Palma and co-workers instead used a genome wide threshold of significance at $P \leq 5.0 \cdot 10^{-8}$ to avoid including false positive associations. Their GWAS identified a genomic region of about 250 000 bp near the APOE-locus on chromosome 19 that associated significantly with the disease (Figure 6.16). This locus has also been implicated in previous GWAS studies and, specifically, one allele, APOE-ε4, is overrepresented in people with Alzheimer's disease, suggesting it is an important genetic risk factor. However, although the association is significant, the allele cannot be directly causative or the only risk factor, because it is also sometimes found in people without the disease and it is absent in some people with Alzheimer's. Pérez-Palma and co-workers therefore also looked for minor effect size contributions from other genes that could help to explain the remaining genetic risk. Their analysis

identified functional relationships as alleles involved in the glutamate signaling sub-network were significantly overrepresented among people with Alzheimer's disease. Glutamate is the most prominent neurotransmitter in the human brain and mutations in glutamate receptor genes have been linked to several neurological diseases, including Parkinson's disease and schizophrenia (e.g. Carlsson & Carlsson 1990).

6.3.6 The enigma of missing heritability

When the first draft of the human genome was published in 2001 (International Human Genome Sequencing Consortium 2001; Venter et al. 2001) geneticists and evolutionary biologists were optimistic that we would soon be able to map the genetic architecture of almost any trait of interest, including a range of complex, heritable medical conditions. From GWAS and other techniques a large number of candidate genes have indeed been identified that show associations with a range of

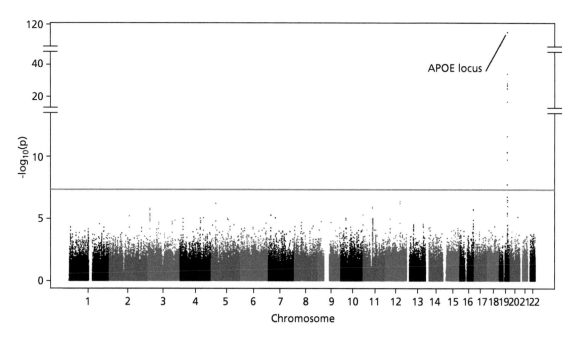

Figure 6.16 Genome wide association study of Alzheimer's disease identifying a candidate region on chromosome 19 that associates significantly with the disease, including the APOE locus. Probability of allelic association (y-axis) is plotted against genomic location with the different chromosomes indicated by change of color pattern of the SNPs. Each dot represents one SNP-allele and the probability that it associates with Alzheimer's disease. The horizontal line indicates the threshold for genome wide significance (5.0×10^{-8}). The null hypothesis of no association is discarded above this threshold. From Pérez-Palma et al. (2014).

different traits. However, scientists soon discovered a worrying discrepancy between the estimated heritability of a trait and the usually much lower proportion of phenotypic variance that could be explained from candidate loci. Moreover, in some cases, SNPs that show statistically significant association with a trait in one study fail to show any association in a replicate study. Thus, there appear to be many false positive candidate genes identified from GWAS and often a disappointingly low proportion of the heritability that can be ascribed to genetic variation at loci that associate with the trait. This has become known as the **missing heritability problem** (Maher 2008).

Several hypotheses have been put forward to try to explain the missing heritability problem (Eichler et al. 2010; Manolio et al. 2010). In short, we may tend to overestimate the heritability of traits or we may be unable to detect a significant proportion of the genes that actually affect a trait, or a combination of both. In human studies it can be difficult to accurately estimate heritability because we cannot conduct transplant or common garden experiments to control for environmental effects. However, the advanced statistical methods available in contemporary twin studies should be quite robust, so inflated heritability estimates are probably not the main cause for the missing heritability. There are, however, several factors that can make it difficult to identify all the causative genes in a GWAS. First, loci with very small phenotypic effects can be hard to detect statistically and may therefore go undetected. Perhaps more importantly, current SNP-marker sets, although large, do not cover the entire genome equally well. Regions of the genome with a repetitive structure are difficult to sequence and few if any reliable SNP-markers are available for such regions. Furthermore, given the large numbers of individual data points required for sufficient power to detect an association in a GWAS, genotype data is often collected using so-called SNP chips, rather than whole genome resequencing, to cheaply and effectively gather data from a population. SNP chips are designed using whole genome data but instead focus on a subset of well-characterized, high quality SNPs. These markers are also often chosen so that they are relatively evenly spaced throughout the genome, such that they are unlikely to be in linkage disequilibrium with one another, but will be in linkage with candidate genes. The disadvantage of this approach is obviously that causative genes may be missed. Another reason for the missing heritability problem is that traits such as genetic diseases may well be caused by mutations that go undetected in a typical SNP survey. They may, for instance, be caused by variation in copy number of a repetitive sequence or by structural changes such as inversions or translocations that cannot be detected by SNP-markers. Moreover, different mutations may be responsible for the same phenotypic effect. A heritable genetic disease such as schizophrenia may have multiple genetic causes and not necessarily the same cause in different families or populations, making it difficult to identify the causative alleles. Finally, epistasis and other interactions can be complex and make it difficult to identify causative genes. For instance, one allele may increase the likelihood of developing a particular disease, but only in certain genetic backgrounds or under certain environmental conditions. For these reasons it seems unlikely that we will be able to identify all the alleles that contribute additively to a complex quantitative trait from GWAS alone. Complementary methods, including whole genome sequencing combined with advanced bioinformatics analysis, will hopefully help us to uncover the missing heritability in the future (e.g. Girirajan 2017).

Study questions

1. You genotype two loci in a population, each with two alleles. The allele frequencies are respectively $p_1 = 0.2$ ($q_1 = 0.8$) and $p_2 = 0.7$ ($q_2 = 0.3$).
   a. Calculate the expected haplotype and genotype frequencies at linkage equilibrium.
   b. Calculate the maximum linkage disequilibrium (D_{max}) that can be found between the pair of loci, assuming a positive association between A_1 and B_1, and between A_2 and B_2.
2. Assume that a novel point mutation takes place at chromosome 1. Discuss whether the mutation would be in linkage disequilibrium with a polymorphic locus on chromosome 2.
3. Review the factors that can cause a pair of loci to be in linkage disequilibrium.

4. Loci that are in linkage disequilibrium will tend to approach linkage equilibrium over time due to recombination. Under otherwise similar conditions, would linkage equilibrium between a pair of loci A and B be approached sooner or later in a small than in a large population, or is this independent of population size?

5. At a point in time $D = 0.25$ between a pair of loci. Assuming that the assumptions of the linkage equilibrium model hold (random mating, infinite population size, no selection, no mutations, no gene flow, and recombination can occur between the pair of loci), calculate the expected level of linkage disequilibrium D_t after 1, 3, and 5 generations when the recombination rate is respectively $c = 0.5$ and $c = 0.02$.

6. What is a supergene?

7. Explain why recombination would usually be severely reduced between an inverted and a non-inverted segment of a chromosome. What are the evolutionary consequences of such reduced recombination?

8. Consider the hypothetical butterfly example in which the focal butterfly benefits from mimicking a red or a blue poisonous model species in Figure 6.9. Discuss whether fitnesses of the different morphs constitute reciprocal sign epistasis in this example or whether they constitute a form of epistasis not covered in the scheme in Figure 6.11.

9. Review the effects of negative and positive epistasis and of reciprocal sign epistasis on adaptive evolution in the case of beneficial and harmful alleles.

10. When a male lion is crossed with a female tiger the resulting offspring is called a liger. This hybrid has, not surprisingly, traits both from the lion and the tiger. However, it is much, much larger than both a lion and tiger. One possible explanation for this phenomenon is epistasis between loci that affects growth and body size. Suggest a way in which epistasis could cause the extreme phenotype of the liger.

11. A pig farmer wants to breed a stock with a longer back to increase meat yield. Mean back length in his stock is 75.0 cm and he chooses the 10 percent with the longest backs for breeding and founding the next generation. The mean back length of the selected individuals is 80.0 cm. After one generation of such selective breeding back length has increased to 78.0 cm in his stock. Use the breeder's equation to calculate the (realized) heritability of back length in his stock.

12. A white supremacist has searched for differences in IQ between groups of people on the internet and found a link to a dubious site claiming that African people on average score 15 points lower on IQ tests than Caucasian people (85 versus 100). At a different site he finds that IQ has been estimated to have a heritability of up to 80 percent in some populations. Based on his internet search he claims that it is scientifically proven that black people are genetically inferior to white people when it comes to cognitive skills. Explain why his reasoning is utterly wrong.

13. Explain the difference between heritability and evolvability.

14. Discuss similarities and differences between QTL and GWAS.

15. Explain the missing heritability problem. What may cause it and how could it be resolved?

CHAPTER 7

Inferring evolutionary processes from DNA sequence data

In chapters 3–6 we developed allelic evolutionary genetic models. These are applicable to genetic markers such as SNPs and microsatellites. However, with the introduction of increasingly cheap and fast methods for generating large numbers of DNA sequences from natural populations we have learned that the traditional allelic models are not always the optimal choice for analyzing such datasets. Most importantly, DNA sequences harbor a lot of information about the evolutionary past that we would miss if we simply treated different sequences as different alleles. In this chapter we shall introduce some important methods and concepts applicable to the analysis of DNA-sequence data. As for the allelic models, we need a reference null model for what pattern of sequence variation to expect under ideal conditions. From there we can compare what we actually find with that expected from our null model and draw inferences about evolutionary and demographic processes that might explain our observed data. The null models for analyzing sequence data are derived from the neutral theory of molecular evolution. Historically, however, the neutral theory has made a large impact on the field of evolutionary genetics that goes beyond serving as a statistical null model. Therefore, we start this chapter by reviewing its important contribution.

7.1 The neutral theory of molecular evolution

7.1.1 The neutral theory of molecular evolution

The neutral theory is not a general theory of evolution. It is also not a competing theory to the neo-Darwinian theory we have outlined in the preceding chapters. It is a theory about evolution at the molecular level—at the level of DNA and protein sequences, and it is fully compatible with the neo-Darwinian view. The neutral theory holds that at the molecular level, most evolutionary changes are caused by genetic drift acting on mutations that are selectively neutral and that the majority of molecular variation found within populations is also neutral. By this we mean that the segregating alleles have no effect on fitness. Natural selection is, however, thought to be common according to neutral theory, in the sense that most non-synonymous mutations are thought to be deleterious. These would therefore rapidly be purged from the population and thus be unimportant in explaining variation within and divergence between populations. Moreover (and despite long debates to the contrary) neutral theory is not concerned with phenotypic evolution. The famous main architect of the neutral theory, Motoo Kimura, explicitly stated that he thought (rare cases of) positive Darwinian selection could explain adaptive phenotypic evolu-

Evolutionary Genetics: Concepts, Analysis, and Practice. Glenn-Peter Sætre & Mark Ravinet, Oxford University Press (2019).
© Glenn-Peter Sætre & Mark Ravinet 2019. DOI: 10.1093/oso/9780198830917.001.0001

tion. However, Kimura argued that at the molecular level, such events are so vanishingly rare that they can largely be ignored (Kimura 1983). Note, however, that although the neutral theory is compatible with the neo-Darwinian theory, this does not mean that it necessarily is correct by default. Indeed, there has been much debate to the contrary (Kern & Hahn 2018). We will discuss the current status of the neutral theory later in this chapter.

The neutral theory was originally introduced by Kimura (1968) and independently described the following year by King & Jukes (1969) as an original and thought-provoking hypothesis to account for the large amount of variation that was being discovered among protein sequences within many different populations and species. As we learned in chapter 4, positive selection tends to lead to fixation of favored variants and elimination of disfavored ones. However, we also learned balancing selection tends to maintain variation. Could it be that balancing selection is responsible for maintaining the large amounts of genetic variation we observe? Kimura tried to calculate the probability that the observed variation in protein sequences was being maintained by balancing selection and concluded that it was minuscule. His argument was that selection at a single locus would tend to reduce population growth below its maximum value because less fit individuals would die or reproduce at a low rate in the process.

Take heterozygote advantage as an example (cf. chapter 4). Because fit heterozygotes necessarily (due to Mendelian segregation) will produce many less fit homozygous offspring, the population suffers a genetic load—i.e. an average fitness which is lower than that of the most fit genotype. Now, imagine that balancing selection is maintaining polymorphisms not only at one locus, but many. If balancing selection was responsible for maintaining all the genetic variation at all the protein coding genes in the genome where we find polymorphisms, Kimura estimated that the total cost of selection would be extremely high—so high that the population could not exist but would instead go extinct. If, on the other hand, the genetic variation we observe is selectively neutral there would be no cost of selection and there would be no problem explaining any level of variation, provided the (neutral) mutation rate is sufficiently high relative to the rate of loss of variation by genetic drift.

7.1.2 Mutation, genetic drift, and the molecular clock

According to neutral theory, molecular polymorphisms are transient. Eventually, (neutral) variants will go to fixation by genetic drift. Hence, over evolutionary time spans populations and species will diverge from each other genetically. Mutations are random errors in the replication and repair machinery of living organisms. It is therefore not unreasonable to assume that, averaged over long time spans, the mutation rate would be fairly constant. Let μ represent the neutral mutation rate. We define it as the average probability that a neutral mutation will occur at any one site along a DNA sequence in an individual during gametogenesis—i.e. one nucleotide mutates to a different one, such as A to G or C to T. Hence, μ is the neutral mutation rate per site per generation. The probability that a mutation occurs at a specific site in an individual is low. For instance, in the human genome the rate of mutation has been estimated to be about 1.2×10^{-8} per nucleotide site per generation (Scally & Durbin 2012). However, in a population effectively consisting of N_e individuals, there are many gene copies around that may mutate. Since each individual in a diploid organism carries two copies of each locus the total number of new mutations in a population per site per generation would be $2N_e\mu$. The number of new neutral mutations is proportional to the population size. More mutations will occur in a large population simply because there are more gene copies around that can mutate.

As we learned in chapter 3, the probability that a neutral allele will go to fixation by genetic drift equals its current frequency in the population. The frequency of a new neutral mutation will be $1/2N_e$— that is, the one mutant out of the $2N_e$ copies present in the population. The probability that a new neutral mutation goes to fixation by genetic drift is thus inversely proportional to population size. Genetic drift is more likely to lead to the fixation of a not so rare new mutant allele in a small population ($1/2N_e$ will be a relatively large number) than to the fixation of a very rare new mutation in a large population ($1/2N_e$ will be a relatively small number).

By combining these two considerations about mutations, genetic drift, and population size we arrive at an elegant key result of the neutral theory which may seem surprising at first: *The expected rate of molecular evolution is constant, independent of population size, and equals the neutral mutation rate μ.* Wait a minute though, didn't we just agree that the mutation rate and the probability of fixation of such mutations depend on population size? Here we make the key distinction between substitutions and mutations. According to the neutral theory, the rate of molecular evolution or the **substitution rate** is the rate at which a population accumulates new fixed variants through mutation and genetic drift along a DNA sequence. So, the rate of molecular evolution depends both on the mutation rate in the population ($2N_e\mu$) and the probability that such new mutations eventually would go to fixation ($1/2N_e$). Hence the expected substitution rate equals:

$$2N_e\mu \cdot \frac{1}{2N_e} = \mu \qquad (7.1)$$

Population size cancels out—so mathematically we can make things simpler and describe mutation rate and substitution rate as one and the same thing. It is important, though, to keep in mind that biologically there are differences between mutation and substitution rate; i.e. mutations are measured at the level of the DNA sequence whereas substitutions are measured in the context of a population. We will see the implications of neglecting this difference shortly. Nonetheless, in a larger population, more mutations would occur but each would have a low probability of becoming fixed by genetic drift. In a smaller population, fewer mutations would occur but each would have a proportionally higher probability of becoming fixed by genetic drift.

In the absence of selection, the substitution rate equals the mutation rate according to the neutral theory. With a constant mutation rate the substitution rate should be constant. Accordingly, genetic differences between populations and species should accumulate at a constant rate. We can use this to estimate the divergence time between populations and species. This is the rationale behind the **molecular clock** (Zuckerkandl & Pauling 1962). Let us say that we compare the nucleotide sequence at a gene of a pair of species and find that the number of nucleotide differences between them is d. If the nucleotide substitution rate in the gene is μ per site per generation, then the expected number of nucleotide differences between the two species would be:

$$E(d) = 2\mu t \qquad (7.2)$$

in which t is the number of generations that have lapsed since they split from their common ancestor. We multiply μt by 2 because it is two lineages that have accumulated nucleotide substitutions since they split from their common ancestor t generations ago. The equation has two unknowns, the mutation rate μ and the number of generations t. We need one to calculate the other—so how would we approach this? With an estimate of the mutation rate μ, we can rearrange equation 7.2 and calculate the divergence time as $t = d/2\mu$. Measures or estimates of mutation rate can be made with several different methods. Parent–offspring trio genome sequencing is the most accurate way to measure mutation rate directly as it allows unique mutations to be identified in offspring that are not present in either parent—i.e. it can determine the mutation rate per generation. Until recently, such a method has been prohibitively expensive for use in anything other than model systems such as humans or *Drosophila*. Although now much more feasible, it still requires a high level of sequencing effort to distinguish real mutations from errors made by the sequencer, as well as access to parents and offspring. However, such direct methods of mutation rates will most likely overestimate the neutral mutation rate. Many new mutations are deleterious and are therefore likely to be removed by selection and not contribute to neutral divergence between populations and species.

Measures of mutation rate can also be made with the d and t parts of equation 7.2—but in order to do this we need an estimate of t. Fossils can be used for this and such an approach is known as a phylogenetic estimate using fossil calibration. Human–chimpanzee divergence is thought to have occurred 5.6–8.3 million years ago; this has been dated using fossil evidence of ancient hominid species (Scally & Durbin 2012). Most reliable fossil calibrations are made using specimens that are from lineages that are clearly more closely related to humans than chimpanzees. For example, both *Ardipithecus* and *Australopithecus* specimens provide evidence that the human lineage had split from the chimpanzee lineage by 3–4 million years ago. However, these specimens are only able to provide a lower limit on estimation of a split time—we know human and chimpanzee divergence must have occurred before this—but we have no clear upper limit, i.e. the true common ancestor almost certainly occurred well before this date. Ideally, we might hope for fossil evidence of a species that exhibited both human and chimpanzee-specific traits, i.e. closely related to the most recent common ancestor of both. Such a specimen would provide a clear upper limit on the time when divergence between the two species must have occurred. The most ancient human-resembling fossil, *Sahelanthropus tchadensis*, appears to have lived around the time of the most recent common ancestor (MRCA) and could even be a candidate for it, but this point remains debated (Brunet et al. 2002). Crucially, it is also important to remember that due to the patchiness of the fossil record, most hominid fossils are unlikely to be direct ancestors of the modern human lineage; instead they are most likely from closely related groups that split from our ancestors after the divergence between the main human and chimpanzee lineages.

If no fossil evidence is available, biogeographical events that might have driven divergence between species can be used, although only if their timing is well resolved. One of the most famous examples is the Isthmus of Panama, which emerged about 2.8 million years ago and which separated the Gulf of Mexico from the Pacific Ocean (O'Dea et al. 2016). This led to divergence between lineages of multiple marine species and was an instrumental calibration point for several early molecular clock studies. Regardless of how we estimate the mutation rate, we can then rearrange equation 7.2 and calculate the divergence time as $t = d/2\mu$.

A large number of studies have used the relationship between the extent of molecular divergence and substitution rates in order to estimate divergence time between pairs of species, and an example is provided in Figure 7.1. Amino acid substitutions in α and β hemoglobin appear to have been fairly constant throughout vertebrate evolutionary history and show a nearly linear relationship with the time of divergence among species and taxonomic groups. Such molecular clocks should however be interpreted with care. The calculation is based on some critical assumptions: that the mutation rate is constant and that substitutions accumulate by

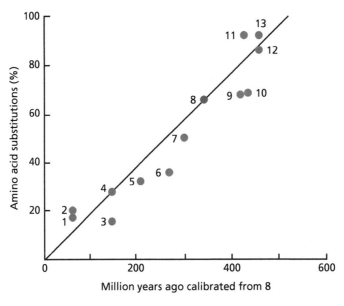

Figure 7.1 Molecular clock. Amino acid substitution rates on α and β hemoglobin in vertebrates show an approximately linear relationship with divergence time as estimated from fossil evidence, using the split between amphibians and terrestrial vertebrates (8) as calibration point. The points refer to the following evolutionary splits/comparisons 1) loris/other placental mammals, 2) primates/other mammals (β hemoglobin), 3) all mammals, 4) kangaroo/placental mammals, 5) monotremes/theria, 6) chicken/mammals, 7) viper/warm-blooded vertebrates, 8) amphibians/terrestrial vertebrates, 9) carp/mammals (α hemoglobin), 10) bony fish/tetrapods, 11) shark/bony vertebrates, 12) all α hemoglobin/all β, γ, and δ hemoglobin, 13) α hemoglobin/β hemoglobin. Redrawn from Runnegar (1982).

random genetic drift and not by selection, that is, it assumes that the neutral theory is correct. If natural selection has affected species divergence at a given gene, or there has been some other violation of the neutral theory, our time estimate would be incorrect. An important empirical question is, therefore, how well the neutral theory approximates molecular evolution and whether the assumptions of the molecular clock model are easily violated. If the clock model was a good approximation we would have an extremely powerful tool for reconstructing the evolutionary past, including timing important events such as speciation. There are however, obviously caveats to the molecular clock model; for example, mutation rates are not constant and they evolve among species and even among human populations (Lynch 2010; Lynch et al. 2016; Narasimhan et al. 2017).

Furthermore, while the molecular clock model assumes a constant mutation rate for a given locus, it does not assume that mutation rates are uniform among loci. Actually, mutation rates vary quite considerably across the genome; in humans, mutation rates are higher in exonic regions (Francioli et al. 2015) and on the male Y chromosome for example (Hodgkinson & Eyre-Walker 2011). Many mutations are not neutral and instead might be weakly deleterious (see next section), meaning it takes selection longer to remove them from a population. As a result, substitution rate estimates can be biased, depending on the timescale on which they are estimated on (Ho et al. 2007, 2015).

Finally, there is the circularity of phylogenetic estimates of substitution rates. To get a substitution rate we need a divergence time and to get a divergence time we need a substitution rate. It is not hard to see how an incorrect calibration or rate can cause considerable uncertainty over estimates of divergence time; indeed the timing of splits between human lineages is debated for precisely this reason.

7.1.3 The generation and population size problem

In the late 1960s and early 1970s several studies set out to test the prediction from neutral theory that the substitution rate should be constant across species and independent of population size using data on amino acid substitutions in proteins such as

hemoglobin (see Figure 7.1) and cytochrome c. By and large, these studies demonstrated that the substitution rates were on average fairly constant across species but not exactly in the way predicted under neutral theory. Each generation, new germline mutations would occur in a population. Hence, according to neutral theory we would predict that organisms with long generation times, such as elephants and apes, would evolve more slowly at the molecular level than organisms with short generation times, such as mice and shrews. However, studies of amino acid substitutions in mammals indicated that the molecular clock is ticking closer to absolute time (years) rather than relative time (number of generations; e.g. Laird et al. 1969; Kohne 1970). The rate of molecular evolution is apparently surprisingly similar across species with very different generation times, which goes against the predictions.

Tomoka Ohta (1973), working as an assistant professor under Kimura, suggested a modified version of the neutral theory to account for this discrepancy. According to her **nearly neutral theory** of molecular evolution, many mutations are not quite neutral but are in fact slightly or weakly deleterious. Species with long generation times such as elephants and apes also tend to have smaller population sizes than species with short generation times, such as mice and shrews. Accordingly, natural selection would be less efficient in removing these slightly deleterious mutations in the former species. In other words, genetic drift would be the dominant force in molecular evolution for a larger proportion of mutations, including slightly deleterious ones. In contrast, in species with short generation times and large population sizes, natural selection would be more effective in purging slightly deleterious mutations, thus slowing down the overall rate of molecular evolution.

7.1.4 The value of the neutral theory as a null model

The generation and population size problem, and Ohta's revised hypothesis to account for it, is but one example showing that neutral theory is too simple to account for all of molecular evolution. Kimura himself recognized that his original formulation was incomplete and credited Ohta for having made an equally important contribution in establishing neutral theory as a way of thinking about molecular evolution. In the late 1960s and 70s, a large number of debates raged between evolutionary geneticists on either side of the selectionist vs neutralist divide. They were convinced that either selection or genetic drift (neutral theory) could explain the large amount of variation we observe in molecular data. Today, such debates seem almost unbelievable, as the actions of both genetic drift and selection are widely recognized as having shaped the genomes of nearly every species studied so far. Selection is clearly an important force and with genomic data there are now thousands of documented examples of how it has shaped diversity and driven molecular evolution. A compelling line of evidence for the importance of selection in shaping the genome is the near ubiquitous positive relationship between recombination rate and levels of genetic diversity (e.g. Nachman 2001; Cutter & Payseur 2013). When selection occurs, it reduces genetic diversity around its target and the extent of this effect is proportional to the probability of recombination. Thus, when recombination rates are low, the effects of selection extend much further into the genome than when it is high, when it affects only a small proportion. As such, diversity is greatest in high recombination rate regions (Figure 7.2).

Such a pattern is not predicted by neutral theory. Kern & Hahn (2018) argue that we should separate

the concepts of neutral evolution and neutral theory. Neutral evolution clearly does occur and can explain diversity and patterns of molecular evolution for parts of the genome during certain time intervals. However, a multitude of demographic and evolutionary processes are molding the genomes of living organisms over long time spans, including mutations, genetic drift, selection, exchange of genes between populations, and changes in population size. As a model, neutral theory simply does not encompass all these processes. In our opinion the main merit of the neutral theory is, therefore, not that of being an accurate description of how molecular evolution occurs. Instead, neutral theory enables us, in some cases, to formulate precise null models for patterns of molecular variation. By comparing observed patterns with these expectations, we can start to make inference about the processes that have affected our study populations. This is not universally accepted; some argue that as a null model, neutral theory is misleading as ruling it out does not necessarily imply positive selection and adaptation (Kern & Hahn 2018).

7.2 Describing and interpreting variation in DNA sequences

7.2.1 DNA sequence versus allelic data

Natural, outbred populations normally exhibit a substantial amount of variation at the DNA sequence level. At several sites along a DNA sequence there may be some form of genetic variation. Consider for example one site; some individuals may be homozygous for an A, others for a G, and yet others still may be heterozygous A/G, derived from a point mutation sometime in the past. In short, there is a single nucleotide polymorphism (SNP) at that site. Estimates from whole genome resequencing of individual humans suggest an average of 3.3 million SNPs—although SNP density is not uniform in the genome (Shen et al. 2013). However, SNPs are not the only form of polymorphism—if we compare a set of sequences from a genome region, we may find that they differ in length. At a specific site, a number of nucleotides may be missing from some sequences relative to

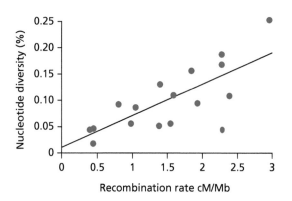

Figure 7.2 Relationship between nucleotide diversity and recombination rate in 17 human genes. The positive relationship suggests that selection has had a stronger effect on reducing genetic diversity in genes with low recombination rates through hitchhiking effects than in genes with high recombination rates. See main text for further explanation. Redrawn from Nachman (2001).

others, and some individuals may be heterozygous for such insertion–deletion polymorphisms, or *indels*. Indel polymorphisms such as this can vary considerably in size but are typically defined as being anywhere from just a single nucleotide to 10 000 bases in length. They are also quite a common form of polymorphism; for example, with 415 436 indels in the human genome, one occurs every 7.2 Kb (Mills et al. 2006). What causes indels? In a similar way to SNPs, mutations such as replication slippage, unequal crossing over, or transposition (see chapter 2) have caused deletions in some individuals and/or insertions in others. Regardless of the exact cause, like point mutations, the resulting variants evolve by selection and/or genetic drift. Despite their abundance and their potential functional impact (~36 percent occur in annotated genes), indels are often overlooked in favor of SNPs, particularly in population genomic studies. One rationale for this is that SNPs and indels are often caused by different mutation processes and may therefore not be directly comparable. It can also be difficult to reliably call indel polymorphisms using standard variant calling methods. For the rest of this chapter we will therefore focus on SNPs in the context of sequence variation.

Due to the large amount of variation natural populations exhibit, very many alleles (different sequences) segregate in a population if we consider a DNA sequence as a locus. In chapter 3, we introduced biallelic models of allele frequency and the statistics we can derive from them such as F_{ST} and F_{IS}. In such models, heterozygosity is an important measure of genetic variation. However, with very many alleles present, most individuals would be heterozygous. Indeed, if we considered an entire chromosome or genome as a locus, all individuals would carry different alleles! Although this is an extreme example, re-defined or different models are needed to describe and interpret sequence variation in meaningful ways.

Imagine we sample individuals from a population and sequence their DNA. If there is variation present, individual DNA sequences drawn from this population would obviously differ from one another. However, sequences would also vary in the extent to which they are different from one another. For example, we may sample three

sequences from this population. Sequence 1 may differ from sequence 2 by only one mutation, whereas sequence 3 may have four mutations not shared by either sequence 1 or 2. Thinking in terms of evolution and the fact that the probability of mutations occurring increases with time, a likely explanation would be that sequence 1 and 2 diverged from one another more recently in their evolutionary history than either of them did from sequence 3. In this way, we can view sequence data as a composite of different polymorphisms which may contain more information about the evolutionary history of a population than if we had considered each variant separately under a simple allelic model, in which all alleles are considered equivalent (either different or similar). As we will see in chapter 9, this feature of sequence data has been successfully leveraged for phylogenetic inference and is essential for resolving evolutionary relationships among taxa.

Finally, evolution proceeds rapidly at the level of the DNA sequence, as evident from the large amount of variation found in most populations and from the fact that genetic differences exist between populations. However, this presents a challenge for developing models of DNA sequence evolution. As you will recall from chapters 3 and 4, the fixation or loss of alleles through evolutionary forces such as genetic drift and selection are fundamental for allelic models of population genetics. In reality, however, there is a probability that any given allele may experience a mutation and become a new allele prior to its fixation or loss. Standard allelic models do not account for such mutations—which at the DNA sequence level may not be a very realistic portrayal of how molecular evolution plays out. In short, we need to apply different models and tools for conducting accurate population genetic analysis of DNA-sequence data.

We add a brief note of caution for readers here. Although different models underlie allelic and sequence-based methods for calculating population genetic statistics, in practice both approaches are now often used alongside one another. In the earlier days of evolutionary genetics, the distinction between models used was clearer because of the differences in the methods used to obtain genetic data. For example, studies using Sanger sequencing of nuclear or mitochondrial loci would clearly use statistics for DNA-sequence data whereas allelic models would be used in studies where microsatellites or other genetic markers were applied. However, high-throughput sequencing platforms blur this line as they allow alleles at SNPs to be called using sequence data (see chapter 1) but also make it possible to reconstruct the DNA sequence of an individual. A typical study using whole-genome resequencing of multiple individuals might therefore include estimates of a statistic such as F_{ST} based on allelic variation at individual SNPs. The same study could, however, also incorporate a statistic such as nucleotide diversity, which requires sequence data (see next section).

7.2.2 Descriptive statistics of DNA sequence variation

There are two important descriptive statistic parameters that quantify the amount of genetic variation present in DNA sequences sampled from a population. The first is Π (pi), defined as the average number of nucleotide differences between pairs of sequences randomly drawn from the population. We can easily estimate Π from sequences obtained from a population, assuming that these constitute a random subset of the sequences in the population, using the following formula:

$$\Pi = \frac{1}{[n(n-1)]/2} \sum_{i<j} \Pi_{ij} \quad (7.3)$$

Here Π_{ij} is the number of nucleotide differences between sequence i and j and $[n(n-1)]/2$ is the number of possible sequence pairs we can make from our sample of n sequences. As an example, let us say our sample consists of $N = 3$ sequences, each 47 base pairs long (Figure 7.3). As evident from the figure, there are two SNPs with C/T polymorphisms

at position 10 and 17, otherwise the sequences are identical.

According to equation 7.3 we must count the pairwise number of nucleotide differences of all the possible pairs and sum these, and then divide by the number of possible sequence pairs. There is one difference between sequence 1 and 2 in Figure 7.3, two differences between sequence 1 and 3, and one difference between 2 and 3, and that is all the possible sequence pairs. Hence, $\Pi = (1 + 2 + 1)/3 = 1.33$.

Note that the value of our estimated Π does not depend on the sample size of sequences we include in any systematic way, since it is the average number of pairwise nucleotide differences between sequence pairs drawn from a population. But of course, the more sequences we include, the more precise the estimate of the actual population average we get. Note also that Π weights polymorphisms according to how common or rare they are. A rare polymorphism would increase Π only slightly, since most sequence pairs would be similar at this site, whereas polymorphic sites where the two variants are about equally common would increase Π substantially, since many sequence pairs would be different at this site.

However, one issue is that the value of Π obviously depends on the length of the sequence. A longer sequence is likely to contain more polymorphisms than a short one. It will often be the case that we want to compare nucleotide variation of sequences of different lengths. We can standardize our parameter by dividing Π by sequence length L in number of base pairs. This standardized parameter is referred to as **nucleotide diversity** π, defined as the average number of nucleotide differences per site between randomly chosen sequence pairs drawn from the population (Nei & Li 1979). In our example from Figure 7.3 we have that $\pi = \Pi/L = 1.33/47 = 0.028$.

Figure 7.3 Genetic polymorphisms among three DNA sequences drawn from a population. For sequence 2 and 3 only sites that differ from sequence 1 are shown. There are two segregating sites, at position 10 and 17, respectively. The three sequences are all different; hence, there are three different alleles present.

The second common descriptive statistic parameter for genetic variation within a population is S, the number of **segregating sites**. This is simply the number of polymorphic sites in our sample of sequences. In the example in Figure 7.3 there were two polymorphic sites, hence $S = 2$. As with Π, we can standardize S by taking sequence length into account. Hence, the number of segregating sites per site is $s = S/L$.

Note that S differs from Π in two respects. First, unlike Π, S does depend on sample size. The more sequences we draw from the population the more polymorphic (segregating) sites we are likely to find. A rare polymorphism is unlikely to be included in a small sample, but likely to be included in a large one. When gradually including more and more sequences, S would increase fast at first and asymptotically level off towards the actual number of polymorphisms present in the population. Second, also unlike Π, all polymorphisms are weighted equally. Any polymorphic site would increase S by 1, regardless of its frequency in the population we sampled. We shall soon see that these different properties of Π and S (or rather π and s) can be used to infer evolutionary processes that may have occurred in our study population. However, first we need to calculate the expected values of π and s under some ideal assumptions. To do that we use the neutral theory of molecular evolution as our starting point.

7.2.3 Coalescent theory and the expected pattern of sequence variation

According to neutral theory, variation at the DNA sequence level is molded solely by mutation and genetic drift. **Coalescent theory** is a powerful tool for estimating population genetic parameters under neutrality (Kingman 1982a; 1982b; 2000). Intuitively, when thinking of evolution under the coalescent, we do so backwards in time. This is best illustrated by the concept of a coalescent event, which is simply the point backwards in time where two alleles can be traced back to their most recent common ancestor (MRCA). In other words, their evolutionary histories coalesce at this point in time, and after this we consider them a single lineage (Figure 7.4). Let us use a simple coalescent model to demonstrate how genetic variation can evolve under neutrality. We sample DNA sequences from a diploid population with effective size N_e. We shall specify a particular null model called the **infinite site model**. According to the infinite site model our idealized assumptions are as follows:

1. Each new mutation on the sequence occurs at a site that has not mutated before (i.e. no recurrent mutations have occurred).
2. No selection has occurred (i.e. we assume neutrality).
3. No gene flow has occurred.
4. Mating has remained random.
5. The population size has remained constant so that the population is at an equilibrium with respect to mutation and genetic drift.
6. Generations are non-overlapping.
7. Recombination has not reshuffled any genetic polymorphisms along our DNA sequence.

Under these assumptions, if we sample one random DNA sequence from the population we are confident that it is derived from a parent from the previous generation. If we sample two random sequences, the probability that they are derived from the same parent the previous generation = $1/2N_e$, since there are $2N_e$ sequences present each generation in a diploid population. If it is the case that the two sequences share a most recent common ancestor (MRCA) in the previous generation, they coalesce one generation back. Otherwise, the two sequences are not derived from the same parent the previous generation with the probability $1-1/2N_e$. Instead they may coalesce two generations ago. The probability for that would be $(1-1/2N_e) \cdot 1/2N_e$. More generally, we can extend the probability of two sequences coalescing t generations ago = $(1-1/2N_e)^{t-1} \cdot 1/2N_e$.

Expected coalescence time of two DNA sequences follows a geometric distribution with the neat property that the mean is simply $2N_e$ generations, but of course with variation around that mean value (Figure 7.5).

You can see from these formulae that effective population size (N_e) is extremely important for determining the rate of coalescent events—i.e. whether two sequences share a common ancestor in the previous generation. If N_e is large, two randomly chosen

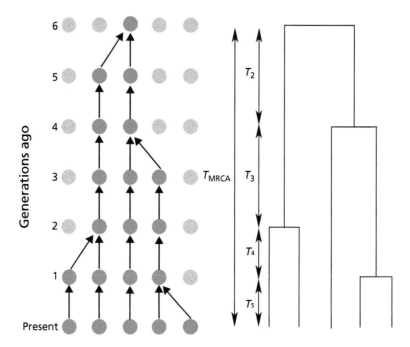

Figure 7.4 Illustration of the coalescent process and genealogy. Left panel: A schematic representation of DNA inheritance. Each row represents a single generation and each circle denotes a DNA sequence. All five sequences sampled in the present generation are indicated with blue circles. The parental sequences that lead to the present sample are also indicated in blue, whereas sequences in the previous generation that did not give rise to any sequence in the following generations are indicated in gray. A pair of sequences coalesce backwards in time because they share an ancestor some generations ago. All the present-day sequences in this example coalesce back to their most recent common ancestor (MRCA) six generations ago. Right panel: The genealogy of the five sequences sampled in the present generation. Time between different coalescence events correspond to T5, T4, T3, and T2.

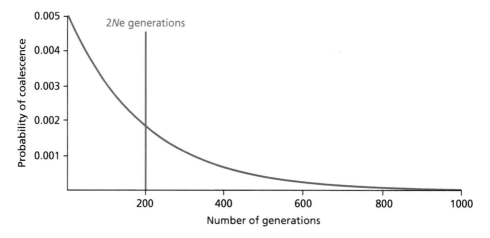

Figure 7.5 Probability distribution of coalescence time for pairs of sequences drawn from a population of $N_e = 100$ individuals according to the infinite site null model. Expected coalescence time follows a geometric distribution (blue line). The average coalescence time for all the sequence pairs (red line) is $2N_e = 200$ generations.

sequences have a very low probability of coalescing in the previous generation. Conversely, if N_e is small, the probability of a most recent common ancestor one generation back is much higher.

Now we introduce neutral mutations. As before, the neutral mutation rate is assumed to be μ mutations per site per generation. If we compare two random sequences from a population we would thus expect that each of them has accumulated μ mutations per site over a time span of $2N_e$ generations since they (on average) shared a common ancestor—in other words, we would expect $2N_e\mu$ mutations per site. Hence, the expected difference between two pairs of sequences randomly drawn from a population—i.e. the average difference between them—would be $2 \cdot 2N_e\mu = 4N_e\mu$ per site. Why do we multiply $2N_e\mu$ by a factor of 2? This is because mutations accumulate on both branches since the sequences shared a common ancestor (Figure 7.6). Again, effective population size is a crucial parameter here—if it is large, the time to an MRCA for two sequences is high and so more mutations are expected to accumulate.

This product ($4N_e\mu$) appears in a number of population genetic formulae and is extremely

important. We refer to it as the **population mutation rate parameter θ** (theta).

$$\theta = 4N_e\mu \tag{7.4}$$

One thing to note here is the similarity between our derivation of the theoretical parameter θ using coalescent theory and our definition of the descriptive statistic of sequence variation, π (nucleotide diversity). Under ideal conditions, i.e. the infinite site model and neutrality, we expect that the average number of nucleotide differences per site between pairs of sequences drawn from a population (π) should equal θ, or in mathematical terms:

$$E(\pi) = \theta = 4N_e\mu \tag{7.5}$$

The descriptive parameter π, the average number of nucleotide differences per site between pairs of sequences drawn from a population (nucleotide diversity), is therefore an estimator of the theoretical population mutation rate parameter θ (Tajima 1989).

Similarly, it should come as no surprise that our other descriptive measure of sequence variation, the number of segregating sites per site s, is also related to θ in a simple way under ideal conditions (i.e. the infinite site model). However, recall that the value of s, unlike π, depends on the number of sequences in our sample. We must therefore scale the expected relationship between θ and s according to sample size. Watterson (1975) showed that the expected relationship between s and θ is:

$$E(s) = a\theta \tag{7.6}$$

where $a = 1 + \frac{1}{2} + \ldots + 1/(n-1)$ and n is the number of sequences in our sample. The elegance of these results is that under the assumptions of the infinite site model the descriptive parameters π and s would only be affected by two factors, the neutral mutation rate and the effective population size. Consequently, θ estimated from the two parameters should be identical. If θ estimated from π is higher or lower than that estimated from s, it is potentially a sign that an assumption of our null infinite sites model has been violated, perhaps by selection.

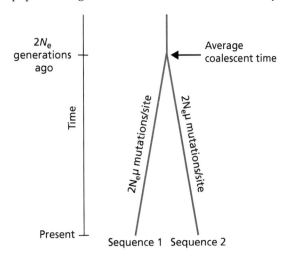

Figure 7.6 Derivation of the population mutation rate parameter $\theta = 4N_e\mu$. On average, two random sequences sampled from a population of effective size N_e would coalesce $2N_e$ generation ago under ideal conditions (i.e. according to the infinite site model). With a neutral mutation rate of μ per site per generation each of the sequences would thus have accumulated $2N_e\mu$ mutations per site since they diverged. Hence, the expected number of nucleotide differences between such a pair of sequences equals $2 \cdot 2N_e\mu = 4N_e\mu$ per site.

7.2.4 Tajima's *D* test

Tajima's *D* test (Tajima 1989) is a statistic that allows us to actually compare the two independent estimates of θ that we introduced in the previous section. Based on the **allele frequency spectrum**, the distribution of allele frequencies of the polymorphic sites (SNPs) along a DNA sequence, Tajima's D uses both π and s and is formulated as follows:

$$D = \frac{\pi - s/a}{\sqrt{V(\pi - s/a)}} = \frac{\theta_T - \theta_W}{\sqrt{V(\theta_T - \theta_W)}} \quad (7.7)$$

where θ_T is Tajima's estimator of θ based on nucleotide diversity, θ_W is Watterson's estimator based on number of segregating sites (Watterson 1975, Tajima 1989), a is as defined in equation 7.6, and V is variance. Hence, the statistic is simply the difference between θ estimates derived from π and s, divided by a variance term that normalizes the test statistic. It is therefore a way for us to test whether our null model has been violated (see previous section). Note that the sign of Tajima's D only depends on which of the θ estimates is the larger. Below we shall explore how different processes can alter the sign and value of Tajima's D.

7.2.5 Natural selection and the sign of Tajima's *D*

Let us assume that all the assumptions of the infinite site model are true except that natural selection is occurring or has occurred at the locus we investigate. How would natural selection affect the sign of Tajima's D? Here we will appeal to your intuition—and it is possible to interpret this test from intuition, despite its apparent difficulty. The trick is to imagine what an evolutionary process would do with genetic variation at the locus, how that would affect the relative size of π and s, and hence, the sign of Tajima's D as a result.

Let us start with purifying (negative) selection—you will recall that this is selection against deleterious alleles introduced by mutation. Say that purifying selection occurs at several sites within our locus, i.e. along our sequence. What would happen? Well because deleterious variants are continually removed by purifying selection, they would remain at a low frequency in the population. How would that affect π and s? Well, the presence of these rare variants would have a large impact on s, inflating it to a higher number (recall that all polymorphisms add equally to the value of s

irrespective of how common or rare the variants are). However, the impact on π would be much smaller (recall that rare variants increase the value of π by only a small amount). Accordingly, our estimate of θ from s (θ_W) would be larger than that from π (θ_T)—Tajima's D would be negative. This example also nicely illustrates that Tajima's D is essentially a statistic summarizing the observed allele frequency spectrum.

Is the effect of positive, directional selection a bit trickier to understand? Say that one mutation occurred at our sequence sometime in the past that increased the fitness of its bearer. What would happen with neutral polymorphisms in the vicinity of the selected site? Since we assume that no recombination is occurring according to the infinite site model, the variants that happened to be on the sequence containing the positive mutation would go to fixation as well. They would "hitchhike" along with the positively selected site. This process is known as a **selective sweep** and is characterized by a loss of variation in the vicinity of the site under selection (Figure 7.7). The extent of this loss of diversity around the target of selection is determined by physical distance along a chromosome and also the probability of recombination—we will deal with this in more detail in sections 7.3 and 7.4. How does this relate to Tajima's D? If all variation is

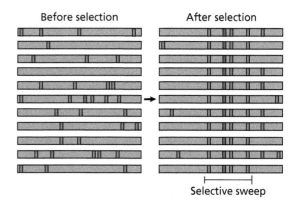

Before selection   After selection

Selective sweep

Figure 7.7 A selective sweep. A positive mutation occurs (red) and goes to fixation by natural selection. Neutral variants (blue) close to the site under selection will also go to fixation by hitchhiking, causing reduced genetic variation in this region. As physical distance from the site under selection increases, recombination will proportionally act to reduce any association between neutral polymorphisms and the positively selected site.

lost due to fixation, both π and s would be zero. However, it is much more likely that we are measuring the population some time after the selection event has occurred. In this case, new neutral mutations would have started to appear on some of the sequences that were originally driven to fixation by the selective sweep. By definition new mutations have a low frequency when they first appear, and if drift does act to increase their frequency, this will take some time. Accordingly, we should expect a time window after a selective sweep has occurred, but before any mutation-drift balance has been reached, in which we (again) would have an excess of rare variants—i.e. variants at low frequencies occurring on several sequences. This means that s/a is large relative to π and so Tajima's D will again be negative.

So, both purifying, negative selection and positive selection (due to the selective sweeps) can yield negative Tajima's D. In order for selection to produce a positive Tajima's D the estimate of θ from π (θ_T) has to be larger than that from s (θ_W). In other words, alleles need to be maintained at an intermediate frequency within a population. Balancing selection, through heterozygote advantage and also negative frequency-dependent selection (see chapter 4), would have that effect.

Balancing selection commonly acts to maintain diversity in genes involved in immune response but it can act to favor pathogens and parasites too. In humans, exposure to antigens from disease pathogens or parasites will result in the production of antibodies that allow adaptive acquired immunity. Therefore, initial exposure to a pathogen can lead to an improved disease resistance in the event of subsequent exposures. From the perspective of a pathogen or parasite, alternative alleles at genes encoding antigen proteins will have a fitness advantage due to negative frequency-dependent selection—i.e. they are rarely encountered by the target immune system and so infection will be easier. Amambua-Ngwa et al. (2012) used allele frequency spectrum statistics such as Tajima's D to look for signatures of balancing selection in the genome of *Plasmodium falciparum*, the bloodborne protozoan parasite that causes the majority of malaria cases in humans. They calculated Tajima's D for the 51 percent of the genes (2853 genes out of ~5560 total) in the

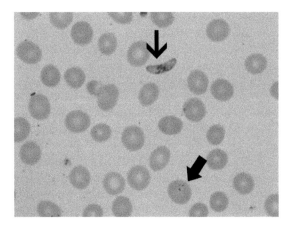

Figure 7.8 The malaria parasite *Plasmodium falciparum* has a complex life cycle involving two hosts, *Anopheles* mosquitoes and humans. The picture shows two life stages of the parasite. Above, marked with a thin arrow is a mature gametocyte that can enter a mosquito that ingests it during a blood meal. Below, marked with a bold arrow, we see a parasite that has recently entered a red blood cell. Parasite genes encoding antigen proteins that are expressed at the stage when the parasite enters a red blood cell appears to be strongly affected by negative frequency-dependent (balancing) selection. Photo licensed under Creative Commons.

P. falciparum genome that contained polymorphisms. Mean Tajima's D was −1.00 but 11.8 percent (337) of the genes had a positive value of the test statistic. Several of the genes with high Tajima's D values were involved in the production of antigen proteins. Furthermore, the highest values were seen in genes with increased expression at the point in the *P. falciparum* lifecycle when it invades red blood cells (Figure 7.8).

7.2.6 Demographic changes and the sign of Tajima's *D*

Besides selection, change in effective population size is an important factor that can alter the sign of Tajima's D. An important assumption of the infinite site model is that the population is at equilibrium with respect to mutation and genetic drift (i.e. constant population size). This will not be the case following a relatively recent change in population size. A population **bottleneck**—i.e. a significant reduction in population size—would change the allele frequency spectrum of a sample of sequences. Rare polymorphisms are likely to be lost, simply because of an increased rate of genetic drift in the now smaller population, whereas sites in which both variants are at intermediate frequencies are likely to remain. Accordingly, the estimate of θ from π will tend to be larger than that from s following a bottleneck, and Tajima's D would be positive. The opposite of a bottleneck is a **population expansion**, a significant increase in population size. Following an expansion, the population mutation rate would increase, since there would be more individuals around in which mutations could occur. Such mutations would be rare at first before mutation–drift equilibrium is restored (i.e. when the rate of mutations entering the population is equal to those lost by drift). Hence, following a population expansion we expect an excess of rare polymorphisms, $\theta_T < \theta_W$ and thus a negative Tajima's D.

Clearly, both selection and demographic changes can affect the Tajima's D test in similar ways—how then can we distinguish between the two? One way is to use multiple loci—something that has become much easier with the availability of genomic datasets. Changes in population size would have genome-wide effects on the rate of genetic drift, whereas selection typically only affects one or a few genes. Two divergent stickleback species, the three-spine (*Gasterosteus aculeatus*) and the Japan Sea stickleback (*G. nipponicus*) occur in the Japanese archipelago (Figure 7.9). Unlike the majority of other stickleback species pairs, the Japanese species pair is unique; strong reproductive isolation has evolved via hybrid sterility—most likely as a result of the evolution of a new sex chromosome system in the Japan Sea stickleback. The leading hypothesis

for how such high divergence arose between the two species has been that *G. nipponicus* was spatially isolated throughout the Pleistocene, when sea level change meant that the Sea of Japan was at least partially enclosed by land. Ravinet et al. (2018a) used whole-genome resequencing data from 10 individuals of each species to investigate how divergence unfolded in the Japanese stickleback system. A striking finding was a highly negative genome-wide skew in the mean value of Tajima's D, estimated from 10 Kb genome windows in *G. nipponicus* compared to *G. aculeatus* also collected in Japan (Figure 7.10). Taken together alongside the results of several coalescent based estimates of past effective population size, Ravinet et al. (2018a) concluded that *G. nipponicus* had likely undergone a bottleneck in the last 400 000 years, after which it

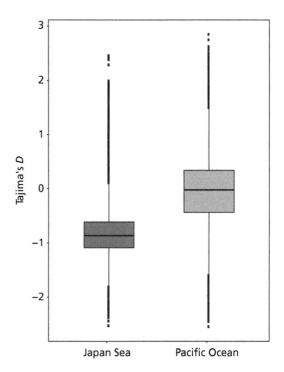

Figure 7.10 Boxplot showing the genome-wide distribution of Tajima's D in Japan Sea stickleback (*Gasterosteus nipponicus*) and Pacific Ocean threespine stickleback (*G. aculeatus*). The black vertical lines show the median values and the colored boxes span the 0.25 and 0.75 quartiles of observed values around those median values. Dots represent outliers. The genome-wide signal of negative Tajima's D values in the former species is consistent with a population expansion which has altered the allele frequency spectrum. Data extracted from Ravinet et al. (2018a).

Figure 7.9 Above: Pacific Ocean threespine stickleback (*Gasterosteus aculeatus*) and below: Japan Sea stickleback (*G. nipponicus*) are two divergent stickleback species that co-occur in marine waters in the Japanese archipelago. Photo by Mark Ravinet.

experienced a population expansion which explains the genome-wide pattern of negative Tajima's D.

7.2.7 Descriptive statistics for sequence divergence between populations

In the same way that we need to be able to describe sequence variation within a population with statistics such as π and s, we need statistics that describe sequence differences between populations. A commonly used measure of divergence between population X and Y is the average number of pairwise nucleotide differences between them, D_{XY}. The calculation of this statistic is thus the same as for Π, except that we count the number of differences between one sequence drawn from population X and one from population Y and calculate the average for all the sequence pairs sampled from the two populations. Or more precisely:

$$D_{XY} = \sum_{ij} X_i Y_j D_{ij} \qquad (7.8)$$

where in population X and Y, X_i is the frequency of the ith sequence in population X, Y_j the frequency of the jth sequence in population Y, and D_{ij} is the number of nucleotide differences between the ith sequence from X and the jth sequence from Y (Nei & Li 1979; Nei 1987). As with Π we can standardize D_{XY} by dividing by sequence length to enable comparison of divergence between different sequences or different regions of the genome.

However, as mentioned in section 7.2.1, F_{ST} is also a commonly used statistic to compare divergence between populations in current population genomic studies. F_{ST} is calculated at the level of individual SNPs and averaged across segments of a chromosome of appropriate length (see section 7.4.2) whereas D_{XY} is calculated at the DNA sequence level. It is often useful to use both a relative measure of divergence such as F_{ST} and an absolute measure such as D_{XY} in population comparisons, because relative and absolute measures can be affected differently by various demographic and evolutionary processes (see chapter 8). Therefore, by using both relative and absolute measures of divergence we are better able to interpret what factors may have affected population differences at the different loci or genomic regions we are comparing.

7.3 Neutrality tests

7.3.1 The HKA test (Hudson–Kreitman–Aguadé test)

If molecular evolution is due to selectively neutral processes (genetic drift acting on neutral mutants) we would expect the level of genetic variation within a population to be positively correlated with the level of genetic differentiation between populations (or species) at a given locus. A highly mutable locus (i.e. one with a high mutation rate) would necessarily have many polymorphisms within a population whereas a locus with a low mutation rate would have few polymorphisms. Accordingly, two isolated populations (or species) would on average also be genetically more different at loci with a high mutation rate relative to those with lower rates. This is because more neutral mutations have occurred at the highly mutable loci, increasing the probability that they would eventually become fixed, rather than lost from a population by genetic drift. The **HKA test** (Hudson et al. 1987) uses this simple prediction to detect deviations from neutral theory. The test statistic of the HKA test is as follows:

$$\chi^2 = \sum_{i=1}^{m} \frac{\left[S_{Xi} - \hat{E}(S_{Xi})\right]^2}{V(S_{Xi})} + \sum_{i=1}^{m} \frac{\left[S_{Yi} - \hat{E}(S_{Yi})\right]^2}{V(S_{Yi})} \\ + \sum_{i=1}^{m} \frac{\left[D_{XYi} - \hat{E}(D_{XYi})\right]^2}{V(D_{XYi})} \qquad (7.9)$$

The formula may look frightening at first sight, but it is actually very similar to the chi-square goodness of fit test we introduced in chapter 3. Let us dissect it and try to understand what the test actually does. Let us take the first part on the right side of the equation first. It says we take the difference between the observed and expected (under neutral evolution) number of segregating sites in population X at locus i, square it and divide by a variance term. The sum sign says that we do this for all the m loci we include in our analysis and just add all the resulting values together. The second part is identical to the first except that we now deal with population Y. That is, we sum the squared differences between the observed and expected number of segregating sites at all the included loci in population 2 as well and divide by a variance term. The last term deals with the differences between the two populations or species. D_{XYi} is the average number of nucleotide differences at locus i between pairs of sequences randomly drawn from population X and Y. As for the first two terms we sum the squared difference between observed and expected number of differences for all the loci and divide by a variance term.

If our loci really have evolved only under neutrality—i.e. all mutations are neutral—the observed and expected number of segregating sites in the two populations, as well as the observed and expected difference between them (D_{XY}), should be the same and our test statistic would be zero. Any deviation from neutrality, say a higher species divergence than expected due to rapid fixations caused by positive selection, would give a positive test statistic (since the differences between observed and expected values are squared). Conveniently, the test statistic yields a chi-square distribution of its values. Hence, we can use the chi-square distribution to approximate the p-value (see section 3.1 in chapter 3).

Whitt et al. (2002) used the HKA test to investigate if genes involved in the starch metabolic pathway in cultivated maize (*Zea mays*) showed signatures of selection (Figure 7.11). Maize is very rich in starch compared to its closest wild relatives and it is likely that there has been artificial selection on this trait in connection with its cultivation. They surveyed genetic variation at six genes known to be involved in starch metabolism, as well as eleven random genes which they had no reason to assume had been affected by recent selection. To conduct the HKA test Whitt and co-workers sequenced the same genes from the closest wild relative of cultivated maize, *Z. mays* ssp. *parviglumis*. The test was highly significant. Closer inspection of the data showed that cultivated maize showed strongly reduced genetic variation at the starch genes, both relative to the eleven random genes and to wild maize. A likely interpretation is that artificial selection for grain yield has favored certain alleles at the starch genes, resulting in selective sweeps that have reduced variation at closely linked sites within these genes. This example also nicely highlights that the effects of selection in the genome are often observed beyond the target of selection, due to

Figure 7.11 Cultivated maize shows signatures of positive selection in genes involved in starch metabolism according to an HKA test, as reported by Whitt et al. (2002). Photo courtesy of Jan Louček.

recombination rate variation and physical linkage along a chromosome—i.e. the action of selection at linked sites can shape genomic variation. This has particular consequences for the interpretation of genome scan data and we will deal with it in detail in section 7.4 and also in chapter 8.

7.3.2 The McDonald–Kreitman (MK) test and its extensions

For protein coding DNA sequences there exist a series of neutrality tests that compare the pattern of variation of mutations that alter the amino acid sequence (i.e. non-synonymous or replacement mutations) of the resulting protein and those which do not (synonymous or silent mutations) in and between populations or species. One of the most famous of these is the **McDonald–Kreitman (MK) test**. The MK test is simply a comparison of the ratio of replacement to silent polymorphisms within the populations with the corresponding ratio of fixations between the populations. Since replacement mutations alter the amino acid sequence and thus protein structure, they are likely to have some kind of functional impact. We can therefore regard tests such as the MK test as quite powerful because non-synonymous mutations are likely targets of positive selection if the process is common in molecular evolution.

Let's imagine we are examining variation at a single gene between two species. If we think in terms of neutral evolution, what would we expect the relative frequency of replacement and silent polymorphisms within populations to be compared to replacement and silent fixations between differentiated populations or species? Well, again they should be positively correlated. Many polymorphisms within a population should lead to many fixations between differentiated populations whether they change the amino acid sequence or not—since they have no outcome on fitness according to neutral theory. However, if selection is a common process at the molecular level we should expect a different pattern.

First of all, replacement mutations are much more likely to be targets of purifying selection than silent mutations. The rationale for this is simple; if a mutation alters a protein that has otherwise survived successfully throughout evolutionary history, this is more likely to result in a less functional protein that will quickly be removed by selection. On the other hand, a mutation that leaves the protein unchanged—i.e. a silent mutation—is likely to be selectively neutral. Hence, at any time we would expect to find more silent polymorphisms in a population than replacement ones. However, not all replacement mutations will be deleterious and under neutral theory the few that remain in the population—i.e. are not removed by purifying selection—are expected to be neutral. We would also expect to find a corresponding rarity of putatively neutral replacement fixations between differentiated populations or species relative to silent ones—again because of the fitness consequences of these mutations. In other words, neutral theory predicts that the ratio of replacement to silent polymorphisms within populations should be the same as that of fixations between populations.

What if positive selection has been important in the evolutionary history of the focal gene we are considering? Well, now and then a replacement mutation with a positive effect on the fitness of an individual carrying that gene will occur and will be rapidly fixed by selection. Over evolutionary time such events would give rise to a very different ratio of replacement to silent polymorphisms within populations versus fixations between populations/species than that predicted from the hypothesis that most mutations are neutral. At the within-population level, replacement polymorphisms should still be rare (since negative mutations are still more likely than positive ones). However, replacement fixations between populations/species are expected to be relatively common in the positive selection scenario—simply because selection in the two different lineages will have fixed different advantageous mutations over time. In other words, the ratio of replacement to silent fixations between populations is expected to be higher than that for polymorphisms within the populations if positive selection has affected the molecular evolution of our focal gene. As we saw previously though, according to the neutral mutation hypothesis these ratios should be the same. This means we can use the MK test to look for deviations from neutrality; all this requires is the calculation of the simplest contingency table statistics such as a Fisher's exact probability test.

The very first McDonald–Kreitman test published was conducted on the gene G6pd that codes for an enzyme involved in sugar metabolism in three *Drosophila* fruit fly species (McDonald & Kreitman 1991). The ratio of replacement to silent polymorphisms at this gene within the three species was found to be 2/42, whereas the ratio of replacement to silent fixations between them was 7/17. In other words, the relative frequency of replacement fixations was much higher than predicted from the neutral theory. Using a Fisher's exact test, the probability that this result could emerge under neutral evolution is $P = 0.0073$ which is much lower than the 0.05 limit. Hence, in these particular species, at this particular gene, adaptive molecular evolution appears to have played a significant role and the neutral theory is rejected.

Since the MK test has been so focal to evolutionary genetics since its introduction, a number of extensions have been proposed. The most well-known of these is the Neutrality Index (N_I – Rand & Kann 1996), which is calculated as follows:

$$N_I = \frac{P_N / P_S}{D_N / D_S} \qquad (7.10)$$

Where P_N and P_S are non-synonymous and synonymous polymorphisms and D_N and D_S are non-synonymous and synonymous substitutions respectively. The N_I is essentially a measure of how strongly the different types of substitutions and polymorphisms are associated with one another. As explained above, under a neutral model, each of these two ratios should be identical—i.e. $N_I = 1$. However, deviations from the neutral expectation will shift the value of N_I around this value. For example, $N_I > 1$ means that there are more non-synonymous polymorphisms than fixed whereas an $N_I < 1$ means a greater number of non-synonymous substitutions are fixed, suggesting positive selection. In their paper introducing N_I, Rand & Kann (1996) calculated the index for human and *Drosophila* mitochondrial genes and found the mean $N_I > 1$—suggesting there is an excess of non-synonymous polymorphism relative to fixed non-synonymous substitutions. The authors suggested this can be explained by a nearly neutral model where non-synonymous substitutions are weakly deleterious and can segregate within a species long enough for us to observe them prior to their removal by purifying selection, but not long enough to contribute to overall mitochondrial divergence between species.

However, the neutrality index cannot be calculated if fixed non-synonymous substitutions or synonymous polymorphisms are absent from a gene. Typically, researchers can derive a mean N_I from across genes where it can be calculated (e.g. Bazin et al. 2006) but this introduces considerable bias to the estimate and meaning there is a risk of wrongly rejecting or accepting neutrality (Stoletzki & Eyre-Walker 2010). To account for such bias, Stoletzki & Eyre-Walker 2010 proposed a modification to the Neutrality index, the direction of selection (*DoS*) statistic:

$$DoS = \frac{D_N}{D_N + D_S} - \frac{P_N}{P_N + P_S} \qquad (7.11)$$

The *DoS* is very simple—it is the difference between the proportion of non-synonymous fixed differences and the proportion of non-synonymous polymorphisms. If mutations in a gene are completely neutral, *DoS* will be zero whereas if some mutations are adaptive and contribute to species divergence, it will be positive. Accordingly, a negative *DoS* suggests a relative excess of replacement polymorphism.

A good example of the application of *DoS* to molecular data comes from studies seeking to test whether positive selection drives higher divergence on sex chromosomes relative to autosomes. The molecular evolution of sex chromosomes is well studied and higher sequence divergence between genes on the sex chromosomes relative to the autosomes has been observed in a number of taxa (Vicoso & Charlesworth 2006). In XY (e.g. humans, *Drosophila*) and ZW (e.g. birds, butterflies) sex chromosome systems, this phenomenon is referred to as the "faster X/Z effect." Two major, but opposing, explanations have been suggested to account for the faster X/Z effect. The first is that selection is more efficient on the X/Z chromosome because in the heterogametic sex (e.g. XY males in humans, ZW females in birds), beneficial recessive mutations are more apparent to selection. The smaller sex chromosome (Y and W) contains fewer genes than the large one (X and Z). Therefore, in the majority of genes on X/Z the heterogametic sex only has one gene copy since it is absent on Y/W—the heterogametic sex is **hemizygous**. A recessive beneficial allele would therefore be targeted by positive selection in the heterogametic sex. In contrast, if such mutations occur on an autosome, selection would only be able to act on them once they occur in an individual homozygous for the beneficial recessive allele—in heterozygotes the allele would be masked by dominance (Mank et al. 2010). Therefore, beneficial recessive mutations are more prone to be lost by genetic drift when they occur on autosomes than on X or Z.

However, the alternative explanation for the faster X/W effect is that the effective population size (N_e) of the sex chromosome is smaller than the rest of the genome. For every four copies of an autosome in a population there are only three copies of the Z or X. This is simply because the heterogametic sex always carries a different sex chromosome—the Y or W. Since N_e is smaller for the sex chromosomes, selection is less efficient and drift plays a more important role; therefore, higher rates of fixed non-synonymous substitutions on the X or Z can be explained by the fixation of slightly deleterious mutations that would be purged by purifying selection elsewhere in the genome (Mank et al. 2010; Vicoso & Charlesworth 2006). Sackton et al. (2014) used whole genome resequencing to test these predictions in an invertebrate ZW system—the silkmoth, *Bombyx huttoni*. They found *DoS* was significantly more positive in Z-linked genes compared to those occurring in the autosomes (Figure 7.12). The authors also showed that female biased genes, that is, genes with a greater level of expression in the ZW female, have more positive *DoS* than genes without any expression bias between sexes. Positive *DoS* values in Z-linked genes therefore point to an

important role for positive selection in fixing beneficial mutations on the Z.

7.4 Genome scans

7.4.1 The logic of genome scans

A **genome scan** is a colloquial term for a set of research methods in which the entire genome, or a significant proportion of it, is searched systematically using various statistical procedures and interpreted for one or more purposes. In chapter 6 we introduced one application of genome scans, namely the search for associations between allelic variation and phenotypic traits, such as alleles associated with a genetic disease using genome wide association studies (GWAS). However, genome scans have many other applications in evolutionary genetics. They can be used to investigate the demographic history of

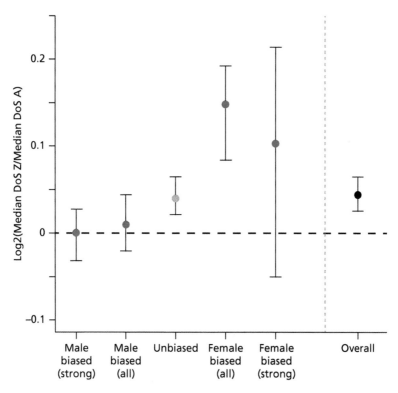

Figure 7.12 Faster-Z in the silkmoth *Bombyx huttoni*. Overall, the Z:A ratio of median scaled *DoS*-score is positive, consistent with more efficient positive selection on the Z chromosome (black right dot). Genes with a higher level of expression in females (red) have more positive *DoS*-score on Z compared to genes with unbiased expression (green) and genes with male biased expression (blue). Error bars represent 95% confidence intervals. Redrawn from Sackton et al. (2014).

populations. The genome wide signal of negative Tajima's D in Japan Sea sticklebacks (Ravinet et al. 2018a) provides one example. However, a particularly important application of genome scans is to identify candidate genes or genomic regions that might underlie adaptive evolution.

A genome scan is an example of a bottom-up approach and can be used to identify signatures of selection without prior knowledge of the genes and, in some cases, the phenotypes that might be divergent between populations. The method is simple; we first take two populations that differ in some way, perhaps they inhabit different environments or feed on different diets. We then estimate statistics such as F_{ST} and D_{XY} between populations and π and Tajima's D within populations, for a large number of genes or loci from across the genome; loci with extreme values, in the tail of the distribution of our test statistic, are treated as being potentially under selection. This is not a new approach; Lewontin & Krakauer (1973) first used variation in F_{ST} among human genes to identify those potentially under selection, arguing that genetic drift should affect all genes equally and that high values of F_{ST} are likely caused by positive selection. Selection does certainly influence the distribution of F_{ST}, but there is a large amount of random variation in F_{ST} even among neutral loci, largely due to the fact genetic drift is a stochastic process (Beaumont & Nichols 1996; Beaumont & Balding 2004; Storz 2005). Genome scans therefore perform best when large numbers of loci are used, making it possible to get a more accurate estimate of the neutral baseline distribution of F_{ST} (Beaumont & Nichols 1996; Storz 2005). In the recent past, use of multilocus F_{ST} approaches was limited in all but a few model species using relatively few loci. However, with cheaper and more cost-effective high-throughput sequencing, alongside a wide variety of methodologies to capture thousands of loci, genome scans of variation in F_{ST} and other statistics have become an extremely popular tool in evolutionary genetics. We will explore how the method has evolved for large datasets in the next section.

The evolution of modern humans largely points to an **out-of-Africa model**, where *Homo sapiens* evolved and diversified in Africa but with modern populations, such as Europeans and East Asians

diverging with repeated migrations out of the African continent. We will see in chapter 9 that this model is somewhat of an oversimplification but for now we can keep in mind that it has been the consensus for modern human evolution since the 1980s. In an early genome scan study, Kayser et al. (2003) argued that since migrating from Africa, European populations are likely to have experienced selection for adaptations to new environments and habitats. To test this, they amplified 332 microsatellite loci in 48 Europeans and 47 Africans and then calculated F_{ST} at each locus, detecting 15 loci (4.5 percent) with extreme F_{ST} values above the average of 0.043. However, in order to confirm that such high values were not merely a result of chance (i.e. via genetic drift), Kayser et al. (2003) used the then newly published human genome sequence to identify and amplify microsatellite loci within 50 Kb of their candidate loci. These close-proximity loci had an average of F_{ST} of 0.171—so they too showed high levels of genetic differentiation. Examining one of the loci with the highest F_{ST}, they found it occurred 900 bp from the E2F transcription factor 6 (*E2F6*) gene. They then amplified a further 5 microsatellites in the region of this gene and found only those in close proximity to the gene had an F_{ST} of above 0.3. The role of *E2F6* and why it might be a target of divergent selection between Africans and Europeans is unclear; however, it is thought to be involved in cancer cell proliferation and mammalian development. Nonetheless, the study by Kayser et al. (2003) gives a good demonstration of how a genome scan can form the basis of further investigation of the roles and functions of candidate genes.

7.4.2 Sliding window genome scans

In one of the first published multilocus F_{ST} studies, Lewontin & Krakauer (1973) surveyed genetic differentiation at nine loci in humans. Even until the early 2010s, studies looking for divergent natural selection in wild populations of non-model organisms used tens or at most a few hundred loci (e.g. Wilding et al. 2001; Grahame et al. 2006; Michel et al. 2010). The genomic revolution with increasingly cheap and accessible methods for sequencing genome-wide variation has made it possible to compare genetic variation and population differentiation at thousands of loci and in some cases across the entire genome. For example, genome scan studies using whole genome resequencing routinely produce millions of SNP polymorphisms at which to examine differentiation (e.g. Poelstra et al. 2014; Lamichhaney et al. 2015). The methods have therefore had to adapt to being truly conducted at the genomic scale.

In general, the principle of a genome scan with millions of SNPs is very similar to a multilocus comparison with just 10 markers, although the former will obviously have a great deal more power. However, one issue introduced by using millions of loci is that there is a great deal of random variation. Since genome scans require the identification of outliers from the genome-wide background of differentiation, many loci can lead to a great deal of noise into the data, making it difficult to properly discern the true signal.

A second issue introduced by using many genome-wide loci is the fact that estimates of population differentiation between some loci are not independent. This occurs because of physical linkage and linkage disequilibrium among loci; if two loci occur physically close to one another on a chromosome, they are more likely to share a recent genealogical history. This is best illustrated by a positive selective sweep—selection at a single locus will also drive alleles in close proximity towards fixation, with a lower probability as physical distance (and thus the probability of recombination) increases. If we genotype variation at both the selected locus and those in close proximity, we might count them as multiple outliers, when in fact only a single locus has undergone selection.

How then do evolutionary geneticists overcome these two issues? A common method is to use a **sliding window** approach (Figure 7.13). Sequence reads are aligned to the reference genome and genotypes called with respect to their physical position. We then traverse each chromosome in

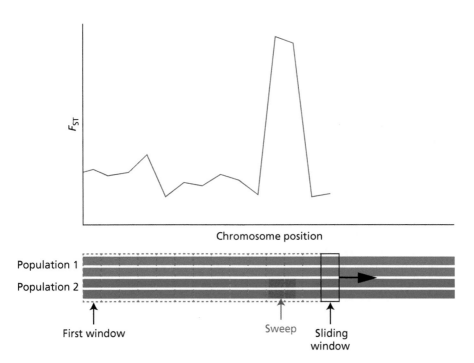

Figure 7.13 Principle of sliding window genome scan. Chromosomes from the populations we compare are aligned. Here only two chromosomes from each of two populations are shown (orange and red respectively). Population genetic statistics, such as F_{ST} are calculated in each of several windows traversing the chromosome. In this example, non-overlapping sliding windows are used. In this hypothetical example, population 2 has experienced a selective sweep (blue part of chromosomes) which translates into peak average F_{ST}-values in the windows that cover the sweep.

windows of a predefined physical distance, calculating parameters of interest (such as π and Tajima's D within each population, and F_{ST} and D_{XY} between the populations) in each window. Consider the largest human chromosome, chromosome 1, which is 248 956 422 base pairs long (i.e. 248.9 Mb). If we set our window size to 100 000 base pairs (100 Kb), we will estimate our statistics for 2490 windows (the last window encompasses the end of the chromosome). Similarly, if we set our window size to 500 Kb, we will examine 498 windows. This form of sliding window approach produces non-overlapping sliding windows—i.e. all of our estimates are conducted on windows that we presume are independent of one another. Alternatively, we might choose to estimate statistics on windows that overlap one another with a small "step" in between them. For example, we could traverse the human chromosome 1 with 100 Kb windows but with a 25 Kb step between them. This means we first estimate a statistic such as F_{ST} for the window 0–100 Kb, then 25 Kb to 125 Kb, then 50 Kb to 150 Kb and so on. This would result in a total of 9959 overlapping sliding windows. It is important to remember that such estimates are not independent of one another. Often, an overlapping approach is used to smooth the signal of the mean value of our chosen statistic across a chromosome and is particularly popular for visualization.

Choosing the size of genome windows for analysis is not just arbitrary; there are several factors to take into consideration. Most importantly, there is a need to choose a window size that reflects recombination rate and the decay of linkage disequilibrium (LD). When SNPs are in close physical proximity to one another, they will have high LD but eventually, at a specific physical distance, LD will drop to a background level. For example, when examining genomic differentiation in Italian sparrows (*Passer italiae*), Elgvin et al. (2017) found that LD decayed to the background rate at around 100 Kb. Therefore, using a 100 Kb sliding window would produce largely independent estimates of F_{ST} and other statistics along the chromosome. Setting a window size too high might also have the effect of averaging out differentiation to the point where important signals are missed—i.e. the window size is simply too coarse. Often, it is best to use a window

size that provides independent estimates of statistics and then if a region of interest is found, use a smaller scale.

Once we have a series of estimates of our statistic of interest from across the genome, we can use our windowed estimates to perform a genome scan. This is exactly the same as we described in section 7.4.1, except now we are using windows instead of individual loci. Windows of increased high divergence relative to the rest of the genome are potentially a result of selection and warrant further investigation for the presence of candidate genes that might underlie adaptive divergence and local adaptation among populations or species. Genome scans based on whole-genome resequencing can be incredibly informative and have pointed towards a large number of interesting and exciting genes and processes involved in recent evolution among species.

In Europe, two forms of the common crow occur and hybridize, the distinctive completely black carrion crow (*Corvus corone corone*) and the two-toned black and gray hooded crow (*C. corone cornix*) (Figure 7.14). Both forms of crow meet in a narrow hybrid zone that extends over much of Central Europe but nuclear loci show very little genetic differentiation between them (Wolf et al. 2010). This is surprising, as there is evidence for barriers to gene exchange via assortative mating in this hybrid zone; i.e. matings between birds with similar plumage color are more common than those with different plumage (Randler 2007). Given that nuclear genetic differentiation between the two forms is low, a genome scan might prove helpful for understanding the genetic basis of plumage coloration. With this in mind, Poelstra et al. (2014) resequenced the genomes of 60 individuals—30 of each form—in order to examine patterns of differentiation. Genome-wide F_{ST} between carrion and hooded crows was low, except for a single 1.95 Mb region on crow chromosome 18. Intriguingly, Poelstra et al. (2014) found only 82 fixed SNPs between the two crow forms (from a total of 5.27 million) and 81 of these occurred within this highly differentiated region. Clustering analysis of the SNPs in this region clearly grouped the two crow forms, whereas the same analysis on non-differentiated regions could not distinguish them. The highly differentiated

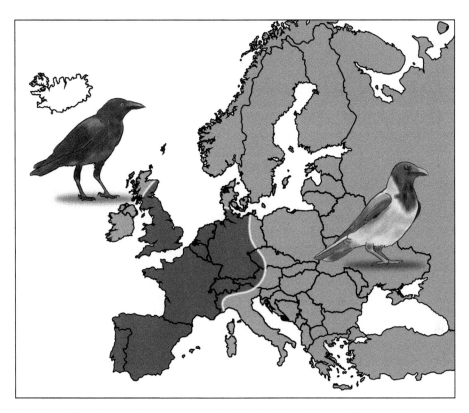

Figure 7.14 Distribution of the carrion crow (*Corvus corone corone*) and the hooded crow (*C. c. cornix*) in Europe. The two crow forms meet in a narrow hybrid zone that extends through Central Europe and across the British Isles. Genome scan analysis shows that the two crows are genetically very similar except in a single region on crow chromosome 18 where F_{ST} peaks to very high values (Poelstra et al. 2014). Genes involved in melanogenesis are found in this genomic region.

region also contained genes involved in the regulation of melanogenesis, that is, candidate genes for the striking difference in coloration between the crows.

7.4.3 Haplotype tests and selective sweeps

Most of the methods for detecting selection we have discussed in this chapter have either relied on genetic differentiation between populations (i.e. F_{ST}) or are based on the expected allele frequency distribution within populations under different selection scenarios (i.e. Tajima's D). In a genome scan context, both of these classes of methods seek to identify signatures of the same phenomenon—selective sweeps occurring as a result of positive selection. However, as we saw earlier, such sweeps produce strong signatures of linkage disequilibrium (LD) and the methods we have examined so far do not incorporate this. A newer set of methods,

developed for genomic data and known as **haplotype statistics**, do incorporate LD. The principle is quite simple; when selection occurs on a locus and drives it to fixation, surrounding variation hitches along for a ride and is also driven to high frequency (see Figure 7.7). In other words, selection fixes a haplotype, not just its target. The physical extent of a haplotype in the genome is proportional to the probability of recombination—i.e. variants in close proximity to the target of selection will be swept alongside it but the probability of being swept decreases as the chance of recombination increases. The temporal extent of a haplotype is also determined by recombination around the target of selection, high recombination will result in a rapid breakdown of reduced diversity around a selected site, whereas low recombination has the opposite effect. Since the signature of a selective sweep decays over time, this means that we are unlikely to be able to detect selection a long time after it has occurred. The chances of observing a sweep just after it has reached fixation, however, is vanishingly small. Much more likely, we will observe signatures of either incomplete sweeps—i.e. those that have yet to reach fixation—or sweeps that reached fixation sometime in the past.

Clearly then, the extent of a haplotype conveys some information about the action of selection in the genome. With this in mind, Sabeti et al. (2002) introduced **extended haplotype homozygosity** or EHH. This statistic starts with a focal allele—e.g. one putatively under selection—and then measures the probability that the region surrounding this SNP is identical by descent in any two randomly chosen haplotypes from a population (Vitti et al. 2013). Identity by descent in this case is calculated using the homozygosity of SNPs surrounding the focal allele (Gautier et al. 2016). Therefore, EHH measures the extent of the haplotype surrounding the allele under selection at the population level. This illustrates an important point about selective sweeps—they are not instantaneous and we are unlikely to observe them immediately after fixation. Therefore, since recombination takes place in the population there is a diversity of haplotype lengths surrounding the focal allele. However, we would expect the regions closely flanking a selected allele to remain similar. EHH is highest at the focal allele (i.e. it is set to 1) but decays towards zero with increasing physical distance from the focal allele (Gautier et al. 2016).

Many other haplotype-based tests for selection have been developed based on EHH. The **integrated haplotype score** or iHS is the area under the curve defined by EHH—i.e. from 1 at the focal allele until the point it decays to zero (Voight et al. 2006). To properly apply the iHS, we first use an **outgroup** species (i.e. a closely related species) to polarize alleles at a locus as either ancestral or derived. For example, to polarize alleles in human sequence data, we would use the corresponding position in the chimpanzee genome. If a site is polymorphic for A and T in humans and fixed for T in chimpanzees, we assume that T is the ancestral allele and A is derived. The iHS is then the log of the division of these two areas. A negative iHS suggests selection has favored the derived allele, while a positive iHS

points to the ancestral allele (Weigand & Leese 2018).

So far, the EHH methods we have introduced focus on the haplotype homozygosity within populations. Obviously, these are useful for identifying signatures of local adaptation or recent selection, but what if our aim is to compare populations or species and show that selection has occurred in one and not the other? The **cross-population extended haplotype homozygosity** statistic (xpEHH) therefore compares the lengths of haplotypes between two populations (Sabeti et al. 2007). Similar to the iHS, xpEHH is the log of the division of iHS values for the two populations. This means a positive xpEHH indicates selection occurring in population A, whereas a negative value suggests selection is occurring in population B (Weigand & Leese 2018).

EHH statistics have proven a very useful method for detecting signatures of selective sweeps, particularly incomplete sweeps. The ability to digest lactose, the main carbohydrate found in milk, into adulthood is a major adaptation in recent human history. In the majority of humans, the enzyme lactase-phlorizin hydrolase (LPH) declines after weaning, meaning that they are unable to digest lactose in adulthood. However, in human societies that have developed and maintained a tradition of raising and herding cattle, lactase persistence occurs—i.e. adults are able to properly digest lactose. Interestingly, the lactase persistence trait appears to have arisen at least twice, and possibly more times, in different human populations that have practiced cattle agriculture. All occurrences of lactase persistence involve the *LCT* gene on chromosome 2 that codes for LPH but different SNPs are involved in different populations. For example, in Northern Europeans where lactase persistence is around 80–90 percent, a T at a position ~14 Kb upstream from *LCT* is associated with the trait in approximately 90 percent of cases (Enattah et al. 2002; Poulter et al. 2003). However, lactase persistence also occurs in African populations where farming cattle is common, but where the European causative SNP is in low frequency. Tishkoff et al. (2007) investigated the association between lactase persistence and SNP polymorphisms surrounding the *LCT* gene in 470 people from Kenya, Sudan, and

Tanzania. Intriguingly, the researchers found three SNPs associated with lactase persistence, each of which occurred approximately 14Kb upstream from the *LCT*, although all in different positions to the causative variant in Europeans. Investigating further, Tishkoff and co-workers focused on the most common of these three positions, a G/C polymorphism that had an average frequency of 39 percent in Kenyans and Tanzanians. The researchers measured the extent of homozygosity using EHH at this SNP position in Africans and also at the causative allele in Europeans. In Africans, EHH indicated a 1.8 Mb region of homozygosity surrounding the derived allele associated with lactase persistence. A region of 1.4 Mb was observed in Europeans for their causative SNP. In both cases, the ancestral allele (i.e. not associated with lactase persistence) had an average EHH of around 2 Kb. Increased homozygosity surrounding the SNPs associated with lactase persistence therefore clearly points to a signature of recent positive selection of the derived allele.

The Masaai are another population of pastoralists from Eastern Africa, and as well as showing lactase persistence they have lower levels of cholesterol despite a diet rich in milk and meat (Wagh et al. 2012). However, they were not included in the study by Tishkoff et al. (2007). Wagh et al. (2012) used a genome scan to compare Maasai with another group of people from Kenya, the Luhya, who do not show lactase persistence; Figure 7.15 shows the main results from this study. First, using F_{ST}, they found a very strong signal of differentiation on chromosome 2 at the site of the *LCT* gene involved in lactase persistence. They next used the xpEHH statistic to test the differences in haplotype length between the two populations. The xpEHH was clearly highly positive at *LCT*, indicating a larger haplotype length in the Maasai. Finally, Wagh and co-workers used the iHS statistic to show that the derived allele in the Maasai had a greater extent of haplotype homozygosity than the ancestral allele (lower panel, Figure 7.15). With three different measures, this is good evidence that a positive selective sweep occurred at the region surrounding the *LCT* gene in the Maasai. To further investigate the polymorphisms at *LCT* that might be involved, the researchers sequenced Maasai individuals at

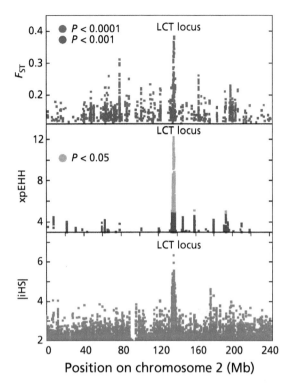

Figure 7.15 Genome scan of chromosome 2 comparing the lactase persistent Maasai population with non-persistent Luhya population. Top panel: F_{ST} shows peak divergence near the LCT locus. Mid panel: Cross population extended haplotype homozygosity (xpEHH) comparing the length of EHH-regions, including the region around the SNP near the *LCT* locus targeted by selection. Maasai have significantly longer EHH than the Luhya as evident from the significant green peak, suggesting positive selection. Bottom panel: Genome scan showing the iHS-statistic which shows that the derived allele in the Maasai has a greater extent of haplotype homozygosity than the ancestral allele. Modified from Wagh et al. (2012).

five loci associated with lactase persistence and found the most common variant from Africans also identified by Tishkoff et al. (2007), the study from our previous example. Both of these examples nicely illustrate the power of genome scans to identify important causative variants underlying adaptive traits.

The power of haplotype statistics for detecting selective sweeps in genome scans is also nicely illustrated in a study of the house sparrow *Passer domesticus* (Ravinet et al. 2018b). The house sparrow is a highly successful human-commensal bird species that occurs on all continents except Antarctica. It is closely associated with human settlements, including cities and farms where it feeds on human food waste. It appears to have become intimately adapted to humans and the new niches that opened up with the Neolithic revolution some 10 000 years ago, to the extent that it now goes locally extinct in abandoned human settlements (Summers-Smith 1988). However, a wild subspecies of the house sparrow, the bactrianus sparrow (*P. d. bactrianus*), resides on the steppes of Central Asia (Summers-Smith 1988; Sætre et al. 2012). The bactrianus sparrow is phenotypically very similar to our house sparrow but it differs markedly in ecology. It is not associated with humans and feeds on natural grass seeds. Also, unlike the human-commensal house sparrow it migrates south during winter. Associated with the difference in ecology, human-commensal house sparrows have evolved a more robust beak and skull than the bactrianus sparrow, presumably an adaptation to cope with the larger and harder seeds of cultivated grass species, such as wheat and barley, compared with wild grass seeds (Riyahi et al. 2013). Coalescent modeling (see chapter 9) using whole genome sequence data dates the split between bactrianus and house sparrow to approximately 10 000 years ago, suggesting that bactrianus sparrows represent a relict population of the ancestral, pre-commensal house sparrow (Ravinet et al. 2018b).

Ravinet and co-workers used haplotype statistics across the genomes of bactrianus and house sparrows to look for signs of recent selection in the house sparrows that may be linked to adaptations to its human-commensal niche. First, using cross-population extended haplotype homozygosity (xpEHH) across the genomes of the two sparrow subspecies, the researchers identified several outlier regions where the haplotype homozygosity extended significantly further in one of the subspecies (Figure 7.16a). Using the highest outlier SNP in each of 59 outlier peaks identified from xpEHH, the researchers then compared site-specific extended haplotype homozygosity (EHH) around each of these outlier SNPs within both house and bactrianus sparrows. EHH-values showed a clear pattern of increased extended haplotype homozygosity in the house but not the bactrianus sparrow, suggesting recent selective sweeps in the house sparrow

genome, but not in bactrianus (Figure 7.16b). One of the highest xpEHH peaks occurred on chromosome 8 (Figure 7.16a). This peak contains two genes. The first gene, COL11A1, is known to be associated with craniofacial development and is therefore a candidate gene for the observed changes in beak and skull morphology in house sparrows. The second gene, AMY2A, is associated with adaptation to a higher starch diet in both humans and dogs following the Neolithic revolution. The Neolithic revolution may therefore have introduced a common selective pressure that has resulted in parallel adaptations in the same gene for three very different taxa—humans, dogs, and potentially also human-commensal house sparrows.

7.4.4 Limits to genome scans

Although popular and relatively simple to apply, genome scans do not come without disadvantages, biases, and limits to what we can learn from them (Hoban et al. 2016; Ahrens et al. 2018). A good researcher should therefore be aware of these limits. Most genome scan methods based on genetic differentiation rely on identifying loci with extreme values of F_{ST} compared to a baseline, i.e. outliers. However, the baseline can be easily distorted by demographic history; for example, a population bottleneck might raise genome-wide F_{ST}, making it very difficult to properly distinguish outliers from the background (Lotterhos & Whitlock 2014). Some demographic processes might also affect individual loci, resulting in a false positive finding of selection when there is none. For example, when a species range expands, a small population often leads the leading edge of the expansion and genetic drift can readily fix individual alleles, giving the impression of selection—this is known as "allele surfing" (Excoffier et al. 2009).

A more pressing problem for genome scan approaches is the fact that they are likely biased towards identifying only certain types of loci involved in adaptation. Genome scans are likely to work well for identifying loci that have a large effect on fitness. However, in reality the genetic basis of most traits probably consists of multiple genes, with relatively small effects (Rockman 2012).

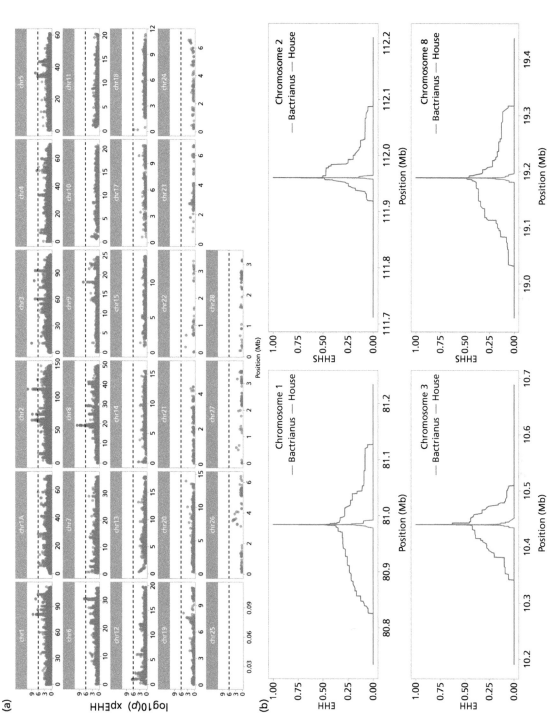

Figure 7.16 Haplotype statistics from genome scans comparing human-commensal house sparrows (*Passer domesticus domesticus*) and wild bactrianus sparrows (*P. d. bactrianus*). A) *P*-values of cross-population extended haplotype homozygosity scores (xpEHH) across the sparrow genome. The genome-wide plot of xpEHH shows clear peaks of divergent haplotype homozygosity occurring throughout the genome, with particularly large peaks on chromosomes 1, 2, 3, and 8. Horizontal dashed line represents threshold of significance ($\log_{10} 1 \times 10^{-6}$), red points indicate outlier SNPs. B) Examination of the four SNPs with highest xpEHH scores across the chromosomes shows a clear signature of site-specific extended haplotype homozygosity (EHH) in the house (blue lines) but not the bactrianus sparrow (green lines). Modified from Ravinet et al. (2018b).

For example, a complicated trait such as human height has evidence of up to 551 alleles associated with it (Field et al. 2016). This suggests that while genome scans are able to easily identify the low-hanging fruit in terms of the genetic basis of traits, we should be aware that there are many signatures of selection they will overlook.

Despite these issues, genome scans are a hugely popular tool for identifying loci under selection, particularly with the rise of cheap and easy sequencing of loci from across the genome (Hoban et al. 2016; Ravinet et al. 2017; Ahrens et al. 2018). However, it is important to keep in mind that they are just one tool in an array of different approaches that evolutionary biologists apply to investigate the genetic basis of adaptation. In the majority of examples we have used in this part of the chapter, genome scans have been used as a first step to point towards a signature of selection at a candidate locus. Experimental work is necessary to properly tie the function of a gene to an adaptation, and further work still is needed to demonstrate that this gene plays a role in adaptation (Ravinet et al. 2017).

Study questions

1. What is the difference between mutation rate and substitution rate?
2. Explain the rationale behind the molecular clock. Review the problems of assuming a molecular clock for timing a speciation event such as between humans and chimpanzees.
3. Explain the difference between Kimura's neutral theory and Ohta's nearly neutral theory.
4. Discuss whether figures such as Figure 7.2 can be used to reject the neutral theory as a general theory for how molecular evolution unfolds.
5. Explain the difference between the parameters Π, S, and D_{xy}.
6. Using coalescent reasoning, explain why nucleotide diversity π is an estimator of the population mutation rate parameter θ.

7. The population mutation rate parameter $\theta = 4N_e\mu$ for an autosomal locus. What would it be for an X-linked locus, a Y-linked locus, and an mtDNA locus?
8. How would Tajima's D be affected respectively by balancing selection, background selection, and positive selection?
9. How would Tajima's D be affected by a population bottleneck and a population expansion respectively? How can we distinguish these patterns from those you described in question 8?
10. You have sequenced the coding sequence of three genes in two species of birds that differ in diet and plumage color: a digestive enzyme, a pigmentation gene, and an mtDNA gene, and made the table below on synonymous and non-synonymous polymorphisms and substitutions. Calculate and interpret the MK-test, Neutrality Index N_I, and the direction of selection DoS. Can you rule out selection for any of the genes?

Gene	P_N	P_S	D_N	D_S
Enzyme	3	62	12	17
Pigmentation	2	15	1	7
mtDNA	18	54	3	17

11. Explain the logic of F_{ST} genome scans and how they can be used to detect loci affected by positive selection. How would you proceed to detect loci experiencing balancing selection using a genome scan approach?
12. Explain the logic of haplotype statistics and explain the difference between EHH, iHS, and xpEHH.
13. Discuss what results we might have gotten if we had used haplotype statistics on the starch genes in maize that we discussed in section 7.3.1.
14. How could Ravinet and co-workers proceed to investigate if there is a causal relationship between the two candidate genes they identified and adaptation to a human-commensal niche in the house sparrow?

CHAPTER 8

Genetics and genomics of speciation

The diversity of life on Earth is something that has long fascinated biologists. When we inspect all this variation more closely, two particularly striking patterns emerge. First, organisms appear as if they have almost been "designed" to live the way they do. However, as we learned in chapters 4 and 5, the theory of natural selection explains this apparent design. The not-so-good solutions get outcompeted by better ones. Evolution thus eventually produces adaptations to the challenges that organisms face throughout their lives. The second striking pattern is that variation appears to be non-randomly distributed. That is, it is clustered into groups of individuals that resemble each other and that are recognizably different from other such clusters. We call these clusters species. To what extent are species biologically meaningful entities in evolution? How do new species originate? What genetic changes occur when one species diverges into two? These are among the many questions we will address in this chapter. We start by discussing what constitutes a species.

8.1 Species concepts

8.1.1 What are species?

It may come as a surprise to many non-biologists, but there is no clear consensus on what constitutes a species among biologists. Yet everyone, young or old, lay-person or professional, has an intuitive idea about what species are. You need only look at a patch of vegetation and count the different types of insects you see—many will be separate species.

Members of a species are usually phenotypically quite similar. However, phenotypic similarity does not always hold true, and there is often a great deal of variation within a species; the two sexes can be quite different, something which is readily apparent in sexually dimorphic bird species such as the superb fairywren we learned about in Chapter 4. Sexual dimorphism aside, individuals within a species are typically phenotypically variable to some extent, often in a geographically structured way like in our own species. Consider for instance the skin and hair color gradient between Northern Europe and the southern latitudes of Africa that encompasses a remarkable range of phenotypic variation. However, phenotypic differences between species often greatly surpass that occurring within species. Members of different species are usually recognizably different, often in a much greater number of traits. Simply by observing organisms in our surroundings we cannot help but notice that biodiversity is structured into relatively discrete clusters. Strolling through a forest in Europe we recognize great tits (*Parus major*) as one cluster of birds that is clearly different from blue tits (*Cyanistes caeruleus*), with no intermediates between them.[1] Likewise, the common hepatica (*Hepatica nobilis*) is an easily recognizable flowering plant species different from, say, the wood anemone (*Anemone nemorosa*). This phenotypic clustering is evidently not just a psychological tendency of the human mind to see categories where none exist, because these phenotypic clusters

---

[1] A caveat that will become important later—great tit × blue tit hybrids are formed on rare occasions.

Evolutionary Genetics: Concepts, Analysis, and Practice. Glenn-Peter Sætre & Mark Ravinet, Oxford University Press (2019).
© Glenn-Peter Sætre & Mark Ravinet 2019. DOI: 10.1093/oso/9780198830917.001.0001

nearly always correspond to clear genetic clusters. There are some exceptions, though. Sometimes we discover clear genetic clusters among individuals that look very much the same to the human eye. We call such groups **cryptic species**, which can be as genetically differentiated as phenotypically different species pairs but with no obvious trait differences (e.g. Pfenninger & Schwenk 2007). Also, sometimes phenotypically distinct groups are extremely similar genetically across the vast majority of their genome, as exemplified by the hooded and carrion crow studied by Poelstra et al. (2014) that we discussed in chapter 7.

Are there other defining characteristics of species? For sexually reproducing animals, they do tend to mate assortatively—i.e. they prefer to mate with members of their own species. Male and female great tits pair up with each other to produce great tit nestlings as do blue tits and interspecific crosses between the two species are extremely rare. However, there are many exceptions to this rule. Hybridization between species is by no means uncommon. In a mixed population of pied flycatchers (*Ficedula hypoleuca*) and collared flycatchers (*F. albicollis*) on Gotland in the Baltic Sea one in 25 breeding pairs are heterospecific and a similar proportion of the breeding birds are hybrids (Alatalo et al. 1990). Yet, the two species remain phenotypically and genetically distinct (Ellegren et al. 2012), so assortative mating alone does not maintain the separation between them. Moreover, assortative mating is certainly not something that only occurs between species—it can also be common within species. In humans, studies have reported assortative mating based on traits such as body height and body mass index (e.g. Silventoinen et al. 2003). Clearly, assortative mating is not a defining characteristic of species.

In sexually reproducing organisms, individuals within a species are typically interfertile whereas interspecific hybrids can be sterile. In high school we learned that hybrids between the horse (*Equus caballus*) and donkey (*E. asinus*) (mule or hinny depending on whether the mother is a horse or a donkey respectively) are sterile and therefore that the horse and the donkey are different species. Interestingly enough, this is not always entirely true. Female mules are occasionally fertile and produce offspring (Yang et al. 2004). Nonetheless,

many interspecific crosses between a wide range of taxa indeed result in sterile hybrids, and some species remain distinct, unable to cross-fertilize at all. Yet again, however, there are countless exceptions. Hybrids between the wolf (*Canis lupus*) and coyote (*C. latrans*) are both fertile and viable, yet the two species remain distinct both genetically and phenotypically except in human disturbed regions, particularly in the eastern parts of North America, where extensive hybridization has resulted in a swarm of intermediates (Stronen et al. 2012). Species often meet in **hybrid zones**, areas where they interbreed with one another but that are geographically restricted by a balance between selection and dispersal—the crows from chapter 7 is a good example of this. Moreover, sterility is by no means a feature unique to interspecific hybrids. Highly inbred individuals can be sterile and some pairs, also in our own species, are reproductively incompatible and struggle to conceive even if neither the male nor female are actually sterile. So again, intraspecific fertility and interspecific sterility seem to be poor defining characteristics of species.

It appears to be difficult to find absolute or general criteria for what distinguishes species except that they are genetically and usually phenotypically different from each other and that they tend to remain distinct even when in contact, although apparently different mechanisms might maintain this distinctiveness.

8.1.2 Typological species definitions

As biologists we may ask questions about species with different purposes in mind. First, we may simply want to label and categorize the variation. Because biodiversity appears to be clustered into species it would be useful to have a definition that allows us to delimit species in a sensible and systematic way. This is not just because categorizing biodiversity is fun: Delimiting species is important for conservation purposes, for instance when establishing legal frameworks for protecting species and for sustainable management. Typological species definitions simply rely on phenotypic and/or genetic clustering. Darwin himself used a typological species concept based on phenotypic resemblance when he wrote about the origin of species. A modern typological definition is the **genetic cluster species concept** which defines species as *groups of individuals that form a genetic (and/or morphologic) cluster with few or no intermediates to other such clusters* (Mallet 1995). Along with the genetic cluster species definition, modern taxonomists have a range of statistical methods at hand that can handle huge sets of phenotypic and genomic data to test for clustering. One potential problem with the genetic cluster species concept is that we can find significant genetic clustering at very fine scales. Closely related, but geographically isolated populations readily show significant genetic clustering. Applying the concept strictly, we may therefore end up with many more species than those we would normally list in a field guide. Mallet (1995) recognized this caveat and therefore pointed out that the genetic cluster definition is particularly useful for classifying organisms that are found in **sympatry** (within the same area). The great tits and blue tits form clear phenotypic and genetic clusters wherever they co-occur and are clearly different species according to this definition. What about the hooded and carrion crow? For the most part they do not co-occur except in a narrow hybrid zone running through Central Europe and across the British Isles. Within this zone there is phenotypic (and we can safely assume genetic) intergradation caused by interbreeding. Yet, the zone of intergradation is very sharp. The two crows could be considered different subspecies according to the genetic cluster species definition, but taxonomists often debate such cases. Indeed, in some field guides the hooded and carrion crow are listed as subspecies belonging to the same species whereas other field guides list them as separate species. Delimiting species based on clustering will necessarily be somewhat arbitrary in some particular cases.

8.1.3 Species concepts that rely on biological properties

As evolutionary biologists, we are interested in how clustering into species comes about as result of evolution. We would hypothesize that there are biological processes at play that tend to keep members of a species similar to one another and processes that prevent different species from blending into each another. If not, biodiversity would be much less structured than what we observe. A second class of species concepts has incorporated hypotheses or statements for *why* biodiversity would become structured into species into its definition. The most popular and widespread of these is the **biological species concept** which defines species as *groups of actually or potentially interbreeding natural populations that are reproductively isolated from other such groups* (Mayr 1942). This is arguably the most famous and well-known species definition of all. The two hypotheses that are incorporated into it are 1) that it is interbreeding within a species that maintains cohesion and 2) that it is reproductive isolation that keeps species apart. When introducing his genetic cluster species definition, Mallet (1995) criticized the biological species concept and argued that it is philosophically problematic to incorporate hypotheses that seek to explain a pattern or phenomenon into the definition of that very same pattern or phenomenon. It is circular in the sense that cause and effect are confused, even if it should turn out that the hypotheses are well supported. Do species exist because of reproductive isolation or does reproductive isolation occur because there are species? We could reasonably argue that defining species is a philosophical question and not a quantitative scientific one. We think Mallet has a valid point so we will examine the support of each of these hypotheses below.

Ernst Mayr, who is usually credited as the founder of the biological species concept (although many of his contemporaries, such as Theodosius Dobzhansky, held similar views on what constitutes a species) had a strong belief in the reality of species as natural units in evolution. He thought interbreeding was a strong force of cohesion within species, even to the extent that it would retard adaptive evolutionary change. Mayr had little or no actual knowledge about how genomes are structured and work, as of course no one had at that time (though we cannot resist mentioning that he contemptuously referred to his contemporary evolutionary geneticists as "bean-bag geneticists" as they mainly played with one and two-locus models of evolution). However, he still argued forcefully that the genomes of species are strongly co-adapted networks of interacting genes, maintained by interbreeding, that could only be broken up under very special circumstances, such as during a founder event (see chapter 5). In other words, Mayr's concept framed gene flow between species as deleterious. He thought reproductive isolation would only build up between geographically isolated populations or otherwise interbreeding would restore cohesion (although he came to moderate this view late in life).[2] Divergence by genetic drift and natural selection in **allopatry** (geographic isolation) could eventually result in differentiated species if the populations remain geographically isolated for a sufficiently long period of time. An indirect effect of evolutionary divergence in allopatry is that the resulting populations no longer would be able to reproduce and bear fertile and viable offspring on secondary contact; they would have become different species. To Mayr reproductive isolation was a key event in organismal evolution, and without such isolation little diversification could occur.

Few if any contemporary biologists share Mayr's original, restricted view on how speciation and evolution occur. We now see countless examples of rapid adaptive evolution taking place as well as divergence between populations developing. Furthermore, these phenomena can occur in the face of gene flow. Geographic isolation is *not* a prerequisite for diversification, although it may play a role in many cases. Gene flow certainly can and does have a homogenizing effect, though, as we have seen in earlier chapters, it can be overridden by sufficiently strong selection. Often populations remain unchanged over long periods of time, but this need not only be due to interbreeding. When ecological conditions are relatively stable, stabilizing selection dominates and little evolutionary change will occur either within or between populations. Thus, current evidence suggests that the first hypothesis incorporated in the biological species concept is insufficient and unsatisfactory. Interbreeding and gene flow do have a homogenizing effect and tend to retard or slow down diversification within a species. However, other factors, including stabilizing selec-

tion, contribute to species cohesion, and, perhaps most importantly, divergence can take place in the face of gene flow provided selection is not too weak to be swamped by it.

What about the second hypothesis, that it is reproductive isolation that keeps species from blending? Well first, by defining species as reproductively isolated entities this hypothesis is untestable because by definition, populations that are not reproductively isolated are not species! Let us therefore relax the criterion. Can we imagine situations where populations remain genetically and phenotypically distinct, to the extent that they meet, say, the criteria of the genetic cluster species definition, if they are not reproductively isolated? The answer is obviously yes. Our field guides for plants and animals would be quite thin if they only included species that are completely reproductively isolated. At least 30 percent of all recognized plant species and 10 percent of animal species have been reported to currently hybridize with one or more species, at least occasionally (Abbot et al. 2013), and these are very likely underestimates. It is also important to move away from the concept of a hybrid as a "dead-end," much like a sterile mule. In many cases these hybrids are fertile and can backcross with one or both of the parent species. When this occurs, it results in gene flow between the species, a process also known as **introgression**.

You may object to this reasoning. Perhaps complete reproductive isolation is a bit too strict a criterion to delimit species. However, could the presence of sufficiently strong partial reproductive isolation still be used to characterize species? Well, yes and no. We argue that by forcing our thinking to fit the biological species concept we come to lump very different mechanisms into a strange, mixed bag which biologists typically refer to as *reproductive barriers*. This can be misleading because the traits and features that often keep species distinct may have little directly to do with reproduction. For instance, many species remain distinct mainly because they are adapted to different ecological niches. In an adaptive landscape they reside on different adaptive peaks (see chapter 4). Such species may sometimes hybridize, even quite extensively, but because hybrids and other individuals of mixed ancestry are phenotypically intermediate they find themselves

---

[2] One of the authors of this book (Sætre), attended a discussion session with Ernst Mayr in Uppsala, Sweden in 2001. There, at the age of 97, he explained that he thought he had been too categorical in his thinking about speciation in earlier days. After reading papers presenting empirical evidence for non-allopatric speciation in cichlid fishes in the Great Rift Lakes in East Africa, he had come to change his mind.

in a fitness valley in the adaptive landscape and are selected against. In other words, hybrids are less fit than either parent in their respective niches. Divergent selection between niches also acts against migrants, as each species is better adapted to its own habitat than that of the other. Darwin's finches provide a good example of such ecological species. This is a group of about 15 bird species that have adapted to different diets and therefore show a remarkable diversity in beak form and function. Many of the species hybridize quite extensively, but they remain distinct from each other in a large part due to the effect of ecological selection (Grant & Grant 2011). In a context outside the biological species concept we would call the ecological selection they experience disruptive and divergent—the different finches are ecologically incompatible. In a seminal and highly influential book on speciation adopting the biological species concept, Coyne & Orr (2004) referred to the same phenomena as reproductive barriers called "ecological prezygotic isolation" when individuals of either species survive poorly in each other's habitat (divergent selection) and "extrinsic postzygotic isolation" when hybrids are ecologically selected against (disruptive selection). Here, pre- and postzygotic refer to selection occurring before and after reproduction and fertilization has occurred. However, apart from this distinction, the mechanism, selection, remains identical. We argue this is unnecessary—why have different definitions for the same phenomenon?

While acknowledging the many contributions of Ernst Mayr to evolutionary biology, we suggest he got his thinking about species, speciation, and evolution slightly wrong in some important respects. Interbreeding is not quite the strong obstacle against divergence he originally thought, neither are species' genomes quite as cohesive, nor is full reproductive isolation the key event in organismal evolution that he believed (Sætre 2013). Therefore, we argue that in terms of studying speciation, it is not necessary to use the biological species concept to define species. For most practical purposes, typological species definitions such as the genetic cluster species definition are much more useful. For instance, the genetic cluster species definition can be used to delimit asexually reproducing species, such as dandelions (*Taraxacum* spp.) and bacteria,

which certainly may show phenotypic and genetic clustering but which are excluded from the biological species concept. Moreover, investigating genetic and phenotypic clustering is much more feasible than testing interfertility of populations that are separated by large geographic distances.

Many modern speciation biologists prefer to think of speciation in sexually reproducing organisms as a continuum—i.e. an accumulation of reproductive isolation over time that starts at relatively low levels of assortative mating or segregation among habitats before building towards the evolution of divergent, separate entities (Hendry et al. 2009; Seehausen et al. 2014). Where we draw the line on this continuum between true and incipient species is basically arbitrary. We can assume a priori that species are real entities in evolution as Mayr did, but this forces us to place our line at the end of the continuum and overlooks the cumulative process that led us to the point of speciation.

On the other hand, more than 75 years of empirical and theoretical research after the biological species concept was founded has taught us that barriers to gene flow, such as assortative mating and hybrid sterility, are important for understanding diversification and speciation in sexually reproducing organisms. Mayr was by no means off target, he was just too categorical in his thinking. Coyne & Orr (2004) described speciation as a cumulative effect of multiple biological factors that restrict gene flow between species and arguably drove our thinking towards a spectrum of isolation at different stages of divergence. Speciation therefore need not be an endpoint or goal. As mentioned above, we can think of speciation as a continuum from undifferentiated populations via all stages of evolutionary divergence to irreversibly reproductively isolated entities. There are different stages of progression towards reproductive isolation and studying these is arguably much more important for understanding evolution than defining the threshold where diverging populations have become "proper species." As specialists in speciation, this is the view that we both agree most strongly with. Hence speciation research, the study of evolutionary forces of cohesion and divergence that eventually enables coexistence of phenotypically and/or genetically clustered (groups of) populations to persist even in close geographic proximity, for the most part investigates the evolution of traits that restrict gene exchange between populations (e.g. Seehausen et al. 2014). In that sense the biological species concept can be said to have had great heuristic value, as it has concentrated our thinking about traits and features of species that generate reproductive isolation. To sum up, the biological species concept hypothesizes that interbreeding maintains cohesion and that reproductive isolation allows divergence. Current evidence suggests that these are important factors, but the concept neglects the importance of selection in maintaining cohesion and in causing and not just allowing evolutionary divergence.

8.1.4 Species concepts that seek to incorporate the historical dimension of speciation

Species are not just the varieties we observe today, they have an evolutionary history. When did species A split from species B and how is it related to species C? Taxonomically, we may want to categorize and name extinct species we find in the fossil record and make qualified hypotheses about their relationship with each other and with extant species. This is largely the focus of phylogenetics, something we will learn about in detail in chapter 9. The biological species concept is pretty much useless for such purposes because it defines species with reference to the time window in which individuals could actually or potentially interbreed. As for the genetic cluster species concept it can be used to delimit fossil species based on phenotypic clustering, but it would say nothing about the evolutionary relationship between them. The **phylogenetic species concept** is an example of a species definition that incorporates the historical dimension of speciation. However, several different versions of this concept actually exist. Broadly, they define species as *monophyletic groups of organisms of common ancestry distinguishable from other such groups of organisms* (e.g. Cracraft 1989, de Queiroz & Donoghue 1990, Baum & Shaw 1995). The key element of this concept is that members of a species share a common evolutionary history separate from other such groups. According to this species concept, in a phylogenetic tree, branching events correspond to speciation events and the tips of the

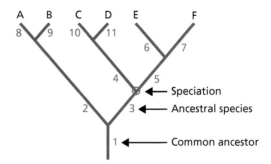

Figure 8.1 The phylogenetic species concept defines species as monophyletic groups of organisms of common ancestry distinguishable from other such groups of organisms. The phylogenetic tree shows the evolutionary relationship among six extant phylogenetic species (A–F) derived from a common ancestor (species 1). Speciation is defined as bifurcation events, as highlighted where ancestral species 3 gives rise to derived species 4 and 5. Using the phylogenetic species concept we can count a total of 11 species during the evolutionary history of this clade, of which 6 are extant.

branches represent extant species (Figure 8.1). We have good statistical procedures for reconstructing phylogenetic trees from morphological or DNA sequence data, and we will learn more about those methods, alongside the origins of tree-based thinking, in the next chapter. Phylogenies are invaluable for taxonomists in determining the evolutionary relationship among organisms. However, for delimiting species it introduces several problems.

First, when using a strict phylogenetic species concept, we typically end up with many more "species" than those we normally would list in a field guide, because bifurcation on the tree occurs at all levels of organization. As an extreme example, a small and isolated population that has become fixed for a single unique mutation by genetic drift would be monophyletic and thus constitute a distinct taxonomic species under this definition. Hence, although phylogenetic trees are useful tools both in taxonomy and evolutionary studies, relatively few biologists rely entirely on evolutionary branching when delimiting species.

Second, not all organisms fit the phylogenetic species concept because they are not derived from bifurcation events. Many plants and some animal taxa are instead derived from hybridization between divergent parent species. Eurasian *Passer* sparrows may serve as an example. In most of Southern Europe and the Near East the breeding

GENETICS AND GENOMICS OF SPECIATION

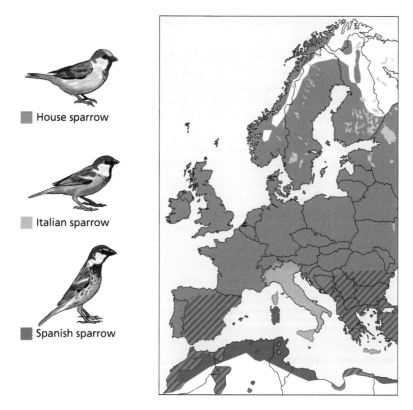

House sparrow

Italian sparrow

Spanish sparrow

Figure 8.2 Distribution of three species of sparrow, the house sparrow (*Passer domesticus*), Italian sparrow (*P. italiae*), and Spanish sparrow (*P. hispaniolensis*). The phenotypically intermediate Italian sparrow has originated through hybridization between the two other species.

range of two distinct sparrows overlap, the house sparrow (*P. domesticus*) and the Spanish sparrow (*P. hispaniolensis*) (Figure 8.2). Males of the two species have different plumage traits. For instance, the head of the male Spanish sparrow is chestnut brown whereas the male house sparrow has a grey crown and nape. Further, the male house sparrow has a smaller black bib and lacks the black streaking on the body flanks characteristic of the Spanish sparrow. In Italy and a few Mediterranean islands, a third sparrow resides, the Italian sparrow (*P. italiae*). Plumage traits of males of this sparrow are a mosaic of the traits found in the two others. Its head resembles the Spanish sparrow, but its back and breast are more similar to the house sparrow—in other words it is phenotypically intermediate (Figure 8.2). Genomic analyses show that the Italian sparrow is of hybrid origin as it is also genetically intermediate between the house and Spanish sparrows (Elgvin et al. 2017).

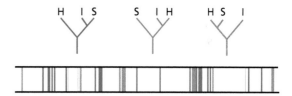

Figure 8.3 Conflicting phylogenetic signals along a stretch of chromosome 1A in three species of sparrows. Phylogenies between house (H), Italian (I), and Spanish sparrows (S) was reconstructed in 100kb non-overlapping windows along the chromosome. Each bar shows if Italian sparrows grouped monophyletically with Spanish sparrows (red), house sparrows (blue), in its own clade (green), or if the phylogeny was unresolved (white) within that window. Constructed from data published by Elgvin et al. (2017).

If we try to reconstruct the phylogeny of these three species from different genomic regions we get conflicting results (Figure 8.3). In certain blocks along a chromosome the Italian sparrow groups phylogenetically with the house sparrow whereas the Spanish sparrow is clearly divergent from the other two. In other blocks the Italian sparrow instead groups with the Spanish sparrow and the house sparrow is the divergent one. There are also some very few and small blocks where the Italian sparrow forms a distinct third group, different from both the house and Spanish sparrow. Finally, for the majority of chromosome blocks the phylogeny cannot be resolved. This is an expected result for a taxon of hybrid origin. An F1 hybrid would be heterozygous at all loci where the parent species are genetically different. However, the genome of a population that originated from hybridization would change over time. In some genomic regions the Italian sparrow has become fixed for house sparrow alleles through selection and/or genetic drift. In other regions it has become fixed for Spanish sparrow alleles. There are also a few regions where the Italian sparrow has diverged from both its parent taxa, probably due to both new mutations and selection that has occurred after the hybridization event. Finally, in some regions alleles from both parent taxa are still segregating in the Italian sparrow making it impossible to resolve the phylogeny. The three taxa do not confirm to the phylogenetic species concept because their evolutionary history differs from the bifurcation model the concept is based on. The trio form a monophyletic group against more distantly related species, however, such as the tree sparrow (*P. montanus*), so we would have to lump them together as one species to fit the phylogenetic species concept, even if they are clearly different and genetically and phenotypically distinct.

Hybridization can make phylogenetic inference difficult even when it does not result in a new species. Introgression of genes between species can generate conflicting phylogenetic signals. The quality of phylogenetic inference can also vary with the number of markers and genome regions used. Inference based on only one locus is often very unreliable as conflicting phylogenies can arise simply due to stochastic processes such as genetic drift. For example, large effective population sizes and short times between species divergence events can mean that genes can coalesce in the common ancestor of a trio of species. This results in **incomplete lineage sorting** where the topology of a tree drawn based on a single gene is different to the true rela-

tionship among species. For example, in a whole genome comparison among humans, chimpanzees, and gorillas, 30 percent of SNP trees showed a monophyletic relationship between either humans and gorillas or gorillas and chimps (Scally et al. 2012). This is despite the fact that the true tree shows a more recent common ancestor between chimpanzees and humans. Furthermore, as we will see in sections 9.1.8 and 9.2.7, conflicting phylogenies across the genome are actually quite effective means of detecting the presence of gene flow. Nonetheless, for a means of defining species the bifurcation model of speciation is too simplistic. This is very much related to our discussion of the biological species concept above. Evolution does not obey the rules of simple models such as staying the same until reproductively isolated and then diverging. Biodiversity is messy and chaotic and species can arise from a number of different processes that are hard to unify under a single definition.

How does the Italian sparrow fit the genetic cluster species definition? Recall that to critically test this, we need to see how a species interacts with others in close proximity. The Italian sparrow lives in sympatry with the Spanish sparrow in South-East Italy where the two species form distinct genetic and phenotypic clusters (Figure 8.4). Sætre et al. (2017) used assignment tests based on a genome-wide set of SNP markers on 327 sympatric birds that had been identified as either Spanish or Italian sparrow based on their phenotypes. All of them were assigned correctly to the Spanish or Italian sparrow cluster with a high probability and none of the birds were F1 hybrids—suggesting no obvious introgression between the two species (Sætre et al. 2017). The Italian sparrow does not live sympatrically with its other parent species, the house sparrow, but they do meet in a narrow contact zone in the Alps, very similar to the contact zone between the carrion and hooded crow discussed earlier and in chapter 7. Genetically and phenotypically there is some intergradation across the narrow contact zone due to hybridization, but also evidence for selection against introgressed house sparrow alleles into the Italian sparrow across the hybrid zone (Trier et al. 2014) and against males with intermediate plumage traits (Bailey et al. 2015). Barriers to gene exchange have evidently developed

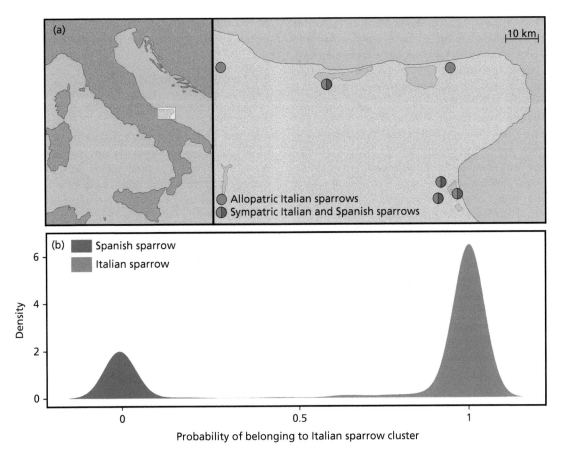

Figure 8.4 Genetic clustering of Italian and Spanish sparrows in an area of sympatry in Southeast Italy. Genotypes of 327 sympatric Italian and Spanish sparrows were assigned as probability of belonging to the Italian sparrow cluster, using a genome-wide set of SNP-markers from protein coding genes. Birds were classified phenotypically as Italian sparrows (blue) or Spanish sparrows (red). Modified from Sætre et al. (2017).

against both parent species but they appear to be stronger against one of the parents.

Based on this lengthy discussion, how do we conclude? Which species concept do we recommend? As a tool for delimiting species we think the genetic cluster species definition is preferable. However, besides genetic and phenotypic clustering, phylogenetic evidence is an important additional line of information which is particularly important when we consider the evolutionary history of a clade. Finally, the scientific impact the biological species concept has had on speciation research should not be neglected as it has drawn our attention to the importance of the evolution of barriers to gene exchange in speciation of sexually reproducing organisms.

8.2 Evolution of barriers to gene exchange

8.2.1 Classifying barriers to gene exchange

During its history, speciation research has often dwelt on debating which of the different species concepts is most appropriate. However, as we outlined in section 8.1.3, most modern evolutionary biologists agree that speciation can be thought of as a continuum, a cumulative process with progression towards increasing divergence and reproductive isolation between lineages or taxa. Yet, where we draw the line between what is and what is not a true species is somewhat arbitrary. Instead, it is more productive to study different points along the speciation continuum and focus on the traits that contribute towards progression along it. A **barrier to gene exchange** is therefore a phenotypic trait that reduces or hinder gene exchange between members of different populations or species. Under this definition, we exclude physical barriers to gene flow. A mountain range or river can prevent gene exchange between populations but it cannot evolve and contribute to the process of divergence in the same way a phenotypic trait might do. Like species definitions, barriers to gene exchange can also be classified in a number of different ways. There is a long tradition in the speciation literature for classifying them according to when during the reproductive cycle they retard gene exchange (e.g. Dobzhansky 1940; Coyne & Orr 2004). Thus, **prezygotic barriers** tend to prevent mating or fertilization from taking place between members of different populations or species, whereas **postzygotic barriers** reduce gene exchange between them due to low fertility and/or viability of hybrids—i.e. after fertilization has taken place. The pre- and postzygotic distinction is often useful. For example, postzygotic developmental incompatibilities can only be exposed to selection after mating has occurred. However, the distinction can be somewhat arbitrary in many cases in the sense that the same mechanism can have both pre- and postzygotic effects. For instance, different species may mate assortatively based on a mating preference for a phenotypic trait that differs between the species. Such preferences will often reduce both the frequency of heterospecific mating (prezygotic barrier) and reduce mating success of intermediate hybrids which are not preferred by either of the parent species (postzygotic barrier). Therefore, we have chosen to classify barriers to gene exchange mainly based on the biological mechanism involved

as this informs us about how different traits and features contributes to speciation (Table 8.1).

8.2.2 Ecological incompatibilities

Gene flow between ecologically differentiated populations can be retarded by low survival of migrants (prezygotic barrier) and/or progeny of mixed ancestry (postzygotic barrier). The rough periwnkle *Littorina saxatilis* nicely illustrates such ecological incompatibilities. These marine snails occupy contrasting habitats exposed to either wave action or crab predation in many places across Europe (Johannesson et al. 2017). Parallel adaptation to the contrasting environments on both local and regional scale has resulted in phenotypically and genetically different populations living in close proximity to each other in alternatingly exposed and less exposed coastal lines. Small "wave" ecotype snails inhabit exposed rocks and experience strong wave action, while thick-shelled, "crab" ecotype snails are larger and experience crab predation on less exposed shores. Gene flow between ecotypes is retarded in large parts because settling snails survive poorly in each other's habitat. Genomic analyses of crab and wave ecotypes show that although overall differentiation between ecotypes is low, some outlier loci are apparent between them (Westram et al. 2014; Ravinet et al. 2016) and that these are shared at different spatial scales too (Westram et al. 2016). Consequently, mating between ecotypes is reduced. Moreover, transplant experiments show that hybrids between *Littorina* ecotypes survive poorly in either habitat and they do not survive better than the parent types even in ecological transition zones where we could assume that the ecological conditions would be intermediate (Rolán-Alvarez et al. 1997). Despite this fact, genomic and phenotypic analyses do suggest that an amount of hybridization occurs in contact zones between the ecotypes (Westram et al. 2018).

In some cases where populations with contrasting ecological adaptations meet there can be little or no negative selection on the parental types but their resulting hybrids may be poorly adapted to either of the parent's niches and an intermediate or hybrid niche may simply not be present at all (postzygotic barrier). As an example, many song birds in the

Table 8.1 Classification of reproductive barriers by biological mechanisms and how they relate to the distinction between pre- and postzygotic barriers.

Mechanism	Pre- or postzygotic
Ecological incompatibility	Prezygotic and/or postzygotic
Assortative mating	Prezygotic and postzygotic
Sexual incompatibility	Prezygotic
Developmental incompatibility	Postzygotic

Northern Hemisphere are migratory, flying south during autumn and returning to their breeding grounds in spring. The route they take between the breeding and wintering grounds can be crucial for their survival. Small animals with high metabolism need to feed during migrations, so they tend to avoid large bodies of open water or unproductive areas such as mountains and deserts. Helbig (1991a, 1991b) investigated migrating behavior and its inheritance in the European blackcap (*Sylvia atricapilla*), a small song bird. In Central Europe there is a migratory divide in this species. No distributional gap exists between populations differing in migratory direction, but they meet in a sharp transition zone in Germany. The eastern population first takes a southeastern (SE) route during autumn migration; recoveries of ringed birds show that many cross the Bosphorus Strait and across Turkey towards the eastern shore of the Mediterranean Sea. Here they shift to a southern direction along the Mediterranean east shore towards their wintering grounds in East Africa (Figure 8.5). In contrast, the western population flies in a southwest (SW) direction to their wintering grounds in the western Mediterranean region without any change in direction. Why is there such a sharp boundary between the SE and SW-migrating population? Helbig showed that the migration route is genetically inherited. By hand-raising nestlings and keeping them in captivity during the period they normally would migrate, he observed that birds became restless and oriented themselves in the direction corresponding to their migration route during this period. Eastern birds first oriented themselves in a SE direction, but after some weeks they changed their preferred direction to the south. In contrast, western birds oriented themselves in a SW direction without any change in preferred direction. He then crossed individuals from the two populations. These hybrids showed restless behavior in an intermediate southern direction. Extrapolating the route these mixed individuals would take, Helbig speculated they would likely have to cross the Alps, the Mediterranean Sea, and parts of the Sahara, a migration route which likely would have been hard to survive. The sharp migratory divide between the two populations may therefore be maintained by an ecological incompatibility between the two populations, linked to their different migratory behavior.

In blackcaps, the low fitness of the intermediate migration route of hybrids has only been inferred via extrapolating from the behavior of captive hybrid nestlings. However, similar sharp migratory divides around major geographical barriers are found in many bird species in the Northern Hemisphere. The Swainson's thrush (*Catharus ustulatus*) in Western Canada is split into populations that respectively migrate east and west of the Rocky Mountains to their wintering grounds. Using light-weight geolocators mounted on the birds, Delmore & Irwin (2014) demonstrated that hybrids between these populations took intermediate migration routes and suffered increased mortality as a result. Therefore, strong selection is likely to act upon these hybrids.

8.2.3 Assortative mating

Assortative mating occurs when like mate with like. "Like" in this sense typically means individuals that are similar with regard to a phenotypic trait such as coloration or body size. In the context of speciation, assortative mating refers to preferential mating among conspecifics (i.e. of the same species) and reduced mating between different species or heterospecifics—in this case it is a prezygotic barrier. When assortative mating results in reduced pairing between parental species and their hybrids or backcrosses, it may also act as a postzygotic barrier.

Assortative mating can arise due to a number of different causes. Often it is directly related to mate choice and mating preferences, such as when a female prefers a male of particular color. Many animal species have elaborate courtship displays and very often members of one sex (usually males) have evolved secondary sexual traits, including pheromones, vocal signals such as bird song, and/or prominent colors and ornaments such as the peacock's tail to attract members of the opposite sex. Divergence in signaling and mate recognition traits directly lead to assortative mating. This constitutes an important barrier to gene exchange either between heterospecifics and/or between hybrids and their parental species.

Behavioral barriers to heterospecific mating have been the subject of much research. Let us take bird song as an example, where males sing to attract females. Bird song often has a strong learning

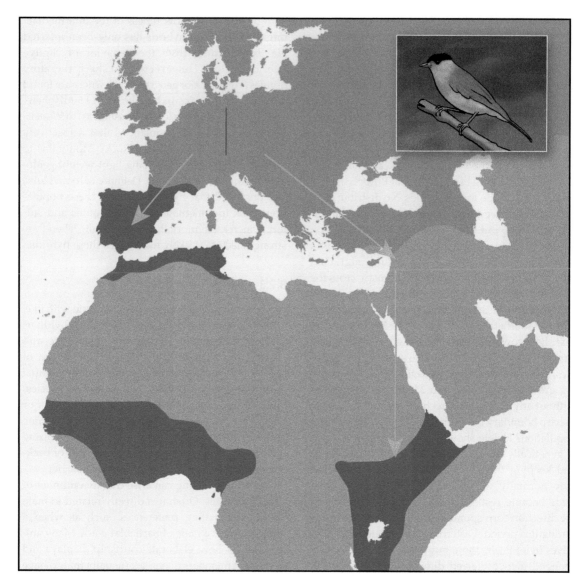

Figure 8.5 A migratory divide in blackcaps (*Sylvia atricapilla*) in Central Europe (red line). Breeding birds (green area) west of the divide migrate in a SW direction in autumn to their wintering grounds (blue area) in Iberia and NW Africa, whereas birds east of the divide migrate SE to the eastern shore of the Mediterranean Sea and then south to their wintering grounds in East Africa. The migration routes are genetically determined and likely constitute a barrier to gene exchange as the intermediate migration route of hybrids would be suboptimal.

component. Reared in acoustic isolation in captivity many songbirds fail to develop their species-specific song. In nature, young birds would be exposed to acoustic stimuli from many different species and not just conspecifics, so how do they learn their species-specific song? In a pioneering study on the swamp sparrow (*Melospiza georgiana*), Marler and

Peters (1977) demonstrated that juvenile male birds, raised in acoustic isolation and tutored only with artificial recordings, tuned into their conspecific song during song learning. The researchers mixed recordings of song from swamp sparrows with recordings from a different but closely related species, the sympatric song sparrow (*M. melodia*), by splicing syllables (i.e. song elements) from the two songs into an array of either swamp sparrow-like or song sparrow-like temporal patterns. Juvenile males chose only to learn songs that contained their own species' syllables and filtered out the foreign syllables in the learning process. It did not matter whether the temporal pattern was swamp or song sparrow-like—in both cases, the males preferentially learned their conspecific song elements. It appears that the birds have some heritable propensities to focus learning. In a later study, the same research group demonstrated that females of the two sparrow species prefer conspecific songs using a similar experimental set up (Searcy et al. 1981). Again, the test stimuli consisted of synthetic songs consisting of arrays of syllables composed from the two species' songs, ordered into a temporal pattern resembling either the swamp sparrow or the song sparrow song. Females of both species responded preferentially (measured as the frequency and intensity of copulation-soliciting displays) to songs containing their own species' syllables as well as their own species' temporal patterns. These studies provide evidence of a heritable display trait (song) and also a matching heritable preference in females.

A parallel to such behavioral barriers can be found in some flowering plants that are pollinated by animals. Plant populations or species can mate assortatively if their flowers attract pollinators of different species due to differences in color, scent, and/or shape of their flowers. For example, the monkey flower *Mimulus lewisii*[3] is pollinated by bees (Figure 8.6). This species has large, open, pink colored flowers. A closely related monkey flower, *M. cardinalis*, has narrow, tubular, red flowers that are very rich in nectar and is pollinated by hummingbirds (Schemske & Bradshaw 1999). By crossing the two species of monkey flowers and conducting a QTL-analysis (F2 intercross, see chapter 6), Schemske and Bradshaw investigated which floral traits were important for the different preferences of bees and hummingbirds as well as the genetic architecture of these traits. Observations from the

[3] The nomenclature of the *Mimulus* genus is currently under revision, as phylogenetic evidence shows that it includes several monophyletic clades. All *Mimulus* species mentioned in this chapter are therefore likely to be placed in the genus *Erythrante* in the near future.

Figure 8.6 Assortative mating between monkey flower species (*Mimulus* spp). Crosspollination between *M. lewisii* (top) and *M. cardinalis* (bottom) occurs very rarely because characteristics of their flowers attract bees and hummingbirds respectively. Top panel: photo courtesy of David Monniaux, bottom panel: photo courtesy of Jason Hollinger.

variable F2-generation showed that bees preferred large flowers, low in concentration of two different color pigments. One QTL-allele, associated with an increase in petal pigment concentration (from pink to red), alone reduced bee visitation rate by as much as 80 percent. Similarly, a single QTL-allele associated with an increase in nectar production doubled visitation rates by hummingbirds. The authors therefore concluded that allelic changes in a few major genes was largely responsible for causing assortative mating between the two monkey flower species due to their large effects on pollinator behavior. Schemske & Bradshaw's (1999) study also represents a good example of a modern speciation research program in which researchers identify traits acting as barriers to gene exchange and then also attempt to identify the genes or genomic regions that underlie them (Seehausen et al. 2014; Ravinet et al. 2017).

That assortative mating can act as both a pre- and postzygotic barrier is nicely illustrated in a study by Naisbit et al. (2001) on mating interactions between two butterfly species, *Heliconius cydno* and *H. melpomene* and their F1 hybrids. These two species coexist in many places in Central and South America. They occasionally interbreed and some crosses yield viable F1 hybrids, albeit with low fertility (developmental incompatibility, see section 8.2.5). The two parent species differ conspicuously in wing color patterns which are linked to predator defense through mimicry of other butterfly species. By mimicking species that are unpalatable to predators, both species have evolved a form of predator avoidance. *H. cydno* is black with white wing markings, whereas *H. melpomene* is black with yellow and red wing markings, as they both mimic different species. However, wing colors and patterns also strongly affect the probability that a male will court a female and is thus important in mate choice. The researchers investigated the likelihood that a male would court a female of the same species as himself, a female of the other species, or a female F1 hybrid produced from crossing the two species in the lab. Males of both species showed strong preference for females of their own species. They would usually reject courtship with a female of the other alternative species. Additionally, hybrid females with intermediate wing patterns were discriminated against by males of both species—the majority of males would not court a hybrid female, such that hybrid females are less fit than females from either of the parent species.

Traits causing assortative mating are often affected by sexual selection. Several comparative studies have reported that clades that experience more intense sexual selection are more species rich than less sexually selected clades, which may indirectly

suggest that sexual selection can be an important driver in speciation (e.g. Owens et al. 1999; Seehausen 2000; Panhuis et al. 2001). However, in a comprehensive review of both theoretical models and empirical studies, Servedio & Boughman (2017) point out that much of the empirical evidence for sexual selection promoting speciation is suggestive rather than conclusive. Moreover, there are situations where sexual selection would oppose rather than promote speciation. For instance, a sexually selected female preference for a male trait is more prone to move between populations via gene flow when it has little or no bearing on other components of fitness than mating success. When the sexual preference migrates into the other population, assortative mating is weakened and the two populations may eventually converge on the same preference and therefore also the same male trait. A related phenomenon is taking place between two subspecies of the red-backed fairywren (*Malurus melanocephalus*) that have come into secondary contact in northeast Australia (Baldassarre et al. 2014). Males of the northwestern subspecies have a black and red plumage, whereas males of the southeastern subspecies are black and orange. Genomic and phenotypic analysis shows that the red color is currently spreading into the range of the orange-colored subspecies. The transition zone between red and orange color is located much further east than the transition zone between other genes that characterize the two subspecies. To sum up, sexual selection appears to contribute to speciation in many cases but this is most likely in concert with natural selection. Mate preferences may need to be linked to some other form of adaptation in order to properly maintain a barrier between species (Servedio & Boughman 2017).

Assortative mating is not always linked to mate choice, however; it can also result more indirectly, for instance through differences in timing of breeding or from different habitat preferences of the populations. A nice example of the latter is provided by the host-race associations seen in the apple maggot fly *Rhagoletis pomonella* (Figure 8.7). Male and female flies of this phytophagous species typically meet and copulate on a fruit. Following copulation, the female deposits her eggs into the fruit body and upon hatching, her larvae feed, causing

Figure 8.7 Adult apple maggot fly (*Rhagoletis pomonella*) The apple-feeding and hawthorn-feeding race of the apple maggot fly mate assortatively because they have a strong preference for aggregating and mating on the same fruit species as the one they lived in as larvae. Photo courtesy of Joseph Berger.

the fruit to decay and drop to the ground before ripening. The insect offspring then eventually leave the fruit and overwinter in the soil as pupae. This *Rhagoletis* fly species used to be mainly associated with hawthorn, a small tree species with an apple-like fruit; indeed, many populations of flies still prefer hawthorn as their primary host. However, an apple feeding race emerged from the hawthorn feeding lineage in the early 1800s after apples were introduced to North America (Bush 1966). Interestingly, flies of both sexes prefer to assemble and mate on the same fruit species as the one they emerged from. The preference for their natal fruit appears to be driven by olfactory cues (Linn et al. 2003); in flight tunnel assays and field tests adult flies preferentially oriented towards chemical blends of their natal fruit. Therefore, it seems likely that in this system assortative mating has arisen indirectly as a result of habitat preference—i.e. flies with the same habitat preference are more likely to meet and reproduce.

It is important to be aware that mate choice, assortative mating, and sexual selection are distinct, albeit sometimes related, phenomena. Mate choice is a behavioral response in which members of one sex choose a mate. Mate choice may or may not involve a preference for a particular trait in the opposite sex. When a female prefers a given trait this *can* lead to assortative mating (like mate with like) and it can generate sexual selection (differential

mating success of males based on variation in that trait). Thus, divergent sexual selection in two populations *can* lead to or contribute to assortative mating. However, assortative mating may also have other causes. For instance, two populations may have diverged phenotypically by other processes than sexual selection. In the *Heliconius* butterflies studied by Naisbit et al. (2001) that we learned about earlier, natural selection for mimicry has been the main driving force for divergence in color pattern. Assortative mating can even occur without any sexual mating preference; the apple- and haw-thorn-feeding races of the apple maggot fly mate assortatively because they preferentially aggregate on different fruit species during mating.

8.2.4 Sexual incompatibility

Gene exchange between divergent populations can be hindered or retarded if they have diverged in reproductive traits in ways which impede successful copulation or fertilization (i.e. a prezygotic barrier). The potential importance of sexual incompatibility has been much discussed in insect speciation where closely related species may differ greatly in sexual organs and related structures that may impede successful copulation between species. For instance, copulation in damselflies (order Odonata, suborder Zygoptera) involves two separate points of contact between the male and female (e.g. Robertson & Paterson 1982). When a male and female damselfly pair up the male grasps the female's thorax with his legs and then clasps her prothorax ("neck") with his anal appendages, yielding the tandem position (Figure 8.8a). The second stage of mating occurs if the female accepts the mating invitation and cooperates. She then bends her abdomen to make contact with his accessory genitalia on the second abdominal segment and so achieves the copula position (Figure 8.8b). Sexual incompatibility between damselfly species may occur at either stage if attachment between the sexes fails or is ineffective (due to divergent morphology), thus disabling successful copulation and transfer of sperm. Robertson and Paterson showed that male anal appendages and corresponding grooves and structures on the female's prothorax differ among closely related species of damselflies and that these

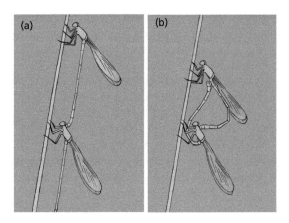

Figure 8.8 Copulation in damselflies involves two points of attachment between the sexes. A) First the male attaches to the prothorax of the female using his anal appendages to form the tandem position. B) If the mating invitation is accepted the female bends her abdomen to make contact with his accessory genitalia on the second abdominal segment and so achieves the copula position. The anatomy of the structures used to make attachment between the sexes differs between closely related species of damselflies and is thought to make successful copulation between species difficult.

structures likely affect the likelihood of successful copulation.

In a number of animal taxa, males are able to mate with and fertilize a female from a different species. However, if the same female then copulates with a conspecific male the majority (if not all) her eggs will be fertilized by the latter male. This phenomenon is referred to as **conspecific sperm precedence**. In chapter 4 we cited a study that found evidence for conspecific sperm precedence among Atlantic salmon and trout (Yeates et al. 2013). Among those two species the fertilization advantage of conspecific sperm appears to be due to proteins in the female ovarian fluid surrounding the eggs that affect sperm from the two species differentially. More generally, chemical communication between cell surface proteins on sperm and egg are important in determining whether or not a sperm cell can bind to and penetrate an egg cell. Therefore, divergence in such proteins can result in sexual incompatibility at the gametic level. Also, in many plant species conspecific pollen takes precedence over heterospecific pollen when females receive pollen from multiple males (e.g. Howard 1999). For instance, in a study on a mixed population of the monkey flower species *Mimulus guttatus* and

M. nasutus, Fishman et al. (2008) found 100 percent precedence of *M. guttatus* pollen over *M. nasutus* pollen in mixed pollination of *M. guttatus*. The researchers identified several loci involved in transmission ratio distortion and suggested that species-specific differences in pollen tube performance likely accumulate gradually through coevolution between pollen and style.

A special kind of selection is thought to often be involved when sexual incompatibility develops between diverging populations, namely selection derived from **sexual conflict**. Males and females will often have conflicting interests over mating, albeit in an evolutionary and not necessarily conscious sense (Trivers 1972, Arnqvist & Rowe 2013; Gavrilets 2014). Males produce many small sperm cells that are relatively cheap to make, whereas females invest a considerable amount of resources into producing a small number of large eggs. This crucial difference between the sexes will often have consequences for how the two sexes can maximize their reproductive success. Males can maximize their reproductive success by mating with many females. Sperm is not a limiting resource, and each of the millions of sperm cells a male produce could in theory fertilize an egg. In contrast, eggs are often a limiting resource for females. They have already invested a considerable amount of resources into each egg, so reproductive success is maximized to a larger extent by being careful that the egg is fertilized by a male of sufficiently high quality to secure survival of the resulting young. There is a great deal of phenotypic evidence to support the existence of sexual conflict; however, we are only just beginning to understand how it can shape diversity at the genome level (Rowe et al. 2018).

Sexual conflict can occur through two genetic routes—within a locus and between loci. First, within-locus conflict occurs when a locus controls a trait expressed in both sexes but where the optimal trait value differs between the sexes. Therefore, optimizing the trait value in one sex leads to a fitness reduction in the other sex. A good example of this kind of sexual conflict is seen in Trinidadian guppies (*Poecilia reticulata*). A striking color pattern on the fins of male guppies helps them to attract mates and so this trait is under strong sexual selection. This comes at a fitness cost—striking, colorful males are also much more obvious to predators; however, the benefits to reproductive fitness outweigh any increased risk of predation. For females, however, there is no such tradeoff—conspicuous coloration is only detrimental and so it is in the best interest of females not to express this trait. How can such a conflict be resolved? One way is that the gene or genes underlying such are traits are expressed differently in the two sexes—i.e. the sexual conflict is resolved via the evolution of sex-specific expression. Wright et al. (2017) examined patterns of gene expression between male and female Trinidadian guppies and found an increased ratio of sex-biased expression between males and females on a non-recombining region of the guppy sex chromosome. Investigating this further, they also found that estimates of Tajima's D were significantly more positive for genes with male-biased expression (Wright et al. 2018). Within-locus sexual conflict essentially causes balancing selection for alleles depending on whether they occur in males or females, thus a higher Tajima's D for genes with male-biased expression suggests sexual conflict was resolved via this mechanism. Within-locus sexual conflict can also be resolved when genes under divergent selection between the sexes become linked to a sex-determining region. In short, this divergent selection will drive a reduction in recombination between sex-biased genes and the sex-determining region; this process is important for the evolution of sex chromosomes. In guppies, whether you are a male is determined by the presence of a sex-determining region on the Y chromosome (Wright et al. 2018). Thus, we might expect that genes with a higher fitness in males, such as those responsible for fin color patterning, will be mainly Y-linked. This is indeed the case and, furthermore, there is good evidence that extent of Y-linkage for color pattern genes varies among habitats with differing degrees of sexual selection. In downstream environments with higher predator abundance, an increased predation risk reduces the coloration of males whereas in the upstream, with fewer predators, males are more colorful. Sexual conflict is therefore higher in the upstream populations at both the phenotypic and genomic level.

Second, between-locus sexual conflict is a little more difficult to define. It occurs between at least two separate loci in each of the sexes. Crucially the conflict arises when a locus that confers a trait beneficial to one sex is detrimental to the other as a result of the interaction between the two sexes (Rowe et al. 2018). A classic example of this is a trait such as mating rate. Imagine that males evolve a high mating rate. This would obviously benefit them but would potentially have a directly deleterious effect on female fitness. A similar conflict has been demonstrated experimentally in *Drosophila melanogaster* where males who are very persistent in pursuing a mate can actually lower the fitness of more attractive females in a population (Long et al. 2009). An expected consequence of such antagonistic behavior is that evolution is likely to favor females that carry a locus resulting in increased resistance to mating attempts from males, which in turn would result in selection for males that are able to overcome the resistance. Consequently, sexual conflict over mating rate will drive rapid coevolution of male and female traits controlling reproduction—essentially a within-species evolutionary arms race. As a by-product of this coevolution, isolated populations will diverge in these traits and therefore potentially lose mutual sexual compatibility if they come into secondary contact. It is for this reason that between-locus sexual conflict is thought to be particularly relevant for the evolution of sexual incompatibilities among species (Rice 1996, Rice & Holland 1997; Gavrilets 2014). A wide range of reproductive traits can be affected by such arms races between the sexes. Anal appendages in male damselflies to lock females in the tandem position versus grooves and structures on the female prothorax to resist locking is one potential example. Likewise, sperm is under strong selection for increased efficiency in fertilizing eggs. However, if more than one sperm penetrates the egg the resulting zygote is usually inviable, so in females there is selection for fertilization barriers that mediate sperm entry. There is ample evidence that fertilization proteins on sperm and egg membranes evolve extremely rapidly in diverging populations, likely because of such coevolutionary arms races driven by between-locus sexual conflict (Palumbi 1998; Firman 2018). Divergence in these proteins likely accounts for conspecific sperm precedence and similar sexual incompatibilities among closely related species.

8.2.5 Developmental incompatibility

Finally, barriers to gene exchange can result from developmental incompatibilities when divergent genomes are combined in hybrids, causing low or non-existing fertility and in some cases complete inviability (postzygotic barrier). As we mentioned at the start of the chapter, a classic example is (the almost complete) sterility of hybrids between horses and donkeys. However, developmental incompatibilities can also affect later generation hybrids, as well as backcrosses between hybrids and either parent species. This is called **hybrid breakdown** and likely plays an important role in retarding gene exchange between many species. By crossing different pairs of African cichlid species in the lab, using species pairs spanning some thousand to several million years of divergence times, and then crossing the resulting F1 hybrids to produce F2 hybrids, Stelkens et al. (2015) demonstrated that F2 hybrids consistently showed the lowest viability compared to both F1 hybrids (overall 21 percent reduction in survival) and non-hybrid, conspecific crosses (43 percent reduction in survival).

You may wonder how developmental incompatibilities could arise if they are products of natural selection. Intuitively one would think that low hybrid fitness stems from heterozygosity in genes where the parent populations have diverged. However, if one of the species has the genotype A_1A_1 at a locus, and the other A_2A_2, at least one of them must have harbored heterozygous individuals during the course of evolution. Shouldn't natural selection remove such unfit heterozygous genotypes and thus make it impossible to evolve from one state to the other? Suppose the ancestral genotype was A_1A_1 and that A_2 was caused by a mutation in one of the populations. As long as A_2 is rare, most carriers of the allele would be heterozygous and, one would guess, have low fitness. Apparently, selection seems unable to explain how one of the species could end up with the A_2A_2 genotype? How then could hybrid incompatibilities evolve?

One of the best-known solutions to this Darwinian problem was for a long time mainly credited to evolutionary geneticists Theodosius Dobzhansky, who conceived the idea (Dobzhansky 1934) and Hermann J. Muller, who refined it (Muller 1940).

However, in the mid-1990s, it was discovered (Orr 1996) that the geneticist William Bateson also conceived the very same idea independently early in the twentieth century (Bateson 1909). However, Bateson's work was forgotten and Dobzhansky and Muller were unaware of it when they developed their own model. These three researchers acknowledged that genes are not expressed in a vacuum—they interact with other genes. The phenotypic effect of the A_2-allele above may depend on the genotype at other loci elsewhere in the genome, i.e. via epistasis (see chapter 6). When two populations diverge, changes are likely to occur at several loci across a genome. The genes within a population therefore become co-adapted. Any harmful mutation is removed by selection, whereas mutations with positive fitness effects on the genomic background of a population will increase in frequency by selection. However, an allele with an overall positive effect in one species' genomic environment need not be universally beneficial.

Let us start with the principle of Bateson, Dobzhansky, and Muller's hypothesis using a simple two-locus model (Figure 8.9). An ancestral population has the genotype A_1A_1 at one locus and B_1B_1 at another, resulting in a multilocus genotype of A_1B_1/A_1B_1. We assume now that some geographic barrier develops that separates the ancestral population into two daughter populations and prevents gene exchange between them. In one of the populations a beneficial mutation occurs at the first locus: A_1 mutates to A_2. The mutant allele will increase in frequency and become fixed to yield the genotype A_2B_1/A_2B_1. Likewise, in the other population, a beneficial mutation occurs at the other locus, B_1 mutates to B_2, and selection eventually fixates the genotype A_1B_2/A_1B_2. Now, what happens if the two populations experience secondary contact and hybridize? Well, crosses between the two parental populations would yield genotypes that have never before been tested by selection. F1 hybrids would have the genotype A_2B_1/A_1B_2. If A_2 and B_2 are incompatible as result of this, the hybrids would have low fitness, despite the fact that both alleles were individually favored by selection in their respective populations. Another way to think of the incompatibility is in terms of epistasis; here the alleles at the two loci show reciprocal sign epistasis (see chapter 6). We

Ancestral population fixed for A_1 and B_1

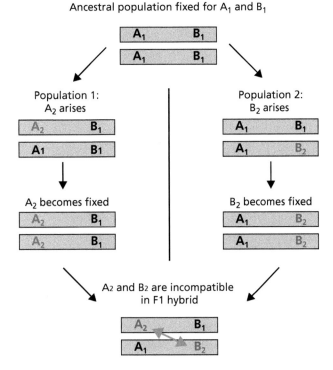

Figure 8.9 The Bateson–Dobzhansky–Muller model for the evolution of developmental incompatibility in hybrids. An ancestral population splits into two geographically isolated populations. Independent mutations (marked with red and blue color respectively) occur at different loci in the two populations and these eventually become fixed in their respective populations by selection. Upon secondary contact incompatible allele combinations that have never before been exposed to selection are brought together in hybrids (marked with green double arrow) yielding low hybrid fitness.

have reciprocal sign epistasis between alleles at two loci when there are two peaks in the adaptive landscape and a fitness valley separating the peaks. In chapter 6 we explained reciprocal sign epistasis using a hypothetical butterfly species that gains protection from predators from mimicking differently colored poisonous model species. Thus, consider a hypothetical environment where our focal species lives together with two poisonous model species, one red and one blue. Color in our focal species is determined by two loci, one controlling the color of the forewing and another locus the hindwing. If our focal butterfly is blue, a mutation causing red forewings would be detrimental on its own. The resulting butterfly would be a poor mimic to both model species. Likewise, a mutation causing red hindwings would be detrimental for the same reason. However, both mutations together would be beneficial—the resulting all red butterfly would

be a good mimic of the red model species. There are two fitness peaks in this adaptive landscape—an all blue and an all red butterfly—and a fitness valley in between consisting of butterflies with mixed colors that are poor mimics to both model species. The Bateson–Dobzhansky–Muller hypothesis represents an evolutionary scenario in which two populations can come to reside on different adaptive peaks in a multilocus adaptive landscape through independent mutations on different loci that becomes fixed in either population, despite the fact that there would be a fitness valley separating the peaks when these mutations are combined in hybrids.

The phenomenon, in which allelic variants at two (or more loci) are incompatible, is referred to as a **Bateson–Dobzhansky–Muller incompatibility (BDMI)**. However, because Bateson's contribution to the model was not rediscovered until 1996, the term Dobzhansky–Muller incompatibility **(DMI)** was already established in the literature. Many biologists still prefer the shorter and simpler DMI and so you are likely to encounter this when reading further.

There is well-documented evidence from model species such as *Drosophila* flies that low fitness in hybrids can be caused by alleles at loci showing reciprocal sign epistasis. As an example, Tang and Presgraves (2015) showed that two genes encoding protein components of the nuclear pore complex cause lethality in hybrid genotypes between *Drosophila simulans* and *D. melanogaster*. Likewise, extensive genome-wide mapping in a hybrid zone between two divergent lineages of the house mouse, *Mus musculus musculus* and *M. m. domesticus*, enabled Turner and Harr (2014) to identify a number of loci that contribute to hybrid sterility through reciprocal sign epistasis. Such incompatibilities are referred to as DMIs (or BDMIs). Some of these examples may certainly have evolved as envisaged in the Bateson–Dobzhansky–Muller model, i.e. due to fixation of alternative alleles during allopatric isolation. However, it is important to note that there are other evolutionary scenarios that would yield similar patterns of fit parental and unfit hybrid genotypes at pairs of loci. One potentially important route is genetic conflict.

Consider for instance the segregation distorter elements we envisaged in chapters 4 and 5. For heterozygous males, a genetic element consisting of a killer allele K_1 at one locus and a receptor allele T_1 at a closely linked target locus will selectively kill sperm with the wild-type alleles K_2T_2 (Feldman & Otto 1991; Larracuente & Presgraves 2012). Such elements are favored at the gametic level because sperm with the distorter gain a transmission advantage by killing competing sperm. However, they are often harmful at the individual level because sperm killing is associated with low male fertility. When the fitness costs at the individual level become severe, genetic modifiers that suppress the element are favored by natural selection, eventually silencing expression of the element, restoring Mendelian inheritance. Now imagine two populations that have lived in geographic isolation. One of the populations has struggled with an aggressive segregation distorter and has evolved an efficient silencing mechanism, whereas the other population has not

had any experience with that element. If the two populations later come together and hybridize many individuals of mixed ancestry would inherit the aggressive segregation distorter from the first population but would also fail to inherit the necessary silencer from that population—they would only inherit a genome region with this crucial trait missing. The segregation distorter would thus literally be brought back to life in individuals of mixed ancestry and yield low fertility. Phadnis & Orr (2009) found that the same X-chromosome linked gene, called *Overdrive*, cause both male sterility and segregation distortion in F1 hybrids between two divergent subspecies of *Drosophila pseudoobscura*, and thus providing direct evidence that genomic conflict can yield developmental incompatibility. Note that in the genetic conflict scenario the two adaptive peaks in the multilocus adaptive landscape separated by a fitness valley where hybrids would fall have come about through evolutionary change happening in only one of the populations— the one that has struggled with a selfish genetic element and thus also has evolved a silencing mechanism.

As we also discussed in chapter 5, there are many genetic elements that are favored by selection acting at the gene level but are selected against at the individual level. A particularly important group of elements is the class I transposable elements (TEs) that proliferate in the genomes of eukaryotes by making copies of themselves that become integrated in the host genome. As with segregation distorters, TEs can have severe fitness consequences for the host, and therefore the host is under strong selection to evolve mechanisms that suppress TE expression or hinder their integration in the genome. Again, hybridization between members of populations that have and have not experienced coevolution with a particular TE can cause severe problems in hybrids that may inherit the TE but not the suppressor mechanism. Erwin et al. (2015) demonstrated that when male *Drosophila virilis* carrying a particular TE family were crossed with females lacking the element, the mother was unable to provide the genome defense via complementary interfering RNAs that normally target and silence the TE in the paternal population. This led to excess TE activation in the germline and hybrid sterility.

Finally, it is also worth noting that ecological incompatibilities can evolve by the very same principle as developmental incompatibilities in the Bateson–Dobzhansky–Muller model. For instance, Arnegard et al. (2014) demonstrated that hybrids between two closely related but ecologically divergent species of threespine stickleback survive poorly due to reciprocal sign epistasis between loci controlling morphological traits involved in species-specific adaptations to different feeding ecologies.

8.2.6 Haldane's rule and the larger X/Z effect

As we have seen, hybrids between two species can have low fertility and/or viability. However, there are also interesting sex differences between hybrids in terms of fitness. The evolutionary geneticist John S. B. Haldane wrote about this in the 1920s and the pattern he described has since been known as **Haldane's rule** (Haldane 1922). In mammals, male hybrids regularly have lower fitness than female hybrids, whereas in birds it is opposite, female hybrids have lower fitness than male hybrids. Sex determination is opposite in mammals and birds. In mammals, males have different sex chromosomes (XY)—i.e. they are heterogametic, whereas females have similar ones (XX). In birds, males have similar sex chromosomes (ZZ), whereas females are the heterogametic sex (ZW). Haldane documented that sex differences in hybrid fitness almost always coincide with their sex-determining system. This extends to other taxa too, beyond birds and mammals. As with birds, butterflies have a ZW-system with female heterogamy, and female hybrids suffer more in terms of reduced fitness than males. Similar to mammals, flies have the XY-system with male heterogamy and male hybrids are less fit than female hybrids. Haldane realized that hybrid fitness somehow depended on whether it has similar or different sex chromosomes. Something makes the sex with dissimilar sex chromosomes more prone to developmental incompatibility.

To understand the mechanism behind Haldane's rule we have to consider Mendelian inheritance (Turelli & Orr 1995). Recall from chapters 3 and 4, some alleles are recessive, some are dominant, and some are partially recessive (or dominant). The Y and the W chromosome typically contain very few genes compared to X and Z respectively. Suppressed recombination on the Y and W leads to a lower effectiveness of selection to remove deleterious mutations on these chromosomes; so as deleterious mutations accumulate gene function is disrupted and gene loss occurs (Bachtrog et al. 2008). Gene degradation on the Y and W means that the heterogametic sex only has one copy of a large number of genes (i.e. X- or Z-linked genes that lack a homolog on Y or W). A male mammalian hybrid, say, would have inherited his Y chromosome from his father (species 1) and an X chromosome from his mother (species 2). For all other chromosomes he would have inherited one each from either species. A female mammalian hybrid in contrast, would have inherited one X from either species. Some of the genes on the X chromosome would likely be incompatible with genes on other chromosomes via epistasis. However, a proportion of these genes would be recessive, or at least partially recessive. These genes would be fully expressed in male hybrids because they only carry a single X chromosome, but buffered by (complete or partial) dominance in females which have two. Haldane's rule can thus be explained by Dobzhansky–Muller incompatibilities between recessive (and partly recessive) sex-linked alleles and genes on other chromosomes (Turelli & Orr 1995).

However, sex chromosomes appear to be important in speciation beyond the prediction from Haldane's rule (e.g. Coyne & Orr 1989; Vicoso & Charlesworth 2006; Presgraves 2008; Ellegren 2011; Payseur et al. 2018). Hybrid sterility is disproportionately caused by genes on the X (and Z) chromosome, beyond what can be explained by Haldane's rule. This phenomenon is known as the **large X-effect**. Although the large X-effect is well established at the phenotypic level, at least in *Drosophila*, the causes of it have been less clear. Part of the explanation may be related to the faster X/Z effect that we discussed in chapter 7. X- and Z-linked loci can evolve faster than autosomal loci because the two groups of loci differ in effective population size, mutation rate, and the efficacy of natural selection in favoring beneficial recessive mutations. Species-specific adaptations may therefore accumulate faster on the sex chromosomes and some of these may be incompatible in the genetic background of the other species. A second class of explanation for the large X-effect involves recurrent bouts of genetic conflict through X-linked segregation distorters such as the *Overdrive*-locus in *Drosophila pseudoobscura* (Phadnis & Orr 2009). A third explanation for the large X-effect involves the regulation of the X chromosome in the male germline. If the meiotic sex-chromosome inactivation machinery is unable to recognize and silence introgressed heterospecific segments on the X chromosome, this may lead to overexpression of X-linked genes that impair spermatogenesis (Lifschytz & Lindsley 1972; Presgraves 2008).

8.3 Modes of speciation

8.3.1 Population structure and allelic models of divergence and speciation

Relationships among populations can be structured in a number of different ways. This has consequences for how populations diverge from one another when subjected to divergent selection. It also plays an important role in determining how easily barriers to gene exchange can accumulate during divergence and eventually lead to speciation. Gene flow has a homogenizing effect on populations and reduces differentiation (Slatkin 1987). Therefore,

population structuring that reduces or eliminates gene flow before any biological barrier to gene exchange has evolved is likely to provide more favorable conditions for divergence and speciation. In short, in the absence of gene flow, reproductive barriers can arise without any opposition. We can broadly distinguish between three or four geographic scenarios that are frequently discussed in the speciation literature (Figure 8.10):

1) The **allopatric** scenario clearly provides the most favorable conditions for speciation. In this scenario a geographical barrier eliminates gene flow so the two populations can diverge independently due to natural selection and/or genetic drift. Geographical barriers might arise due to colonization of newly formed volcanic islands or because of the deepening of a river between populations, for example. Whether or not genetic drift can lead to the evolution of barriers to gene flow is unclear, although it has long been hypothesized that it could play a significant role in the evolution of DMIs. Certainly, drift will contribute to genetic differentiation between populations in geographical isolation but this does not mean it is the cause of incompatibilities. Theories of speciation through genetic drift are beginning to go out of fashion because empirical evidence points

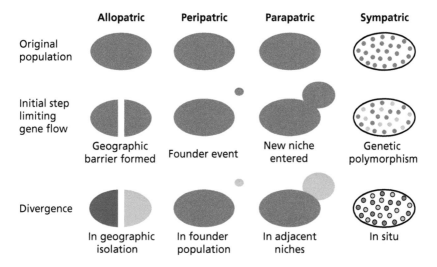

Figure 8.10 Four geographical scenarios in which divergence and eventually speciation can be hypothesized to occur. In the allopatric and peripatric scenarios there is no gene exchange between the diverging populations, in the parapatric scenario gene flow is reduced by population structuring but not absent, and in the sympatric scenario divergence is initiated in one randomly mating population.

to a dominant role of selection in explaining the evolution of barriers to gene exchange (e.g. Seehausen et al. 2014).

2) A variety of the allopatric scenario is the **peripatric** scenario where a small number of individuals from a large population (say a continental population) colonize a disjunct area (say a remote island) through a founder event. This is the scenario we discussed in chapter 5 in which Mayr (1954) hypothesized that the founder population would be likely to diverge particularly rapidly. His idea was that increased rate of genetic drift in the small founder population would tend to break up co-adapted epistatic networks between genes and that this would set the stage for evolution to take new directions, leading to rapid divergence and development of reproductive isolation should the "island population" later come in contact with the "continental population." Evidence for Mayr's hypothesis that founder events play a disproportionate role in speciation is limited and as a result this geographical mode of speciation is no longer considered particularly important. For this reason, we will not explore the peripatric scenario further here.

3) In the **parapatric** scenario geographically structured populations experience contrasting ecological selection. This geographical mode tends to arise when populations colonize newly available habitats or ecologically different habitats adjacent to another. Lake–stream divergence in fish species is a classic example of parapatric speciation that appears to occur in many freshwater fish taxa, including stickleback (Hendry et al. 2002) and cichlids (Seehausen and Wagner 2014). In a parapatric scenario, geographic structuring reduces, but does not eliminate, gene flow. The two populations may meet at a sharp environmental transition zone that is difficult for migrants to cross but where interbreeding takes place. Alternatively, populations might be geographically slightly separated but are still affected by gene flow through the exchange of migrants across a weak geographic barrier. Either way, divergent selection between the habitats drives increased differentiation and local adaptation that in turn increases selection against migrants and hybrids.

4) Finally, in the **sympatric** scenario there is no obvious geographical structuring at all. Instead populations co-occur in the same location and are able to freely interbreed without geographical barriers. Divergence is therefore initiated by disruptive selection acting on a genetic polymorphism that segregates in the population.

These different speciation scenarios, also known as geographical modes, are useful when developing mathematical population genetic models that explore potential mechanisms of speciation. Although they are categorized by geography, this is largely a proxy for how freely diverging populations are able to exchange genes. However, it is important to remember that such models are caricatures of how speciation likely plays out in real world. For instance, real populations may experience alternating phases of allopatry, parapatry, and sympatry during speciation, with different barriers arising at different times. However, by considering each scenario separately as discrete entities, rather than part of a larger process, models can help us in identifying key factors that are antagonistic to progress towards speciation and factors that can promote speciation (Gavrilets 2004).

8.3.2 Non-allopatric speciation

Traditionally, allopatric speciation has been regarded as a null model for understanding how speciation progresses (Coyne & Orr 2004). This is because in the absence of gene flow, two populations are expected to diverge and for barriers to gene exchange to arise as an indirect result. Since speciation in this scenario is essentially an emergent property, scenarios where gene flow is possible are therefore more interesting, as barriers to gene exchange must arise to overcome its homogenizing effects. For this reason, and because speciation with gene flow until recently was considered controversial, a large number of theoretical population genetic studies have investigated the sympatric and/or parapatric scenario and the interplay between selection and gene flow. One study in particular, a classic paper by Felsenstein (1981), clarified that progress towards speciation in the face of gene flow can be seen as a battle between selection and recombination.

Unfortunately, this can be counter-intuitive and challenging to understand. Furthermore, even the simplest population genetic models we are able to imagine very soon become analytically unmanageable (e.g. Gavrilets 2004). We therefore usually have to rely on computer simulations to investigate the evolutionary dynamics of even simple models—something we cannot easily do in the pages of a textbook! Luckily, however, some thought experiments are able to help us to clearly understand the principles of the balance between selection and recombination in a qualitative manner.

We will start with illustrating the sympatric scenario (NB! A short review of section 4.3.2 will help you here). A starting point for divergence that might potentially lead towards eventual speciation is the establishment of a genetic polymorphism that is under disruptive selection. In chapter 4 we analyzed such scenarios using one-locus models. We learned that cases where both homozygotes have higher fitness than the heterozygote are inherently unstable. When fitness is constant A_1 or A_2 would go to fixation depending on initial conditions. Although a polymorphic equilibrium exists in which $w_1^* = w_2^*$, that equilibrium is unstable. However, when selection is also negatively frequency dependent a stable polymorphism where heterozygotes are selected against can arise. For this to occur the inequality $S + M < 0$ must be satisfied, or more plainly, negative frequency-dependent selection must be sufficiently strong to stabilize the system (Udovic 1980).

Let us make a thought experiment where these conditions are met. Imagine an arbitrary chess-board-colored world in black and white. The black and white squares are equally large and equally numerous. Imaginary butterflies have become a theme in this book when we need to explain difficult mechanisms, so we will populate our chess-board-world with a mindless butterfly species, fluttering around erratically (Figure 8.11). We focus on one locus which controls wing color so that A_1A_1 is white, A_2A_2 black, and the heterozygote A_1A_2 is grey. A visually hunting predator is introduced. The black and the white homozygous butterflies are protected by camouflage half the time when they happen to match the background, whereas the grey heterozygote is poorly camouflaged in all squares. Because of this, the heterozygote will be on average

selected against at equilibrium where the two alleles are equally common ($\hat{p} = 0.5$). Except for selection occurring at our focal locus we assume that all the assumptions of the Hardy–Weinberg model are met (random mating, infinite population size, etc.), and to make sure that $S + M < 0$ and the equilibrium is stable we assume that fitness is frequency dependent. This might arise when the predator forms a search image based on common prey. Hence, when the black morph is rare the predator is mainly focused on attacking white prey on black squares and is less easily alerted to the other color mismatch. Hence, black butterflies tend to escape when rare, and vice versa when the white morph is rare.

We clearly have a genetic polymorphism in a randomly breeding population. Yet, we could force ourselves to think of the population in a different way—namely as a black colored and white colored ecotype which hybridize extensively and produce unfit "hybrids" (grey heterozygotes have low survival). Thinking about the model in this way, migrants between squares are also unfit, as they result in a clear color mismatch. Could we imagine a new adaptation that could evolve in the black ecotype and drastically improve its fitness? Say that a dominant mutation occurs in a black individual at a different locus that makes it preferentially sit on black squares when it rests so that the proportion of time it is protected by camouflage increases. Let us assume that this adaptation is controlled by locus B. The ancestral genotype B_1B_1 flutters around randomly with respect to square color whereas carriers of B_2 spend more time in black squares. Clearly, A_2B_2 is a favorable combination. However, because these butterflies mate randomly that combination would easily break up by recombination. B_2 might increase in frequency at first, but what happens when a black butterfly carrying the B_2 allele (say it has the genotype A_2B_2/A_2B_2) mates with a white one (A_1B_1/A_1B_1)? Since mating is random that should not take long! The resulting offspring would in this particular case be A_1B_1/A_2B_2. They would be grey (A_1A_2) and tend to rest on black squares (B_1B_2). Such "hybrids" carrying B_2 would eventually mate with white butterflies (again, mating is random). If the two loci are unlinked, half the gametes of a A_1B_1/A_2B_2 hybrid would be A_1B_2. When A_1B_2 is

Sympatric scenario

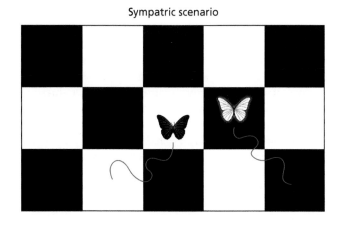

Allopatric scenario

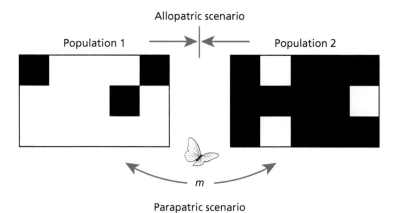

Parapatric scenario

Figure 8.11 Thought experiment illustrating geographic modes of divergence and speciation. In the sympatric scenario a polymorphic butterfly inhabits a two-colored world and experiences disruptive selection related to camouflage. In the allopatric and parapatric scenarios different morphs are favored by selection due to habitat differences (amount of white and black) without gene flow occurring (allopatric) or in the face of moderate gene flow (parapatric).

combined with A_1B_1 at fertilization you have white butterflies flying around that tend to rest on black squares—a bad combination that would be selected against. Likewise, black butterflies readily get the less favorable A_2B_1 combination through interbreeding and recombination. As we can see, recombination is antagonistic to selection in the face of gene flow; it acts to break down associations between alleles which when combined have a greater adaptive advantage. In the case of sympatry and random mating, gene flow is at its maximum; i.e. in our example, migration between "ecotypes" is $m = 0.5$. It is not entirely impossible that favorable B_2-like mutations could increase in frequency and remain

associated with A_2 in the sympatric scenario. For instance, if the two loci are tightly physically linked on the same chromosome, recombination is less likely to break them apart. Similarly, if an inversion occurs that protects the A_2 and B_2 alleles from recombining with A_1 and B_1 (like in the different morphs of the ruff that we discussed in chapter 6), they would also remain associated. Finally, extremely strong selection against heterozygote "hybrids" would mean that recombinant genotypes rarely, if ever, occur. So, from this thought experiment it is easy to see that increased divergence or progress towards speciation by no means is guaranteed in sympatry. In fact, it is quite hard!

We will momentarily return to the other extreme of migration and imagine our butterfly in an allopatric scenario. This will allow us to better reflect on how gene flow and recombination interact. Let us assume that our chessboard is divided, such that the environments the two allopatric butterfly populations inhabit are different from one another. In area 1 the majority of squares are white and in area 2 the majority of squares are black (Figure 8.11). We will assume that this has tipped the balance such that no stable polymorphic equilibrium is able to exist in either population. Therefore, A_1 is overall favored in population 1 and goes to fixation, whereas A_2 goes to fixation in population 2—in other words, we have all white morphs in population 1 and all black morphs in the other. Now recall our favorable mutation at locus B, B_2 that means individuals spend the majority of their time in black squares. Clearly, in this allopatric scenario there is nothing stopping this favorable B_2 mutation from increasing in frequency in population 2. The black individuals carrying that allele would spend even more time in the favorable black environment and consequently survive better and so B_2 would go to fixation. In the allopatric scenario there is no gene flow between the populations so $m = 0$. The two ecotypes do not exchange genes and so recombination is unable to produce maladaptive associations between the A and B locus.

We can also use our imaginary butterflies to model the parapatric scenario. To do this, we just need to let the two ecologically differentially selected populations from our allopatric scenario exchange migrants at varying rates ($0 < m < 0.5$). By adjusting our assumptions slightly, the parapatric scenario becomes mathematically identical to the sympatric scenario above when $m = 0.5$ and to the allopatric scenario when $m = 0$. Theoretical investigations indeed confirm our intuition that reducing m from 0.5 towards 0 makes the conditions for population specific adaptations and speciation progressively more favorable. So, to sum up so far, population genetic models such as those analyzed by Udovic (1980) and Felsenstein (1981) suggest that, compared to the allopatric scenario, gene flow makes ecological divergence and progress towards speciation harder. The greater the level of gene flow, the more difficult progression towards divergence and speciation becomes. However, these models also make clear that when gene flow is occurring, there is a balance between the antagonistic effects of recombination and selection too. Therefore, progression towards speciation can be enhanced by sufficiently strong selection, reduced gene flow, and/or factors that reduce recombination between favorable allele combinations, including tight physical linkage and reductions in recombination—i.e. due to inversions.

However, we can also easily imagine mutations that would facilitate sympatric and parapatric speciation much more easily than the example above. Recall that without the mechanisms just outlined, the B_2 mutation was not a guaranteed success in the sympatric scenario. Now imagine a mutation occurred at a third locus C that mutates the ancestral C_1 allele to a dominant allele C_2 which causes the butterfly to match its wing color with the background when it rests. The mutation would soon recombine away from the color-determining A allele it segregates with but would clearly be beneficial for both black and white butterflies. In hybrids, C_2 would be neutral as these individuals have no background they could match in our thought experiment. C_2 would therefore be favored overall and would go to fixation in the population. The fitness of the homozygotes A_1A_1 and A_2A_2 would increase relative to the heterozygote from the action of the C_2 allele as their survival would increase. The black and white butterflies would be well camouflaged for more than half the time, so the predator would take proportionally more grey ones. A likely additional effect as C_2 becomes common in the population is assortative mating. Black C_2 butterflies spend more time in black squares so when they are ready to mate the nearest potential mate is more likely to be black than white. Likewise, a white butterfly looking for a mate is more often surrounded by other white butterflies on white squares, so like would tend to mate with like. Assortative mating would then act to reduce the frequency of grey hybrids even further. The C_2 allele is an example of what Felsenstein (1981) called a one-allele mechanism as the same allele is facilitating speciation irrespective of genomic background. In other words, a single allele like this is able to generate indirect assortative mating by driving its carriers to mate where they occur.

8.3.3 Genomic "islands of speciation" and the divergence hitchhiking hypothesis

Imagine that we sequence and compare the genomes of our hypothetical butterfly from the chessboard-world. Let us take the parapatric scenario this time. There is overall ecological selection for white butterflies in population 1 and for black butterflies in population 2. However, since butterflies migrate between the two chessboard habitats all three colors can be found in both places. Yet, A_1 will have the highest frequency in population 1 and A_2 in population 2 and the allele frequency difference between the two populations would depend on the strength of selection and the migration rate m. Comparing individuals from the two populations using a genome scan (see chapter 7) the locus that causes the color polymorphism will stand out with a much higher F_{ST} value than polymorphic sites elsewhere in the genome. The divergently selected color alleles are restricted from moving between populations by the action of selection. Meanwhile, regions of the genome not experiencing selection move between populations without being hindered, as recombination rapidly breaks down their initial association with the selected loci. Hence, as we have seen in many of the examples of the preceding chapter, F_{ST} will be elevated in a region around a locus under divergent selection due to hitchhiking, but this signal will decay with physical distance along the chromosome as the probability of recombination increases. This means that in principle, it should be possible to use a measure of genetic differentiation such as F_{ST} to identify barrier loci—i.e. genes involved in reproductive isolation between populations. Thinking in these terms, Wu (2001) suggested that adaptations involving individual genes rather than entire genomes is what characterizes species. This hypothesis, combined with the decreasing costs of sequencing and widespread availability of next-generation sequencing, led to an explosion of speciation genomics studies aiming to use genome scans to understand how divergence between populations progresses.

In an early application of genome scans to speciation research, Turner et al. (2005) suggested that the low baseline level of F_{ST} they observed genome wide when comparing two forms of *Anopheles* mosquitoes resulted from gene flow homogenizing neutral variation. However, a few conspicuous F_{ST} peaks stood out above this neutral background and were referred to as "genomic islands of speciation" as they were assumed to have been caused by disruptive selection on hybrids and divergent selection on migrants between the two forms. Expanding on the metaphor of genomic islands of speciation, Via and West (2008) noted that the effective migration rate (m_e) is reduced locally in the genome around loci under selection compared to the gross migration rate (m) between two populations of pea aphids (*Acyrthosiphon pisum*). They suggested that such regions of locally reduced effective migration could therefore function as safe havens not only for neutral polymorphisms but also for weakly positive mutations. In other words, these mutations would only be able to overcome the antagonistic effect of gene flow when (effective) migration is reduced by selection at closely physically linked loci. In our chessboard-world thought experiment a B_2-like mutation in population 2, causing individuals to spend more time on dark background, could be targeted by selection and rise to high frequency if (and only if) it occurred in the region of reduced effective migration created by selection on A_2. In the process when B_2 increases in frequency the "island" would grow in height and width, meaning that the F_{ST}-peak in our genome scan would become higher (m_e becomes even lower) and a larger region around the two selected loci would experience reduced effective migration. Hence, we could imagine such islands of speciation growing and expanding through the accumulation of co-adapted alleles during non-allopatric speciation, maybe until gene flow is completely cut off. The metaphor of genomic islands of speciation has gained considerable traction since it was first introduced, with some authors suggesting the neutral baseline in the genome represents "sea-level" (Nosil et al. 2009) and that islands will grow to "continents of speciation" (Michel et al. 2010).

Elaborating on the metaphor, Feder et al. (2012) presented a four-phase conceptual model for genome divergence during non-allopatric speciation (i.e. speciation with gene flow). In phase 1 there is direct selection on one or a few loci, like in our virtual example with divergent selection on a color locus in the chessboard-world. Phase 2 is the one we just described where divergence hitchhiking occurs—weakly positively selected loci increase in frequency within divergence islands, causing the islands to expand and grow over time. Eventually speciation may go into phase 3. When multiple loci across the genome are under divergent selection, surrounded by larger and larger islands of speciation, effective migration (m_e) starts to become globally reduced instead of only in localized windows around single loci. This occurs because the correct combinations of alleles at these loci have a higher fitness than recombinant genotypes. Therefore, selection against individuals (i.e. hybrids and migrants) with the wrong combinations creates strong linkage disequilibrium between the loci, even though they are not physically linked in the genome. Now conditions are favorable for *genome hitchhiking*, resulting in a rapid and general uplifting of large "continents" of divergence. In the final phase 4, speciation is essentially completed. No alleles can readily introgress across the barriers to gene exchange that have accumulated. At this stage, diagnostically fixed differences can accumulate between the populations that now have become separate species. It is important to note that the four-phase model is conceptual and that evidence for each of the stages is mixed. Comparisons of species at different stages of divergence, i.e. samples along the speciation continuum, certainly suggest that genomic differentiation increases as progression towards speciation occurs (Martin et al. 2013; Riesch et al. 2017). However, evidence for processes such as divergence hitchhiking is very limited and whether it occurs at all is uncertain (Ravinet et al. 2017).

Genome scans are important tools in speciation genomics and the divergence hitchhiking model for non-allopatric speciation is intriguing and continues to inspire research in a range of study systems (Seehausen et al. 2014; Ravinet et al. 2017). There are a huge number of interesting examples to choose from and we highlight only three here. The yellow monkey flower (*Mimulus guttatus*) is a wildflower that occurs across the western parts of North America. Two ecotypes occur—a perennial form that grows in constantly wet environments and an annual form that experiences drought periods during the summer (Figure 8.12). As well as habitat differences, the two ecotypes differ in growth rate, flowering time, and size, resulting in striking phenotypic differences. Both ecotypes co-occur and overlap across their distribution and readily interbreed. However, work by Lowry & Willis (2010) identified an inversion that is a major QTL contributing to between-ecotype phenotypic variation. A genome scan between the annual and perennial forms confirmed that the inversion region shows high F_{ST} between them compared to the co-linear, non-inverted genome background (Twyford & Friedman 2015). Furthermore, SNPs from this region clearly show population structure between the forms, whereas SNPs from outside the inversion only show geographic structuring, suggesting the suppression of recombination of genes in the inversion allows the perennial and annual ecotypes to coexist in the face of gene flow.

A further example of divergence-with-gene-flow is that of the European rabbit subspecies *Oryctolagus cuniculus algirus* and *O. c. cuniculus* that are found across Spain and Portugal. The subspecies are thought to have diverged as a result of spatial isolation during the repeated glacial and interglacial cycles of the Pleistocene; however, they now co-occur in secondary contact. Using genome-wide SNP data from individuals sampled well away from the contact zone, Carneiro et al. (2014) found multiple regions showing high differentiation across the rabbit genome, despite an overall moderate level of differentiation between the two subspecies. Interestingly, these regions are overrepresented in centromeric regions and on the X chromosome—i.e. regions of reduced recombination rate. However, in the case of the rabbit subspecies, the genetic basis of reproductive isolation is unclear.

This is not true for *Heliconius* butterfly species in Central and South America. Whole genome resequencing of species at different degrees of spatial overlap and divergence show a repeated signature of genetic differentiation at two loci, *HmYb* and *Hmb*, that respectively control yellow and red patterns on the wings of these species (Martin et al. 2013). Since these *Heliconius* butterflies mimic the wing patterns of other species in order to avoid predators, maladaptive combinations of alleles at wing pattern loci will be selected against and thus these loci are likely responsible for a barrier to gene exchange between the species.

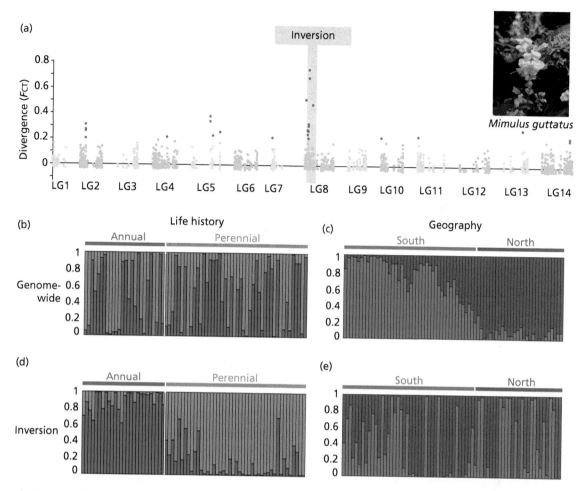

Figure 8.12 An inversion in linkage group 8 (a putative chromosome) appears to prevent recombination from dissociating co-adapted genes in the face of gene flow in yellow monkey flowers (*Mimulus guttatus*) adapted to annual and perennial life histories respectively. Annual and perennial monkey flowers differ with respect to having the inversion or not. A) A genome scan comparing genetic divergence between annual and perennial monkey flowers (F_{CT}, which like F_{ST} is a measure of population divergence) along the 14 linkage groups of the genome. Red dots represent outlier loci showing elevated divergence between the two forms, which to a large extent are concentrated within the non-recombining, inverted region marked in grey. B–E are different STRUCTURE plots. Each bar in each of the panels represents the assignment probabilities of one individual to either of two putative populations. In B and C genome-wide genotypes are used to classify individuals, whereas D and E only include genetic markers from the inverted region. B and D assign individuals to life history (annual and perennial) whereas C and E assign individuals to two geographic populations (south and north). Presence or absence of the inversion assigns individuals to the annual or perennial life histories with high probabilities (panel D), whereas genome-wide there is only some geographic structuring (panel C). Redrawn from Twyford & Friedman (2015). Photo courtesy of Christopher M. Luna.

However, as we learned in chapter 7, genome scans should be analyzed and interpreted carefully. We should certainly not see F_{ST}-peaks or "speciation islands" as the sole evidence for the presence of barrier loci and low baseline F_{ST}-levels as evidence for gene flow without further analyses (Noor & Bennett 2009; Cruickshank & Hahn 2014; Burri et al. 2015; Ravinet et al. 2017; Wolf & Ellegren 2017). Indeed, genome scans are best used in combination with a range of other methods and approaches. Why is it that we need to be so cautious when using genome scans in a speciation context? A range of demographic and evolutionary processes besides divergent selection and gene flow can affect genome divergence and differentiation. The heterogeneity we see in genome scans can therefore be driven by processes that are not related (directly at least) to speciation. We can turn to divergence between allopatric populations as an example to illustrate this. If we perform a genome scan between allopatric populations that experience no gene flow, we might also see the familiar peaks and troughs that in other systems are used to point to barrier loci. This can occur even if these allopatric populations do not experience divergent selection!

Firstly, if two allopatric populations separated very recently, they would have low average F_{ST} at polymorphic sites genome-wide (Cruickshank & Hahn 2014). This is because drift and selection have had only a very short period of time to work and are extremely unlikely to have driven sites to fixation. Moreover, if the effective population sizes of the two populations are large, genetic drift is a weak force of divergence, so F_{ST} will increase only very slowly at neutral loci. Gene flow can act to maintain low F_{ST} at both neutral and weakly selected loci, but a low F_{ST} alone is clearly not sufficient evidence that two populations exchange migrants. Several authors have suggested study designs that incorporate geographical comparisons, i.e. populations in allopatry vs sympatry. A lower F_{ST} between sympatric populations provides evidence gene flow might be occurring. Similarly, evidence of hybridization, admixture, and backcrossing in contact zones between species is also a good means of further supporting the conclusion that gene flow might be taking place. There are also many sophisticated methods for inferring whether gene flow has taken place during the evolutionary history of a species, including ABBA-BABA statistics and demographic inference; we will return to both of these topics in chapter 9.

Secondly, locally elevated F_{ST}-values (peaks) between populations can have causes other than divergent selection. Statistics such as F_{ST} are relative measures of population differentiation. Values of these statistics can therefore be high, either because within-population variation is low at a given locus, or because the between-population difference is large (Charlesworth 1998). These two explanations may also both contribute to relative differentiation—i.e. F_{ST} is high because within-population variation is low *and* because between-population difference is large. This has some serious implications for the

interpretation of F_{ST} because processes other than positive or divergent selection can inflate the statistics. For example, purifying selection to remove deleterious mutations in functional regions of the genome will reduce diversity and so increase F_{ST} (Charlesworth et al. 1997). This is one reason why researchers often compute D_{XY} in addition to F_{ST} in genome scans, as D_{XY} is an absolute measure of population differentiation and will not be affected by current levels of within-population diversity. Clearly, when interpreting multi-locus or genome sequence data it is important to be aware of how different measures of differentiation would be affected differently by various demographic and evolutionary processes (e.g. Noor & Bennett 2009; Cruickshank & Hahn 2014).

The problems with relative measures such as F_{ST} should not give the impression that D_{XY} is a panacea to correctly interpreting population differentiation in genome scans. Absolute divergence can also be influenced by other processes besides divergent positive selection. For example, balancing selection might maintain two haplotypes at a locus for a long time prior to a divergence event between two populations/species. If, after divergence, each population becomes fixed for different haplotypes, a genome scan using D_{XY} might suggest strong, recent positive selection where none has occurred (Guerrero & Hahn 2017). Population structure in the ancestral population prior to divergence might also drive a similar pattern. In contrast, a selective sweep in the ancestral population (i.e. prior to the split between lineages) will reduce nucleotide diversity at a locus, meaning that D_{XY} for this locus will be lower relative to other loci which may have not experienced selection until after the population split occurred. The use of both relative and absolute measures of differentiation has become a matter for considerable debate in speciation genomics. Nonetheless, a positive way forward to identify genes involved in speciation is to combine genome scan data with additional evidence of gene flow, selection, and gene function.

8.3.4 Allopatric and non-allopatric divergence and the study of hybrid zones

Many divergent populations currently have parapatric distributions. For the most part they have non-overlapping ranges but they meet in hybrid zones where interbreeding takes place (Barton & Hewitt 1985; Gompert et al. 2017). Contact zones between *Littorina* ecotypes, blackcap and Swainson's thrush populations migrating in different directions, hooded and carrion crows, marine and freshwater stickleback, and Italian and house sparrows are some examples we have already met. Parapatric distributions with zones of intergradation may arise *in situ* by divergent selection (incipient parapatric speciation). *Littorina* ecotypes that interbreed in ecological transition zones and marine and freshwater stickleback exchanging genes in brackish water are likely examples of such **primary hybrid zones**. However, hybrid zones also frequently arise by secondary contact between populations that have diverged some time in allopatry. For instance, during the last glaciation about 115 000–12 000 years ago glaciers covered large parts of the Northern Hemisphere so temperate habitats such as deciduous forests and their inhabitants became pushed southwards (Hewitt 2000). In Europe, the peninsulas in the Mediterranean region (Iberia, Italy, and the Balkans) acted as geographically isolated refugia for the temperate biota during the cold glacial period, so when the ice finally began to retreat 18–12 000 years ago and the forest and other temperate habitats recolonized the continent, many previously isolated populations of plants and animals came into secondary contact (see also section 9.2.2). Blackcap populations with different migration routes and hooded and carrion crows likely diverged initially in different refugia. If we take an even longer perspective, the Northern Hemisphere has experienced repeated cycles of glaciation and warmer interglacial periods during the last 2 million years (Hewitt 2000). Also, elsewhere on the planet climate has been fluctuating, sea levels have risen and fallen, rivers have altered their courses so that waterways have become cut off and reopened. In short, many organisms would have experienced both periods of restricted gene flow or allopatry and periods of secondary contact and gene flow during their evolutionary history. Hybrid zones are gold mines for speciation research because they allow us to investigate divergence at different stages along the speciation continuum—how barriers to gene exchange may accumulate over time and how selection and hybridization affect the hybrid-

izing populations and their genomes (Gompert et al. 2017).

By investigating the pattern of introgression of traits and alleles across a hybrid zone we may identify traits and genes that are under divergent selection and/or act as barriers to gene exchange (Barton & Gale 1993; Gompert & Buerkle 2016). Many alleles may exhibit a change in allele frequency across the hybrid zone. We call such a change in allele frequency across a geographic gradient a **cline**. But how can we distinguish clines caused by divergent selection from clines caused by neutral diffusion through dispersal and interbreeding? The shape and steepness of a cline will depend on factors such as the amount of gene flow, the pattern of dispersal, how long ago the two populations met, stochastic variation, and, potentially, selection. It is therefore not entirely straightforward to identify loci under selection in cline analysis. Current methods use different statistical procedures to estimate the neutral expectation based on how the majority of loci behave and identify outlier loci that deviate from that pattern. Typically, loci under selection exhibit steeper clines than neutral loci and greater changes in allele frequency across the geographic gradient. In Figure 8.13 we see an example of cline analysis across a hybrid zone between wave- and crab-adapted *Littorina* snails (Westram et al. 2018). Quite a few loci deviate from neutral expectation, which in this case was derived using a simulation approach. Loci identified as experiencing divergent selection in the two habitats were not randomly distributed in the genome but were found to be clustered into three putative inversions which likely protect ecologically important and divergently selected loci from recombining among the snail ecotypes (Faria et al. 2018).

8.3.5 The fate of hybridizing populations

When differentiated populations meet in primary or secondary hybrid zones different outcomes are possible (Figure 8.14). First, extensive hybridization and backcrossing may lead to a **speciation reversal** in which the populations converge and fuse into one population (Figure 8.14a). The collapse of wolves and coyotes into a hybrid swarm in human disturbed regions in eastern North America, which we learned about in section 8.1.1, is a likely

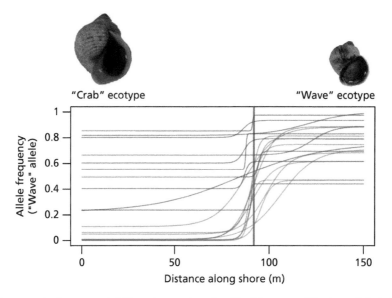

Figure 8.13 Examples of genetic clines across a hybrid zone between "wave" and "crab" ecotypes of *Littorina saxitalis*. Loci exhibiting clines that deviate from neutral expectation are in orange, whereas neutral clines are in gray. The orange vertical bar represents the average cline center. Modified from Westram et al. (2018).

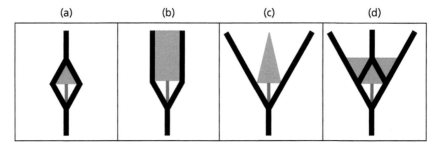

Figure 8.14 Four potential outcomes of hybridization between divergent populations. In all cases we assume that two populations have first diverged in geographical isolation (red line) and start to hybridize and exchange genes upon secondary contact (blue). A) Speciation reversal, in which hybridization causes the populations to converge and fuse back into one lineage. B) A stable hybrid zone is being maintained by a balance between selection within the hybrid zone and gene flow into it. C) Speciation by reinforcement and related phenomena, in which barriers to gene exchange are strengthened over time, leading to reduced hybridization and continued divergence. D) Hybrid speciation, in which a third lineage of hybrid origin emerges through hybridization that develops barriers to gene exchange against both parent species.

example of this phenomenon. The barriers to gene exchange that normally keep these two species distinct appear to be breaking down in this geographic region due to extensive hybridization and backcrossing. Speciation reversals have similarly been reported in whitefish (*Coregonus* spp.) in Swiss lakes (Vonlanthen et al. 2012). Analyzing historic and contemporary fish samples from pre-alpine lakes, Vonlanthen and co-workers concluded that anthropogenic eutrophication has diminished ecological

opportunity in the lakes, causing ecological niches to collapse. This has led to increased hybridization and backcrossing between formerly isolated whitefish species, resulting in a loss of certain species and also a loss of genetic and functional distinctiveness among those that have managed to remain.

Second, hybrid zones can remain fairly stable for long periods of time (Barton & Hewitt 1985) (Figure 8.14b). In many hybrid zones, individuals of mixed ancestry with low fitness are constantly being produced. Consequently, population growth will tend to be lower within the hybrid zone than outside of it. The hybrid zone would therefore act as an ecological sink of low productivity that continues to receive migrants from the more reproductively successful parental populations on either side of the hybrid zone. An equilibrium between selection against hybrids and gene flow may therefore be established.

Third, the hybridizing populations may proceed towards speciation in the face of gene flow (Figure 8.14c). For instance, when hybrids have low fitness and are selected against, natural selection may favor traits that reduce maladaptive hybridization—including traits facilitating assortative mating. This is the process of **reinforcement** of prezygotic barriers to gene exchange (Dobzhansky 1940; Servedio & Noor 2003). We will discuss reinforcement and other processes of coupling between barriers to gene exchange in the next section.

Finally, hybridization can lead to the formation of a third species of mixed ancestry that develops reproductive barriers against both parental species (Figure 8.14d). The Italian sparrow that we learned about in section 8.1.4 is one example of a species that has originated through such hybrid speciation. We will discuss this interesting form of speciation in section 8.4.

8.3.6 Reinforcement and coupling of barriers to gene exchange

In theory, reproductive isolation could result from just one barrier to gene exchange, say a single DMI that causes hybrids to have zero fitness. However, usually gene exchange between species is retarded by the joint action of several barriers. How multiple barriers to gene exchange come to coincide and become coupled is therefore important for understanding speciation (Butlin & Smadja 2018). Reinforcement of prezygotic isolation is a classical theory for how different barriers can become coupled and contribute to progress towards speciation (Dobzhansky 1937, 1940; Servedio & Noor 2003). It was originally formulated in the context of a pair of incipient species that have diverged some time in allopatry. Upon secondary contact hybridization takes place and hybrids have reduced but non-zero

fitness and are thus selected against. According to the reinforcement hypothesis, natural selection might then favor traits in the parental populations that increase assortative mating. Reinforcement is therefore a process where natural selection directly favors progress toward speciation.

Reinforcement likely explains plumage color evolution in sympatric pied flycatcher (*Ficedula hypoleuca*) and collared flycatcher (*F. albicollis*) populations in Central Europe (Sætre et al. 1997). In allopatry, e.g. in Scandinavia, UK, and Spain, male pied flycatchers have a plumage color that suits their name—they have blackish heads and backs which contrasts against their bright white belly and white forehead and wing patches. However, in Central Europe where their breeding range overlaps with the black-and-white collared flycatcher, male pied flycatchers are brownish and resemble female flycatchers (Figure 8.15). The collared flycatcher also appears to have diverged in plumage traits in sympatry. Allopatric collared flycatchers, e.g. in Italy, have on average smaller patches of white on their forehead, neck, and wings than sympatric ones (Figure 8.15). Hybrids between the two species have very low fitness. Female hybrids are completely sterile and male hybrids also have strongly reduced fertility (Sætre et al. 1997; Wiley et al. 2009). Thus, very little gene exchange occurs between the two species (Ellegren et al. 2012). Sætre and co-workers therefore speculated that selection against maladaptive hybridization could have driven female mate preferences in the two species in opposite directions, causing the observed divergence in plumage traits and a resulting reduction in rate of hybridization—i.e. reinforcement. Using mate choice experiments, Sætre and co-workers showed that female mate preferences in the flycatchers indeed have shifted in sympatry in favor of divergent male plumage traits and that the resulting divergence in color helps species recognition and thus increases assortative mating (Sætre et al. 1997).

Hopkins & Rausher (2011, 2012) investigated a similar process in two species of phlox, a type of flowering plant, in a zone of distributional overlap in Texas, USA, namely *Phlox drummondii* and *P. cuspidata*. The two species produce hybrids with strongly reduced fertility when crosspollinated. In sympatry with *P. cuspidata*, the flowers of *P. drummondii*

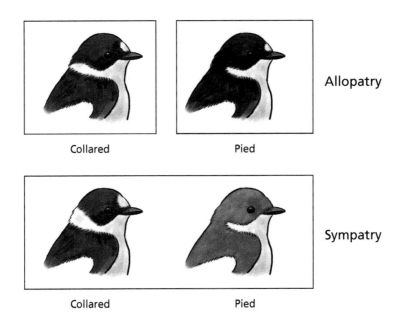

Figure 8.15 Male plumage color in allopatric and sympatric collared flycatcher (*Ficedula albicollis*) and pied flycatcher (*F. hypoleuca*). Males of the two species have diverged in plumage traits in sympatry. Mate choice experiments (Sætre et al. 1997) suggest that the divergence in plumage traits is due to changes in female mate preferences caused by selection against maladaptive hybridization.

exhibit increased pigment intensity compared to when they occur in regions without the other species (i.e. in allopatry). Thus, in allopatry, both species have light-blue flowers, whereas in sympatry *P. drummondii* have dark-red flowers, which are very different from the light-blue flowers of the sympatric *P. cuspidata*. The sympatric color divergence is caused by alleles at two different loci that are absent in allopatric *P. drummondii* but nearly fixed in sympatric ones (Hopkins & Rausher 2011). In common-garden experiments in sympatry the researchers demonstrated strong selection in *P. drummondii* for the derived, red color alleles whereas no selection could be detected in the absence of *P. cuspidata* (Hopkins & Rausher 2012). The authors further demonstrated that reinforcing selection is generated by pollinator behavior. Butterfly pollinators are much less likely to move between the two species (thereby causing crosspollination) when the flowers have different colors (red vs light-blue) than when they are similar (both light blue). A later follow-up study also used population genomic data to demonstrate strong selection on the two color alleles in nature (Hopkins et al. 2014). Thus, crosspollination is strongly reduced in sympatry due to reinforcement selection.

Reinforced assortative mating due to selection against unfit hybrids is, however, not the only route to **coupling of barriers to gene exchange** (Butlin & Smadja 2018). For example, selection against unfit hybrids can increase sexual incompatibility through a similar reinforcement process. Matute (2010) demonstrated significantly stronger gametic incompatibility between sympatric *Drosophila yakuba* females and male *D. santomea* in a hybrid zone on the African island of São Tomé than when allopatric *D. yakuba* females were crossed with male *D. santomea* in the lab. Moreover, prezygotic barriers can also induce selection for additional barriers to gene exchange. In our chessboard-world thought experiment the black and white butterflies are ecologically incompatible in the sense that they survive poorly in each other's optimal habitat—i.e. background color (prezygotic barrier) and that heterozygous grey "hybrids" have lower survival in either habitat (postzygotic barrier). We envisaged scenarios where alleles that would cause either or both morphs to spend more time in their optimal habitat could potentially spread in the face of gene flow. For instance, the dominant C_2 allele would cause both black and white butterflies to spend more time in their optimal habitat as they would match their wing color to the background when resting. This hypothetical adaptation was largely induced by a prezygotic barrier through ecological divergent selection (the differently colored morphs survive poorly in each other's optimal habitat). The resulting adaptation would increase postzygotic isolation as the fitness of grey hybrids would become relatively lower due to the increased survival of the black and white butterflies. Moreover, the same trait would also increase prezygotic isolation because site fidelity or habitat choice would indirectly result in assortative mating.

A most fascinating example of coupling of barriers to gene exchange is provided by a case of parallel parapatric speciation among *Pundamilia* cichlids in the Mwanza Gulf south in Lake Victoria (Seehausen et al. 2008; Meier et al. 2017a; Meier et al. 2018). *Pundamilia* cichlids are found along light gradients ranging from blue-shifted shallow waters to red-shifted deep waters around different islands in the Mwanza Gulf. Two species are found in this environment, but they differ with respect to progress towards speciation at different islands—a shallow-water species where males have blue-backed nuptial colors and a deep-water species where males have red-backed nuptial colors (Figure 8.16). Key alleles that separate the two cichlid species are linked to genes affecting color vision. The shallow-water species has a lower threshold for the detection of red light, whereas the deep-water species has a lower threshold for the detection of blue light. The difference in color vision is mainly caused by different alleles at the long wavelength-sensitive opsin gene but other color vision genes are also involved. Their different visual systems are clearly adaptations to live at different water depths as short-wavelength (blue) light is absorbed more rapidly than long-wavelength light at increasing water depths. Blue and red male nuptial colors are the result of sexually selected female preferences, which apparently contrasts in the two species because the females differ in color vision. At intermediate depths where both species can be found, females mate assortatively—those adapted to red-shifted light prefer red males, whereas those adapted to blue-shifted light prefer blue males.

Figure 8.16 Males of two closely related *Pundamilia* cichlid species from the Mwanza Gulf of Lake Victoria. The two species diverge by selection for different alleles at color vision genes as well as in male nuptial color in blue-shifted shallow waters (top panel) and red-shifted deep waters (bottom panel). Male nuptial colors are sexually selected and diverge in the two species likely due to the different color vision of the differentially adapted females. Hence, at intermediate water depths where both species are found, females mate assortatively based on male nuptial color. Photo courtesy of Ole Seehausen.

Interestingly, the two species appears to be derived from hybridization between two more distantly related cichlid species that occur elsewhere in Lake Victoria, the deep-water adapted *P. pundamilia* and the shallow-water adapted *P. nyererei*. Some of the traits and genes that distinguish these more distantly related cichlids are indeed related to color vision and male nuptial colors, but they also differ in several other traits that are not paralleled in the younger species pairs of hybrid origin (Meier et al.

2018). Apparently, parallel divergent selection linked to color vision at different water depths, coupled with sexual selection for striking male nuptial color, is driving speciation in the younger species pairs.

As we discussed in section 8.3.2 recombination is antagonistic to coupling of barriers to gene exchange in the face of gene flow. However, we also discussed factors that could facilitate coupling, including one-allele mechanisms, physical linkage, and inversions. In the case of *Pundamilia* cichlids assortative mating and divergence in male secondary sexual traits may be an emergent property of their different color vision. If so, *Pundamilia* color vision could be what speciation researchers refers to as a "magic trait" (Gavrilets 2004; Servedio et al. 2011) or multiple-effect trait (Smadja & Butlin 2011). Divergent selection on one trait (color vision in this case) would reduce gene exchange beyond the direct effect of the divergence in the selected trait (assortative mating in addition to ecological divergence in this case).

Sex chromosomes have been a recurrent theme when we have discussed barriers to gene exchange. For instance, we have seen that sexual conflict is resolved in guppies by sex chromosome linkage of male secondary sexual traits. Haldane's rule is explained by exposure of recessive sex-linked incompatibilities in the heterogametic sex. Moreover, hybrid incompatibilities are disproportionally often caused by sex-linked genes (larger X/Z effect). In short, sex chromosomes play an outsized role in speciation (e.g. Qvarnström & Bailey 2009; Payseur et al. 2018). Sex-linkage may also facilitate coupling of different barriers because genes affecting different barriers will often tend to be inherited together on sex chromosomes and are less likely to be broken up by recombination. For example, reinforcement in pied and collared flycatchers appears to be facilitated by sex-linkage of genes affecting hybrid fitness and those affecting assortative mating (Sætre et al. 2003; Sæther et al. 2007). Indeed, despite genomic evidence for introgression occurring across large regions on autosomal chromosomes in sympatric pied and collared flycatchers via partially fertile male hybrids, there is no evidence for introgression of Z-linked alleles (Sætre et al. 2003; Ellegren et al. 2012). Hence, apparently, key genes affecting different barriers to gene exchange essentially do not recombine but are inherited as a unit on the flycatcher Z chromosome.

8.4 The role of hybridization in speciation

8.4.1 Polyploid hybrid speciation

A substantial fraction of speciation events does not occur as a consequence of divergence between lineages but instead involves the reunion of divergent genomes through hybridization. There are two routes to hybrid speciation—homoploid and polyploid (Mallet 2007; Rieseberg & Willis 2007). **Homoploid hybrid speciation (HHS)** occurs without any change in the number of chromosomes. **Polyploid hybrid speciation (PHS)** involves the full duplication of a hybrid genome (**allopolyploidy**). However, polyploidy can also occur within a species (**autopolyploidy**). Of these PHS is the more common and well understood mode of hybrid speciation.

PHS is very common in plants, but the true rate is difficult to estimate (Stebbins 1971; Otto & Whitton 2000; Wood et al. 2009). Often several different polyploid species occur within a genus and disentangling whether there was just one transition to the new ploidy level followed by divergent speciation or whether each polyploid species originated independently is difficult and requires careful analysis in each particular case (e.g. Marcussen et al. 2014). A more recent estimate suggests that 15 percent of angiosperm and 31 percent of fern speciation events are accompanied by ploidy increase (Wood et al. 2009). However, PHS is taxonomically widespread and has been reported from fungi and animals, including vertebrates, although at a significantly lower rate than in plants (McLysaght et al. 2002; Albertin & Marullo 2012; Betto-Colliard et al. 2018).

There are different routes to polyploidy. First, polyploidy may arise from somatic chromosome doubling in vegetative tissue (such as the stem of a plant) that gives rise to polyploid reproductive organs (such as the flower). Second, a stable tetraploid can be established via a so-called "triploid bridge"—triploids may arise through the fusion of an unreduced diploid gamete with a normal haploid gamete. Although such triploids normally would produce aneuploid inviable gametes and thus be sterile, they may occasionally give rise to an interfertile tetraploid by themselves producing unreduced triploid gametes that fuse with a normal haploid gamete (Rieseberg & Willis 2007). However, the most common route to polyploidy is thought to occur through the fusion of unreduced gametes (Mason & Pires 2015). Fusion of diploid gametes from two different species results in tetraploid hybrids that instantaneously exhibit strong reproductive isolation against both parent species. As we discussed in chapter 2, this is because triploid backcrosses between the hybrid and either parent would produce aneuploid, inviable gametes and have very low fertility (Rieseberg & Willis 2007).

However, strong barriers to gene exchange between tetraploid hybrids and the parent populations are not sufficient conditions for successful PHS. Assuming new polyploids are rare in an outcrossing population, they will mate mostly with incompatible parentals. Thus, founding an interfertile population of tetraploid hybrids is very difficult in outcrossing species, a phenomenon known as the **minority cytotype exclusion principle** (Husband 2000). This is probably the reason why PHS is more common in plants than in animals. First, plants are often perennial and temporarily clonal, allowing multigenerational persistence of hybrid lines. Second, plants are often hermaphroditic. Therefore, polyploids may rely on self-fertilization as a means of sexual reproduction when they are still rare. Finally, many plants, being sessile, have much more restricted dispersal than animals. Thus, a local cluster of polyploids can accumulate relatively unhindered by gene flow to overcome the minority cytotype disadvantage. Indeed, polyploidy is strongly associated with asexual reproduction, self-fertilization, and perenniality in plants (Otto & Whitton 2000; Brochmann et al. 2004).

Overcoming the minority cytotype exclusion principle is, however, not the only challenge facing a newly formed polyploid hybrid lineage. It must be able to successfully compete with its diploid parent species or adapt to a different niche in order to establish itself. Polyploidy may in itself facilitate this process, as having a double set of all genes affects gene dosage and gene regulation. Moreover, they would be heterozygote at genes where the

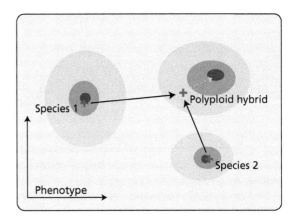

Figure 8.17 Hypothetical adaptive landscape of polyploid hybrid speciation. Blue contours represent isoclines of equal fitness with increasing fitness shown in darker colors. Two species reside near different peaks in the adaptive landscape (red crosses represent the mean phenotypes of the three populations). Polyploid hybrids are produced with a mean phenotype that by chance is adjacent to a third peak in the adaptive landscape. The hybrid population may eventually evolve towards the new adaptive peak by selection (yellow dashed arrow) and become established as a third species of hybrid origin. Factors that can promote polyploid hybrid speciation include barriers to gene exchange against the parent species due to different ploidy level, heterozygote advantage, and differences in gene dosage and gene regulation caused by the difference in ploidy. However, a vacant niche would normally have to be present in the adaptive landscape for the three species to be able to co-exist.

parent species are fixed for different alleles, and so may benefit from heterozygote advantage at some of these genes (Mallet 2007). Therefore, polyploids may differ from their parents in many traits and this may sometimes render them competitive or able to enter an unoccupied niche (Figure 8.17).

8.4.2 Homoploid hybrid speciation

Homoploid hybrid speciation (HHS) occurs without any changes in ploidy level. As we saw in the previous section, a change in ploidy facilitates hybrid speciation because strong barriers to gene exchange are instantly in place between the polyploid hybrid and its diploid parent species and also because polyploidy affects phenotypes through dosage and gene expression effects that would tend to make the hybrid phenotypically different from either parent and not merely intermediate. Homoploid hybrids have fewer such initial advantages and HHS is therefore more difficult to accomplish and likely a much rarer mode of speciation (Mallet 2007). Nevertheless, well supported empirical examples are growing in number, so evidently HHS does occur from time to time. Analysis of particular cases has helped us to identify factors that can facilitate this peculiar mode of speciation. These include transgressive segregation, hybrid traits that cause assortative mating, and sorting of barriers to gene exchange that would normally isolate the parent species, resulting in an interfertile hybrid retaining barriers against both parent species.

When we think of hybrid traits we tend to think that these would be intermediate between the parental traits. However, this is not universally true. Very often hybrid traits show **transgressive segregation**. This means that a phenotypic trait value in a hybrid exceeds trait values found in either parent. A classic example of transgressive segregation is the enormous body size of ligers—a cross between a male lion (*Panthera leo*) and a female tiger (*P. tigris*) that results in hybrid offspring with a body size that greatly exceeds that of either parent species. There are different routes to transgressive segregation. However, one important route is thought to be polygenic inheritance in which alleles at different loci contribute to the trait value but in opposite directions, and where the parent species are fixed for different allele combinations (haplotypes). Consider for instance a trait such as body size where alleles at different loci contribute to larger (+) or smaller size (−). Two parent species of similar size could for instance be fixed for the following combination of + and − alleles at six loci that affect the trait: (+ + + − − −) and (− − − + + +) respectively. Recombination between the six genes in individuals of mixed ancestry can then liberate "transgressive" quantitative variation, more extreme than either parent, for example (+ + + + + +) or (− − − − − −). A large proportion of early generation hybrids may be unfit, but sometimes extreme transgressive hybrids may be able to colonize niches that are not available to either parent. Transgressive segregation appears to have been important in HHS in wild *Helianthus* sunflowers in North America (Rieseberg et al. 2003). Phylogenetic evidence suggests that three out of 11 species in this genus are stable hybrid diploid species that have originated from hybridization between two common and widespread sunflower species, *H. annuus* and *H. petiolaris*

(Rieseberg 1991). The three hybrid species appear to have originated independently through hybridization between the two parental species and have become adapted to extreme habitats where neither of the parent species thrive: *H. anomalus* lives on sand dunes in Utah and northern Arizona, *H. deserticola* inhabits dry, sandy soils in the deserts of Nevada, Utah, and Arizona, whereas *H. paradoxus* thrives in brackish salt marshes in western Texas and New Mexico. Rieseberg and co-workers crossed the parent species to produce variable second-generation backcrosses (BC2) exhibiting a range of transgressive phenotypes. Comparing trait values in the three hybrid species with those artificially created from crossing the parent species the researchers were able to recreate the majority of the extreme traits found in the hybrid species (Rieseberg et al. 2003). In a follow-up study, the research team conducted experiments in which BC2-individuals, the parental species, and one of the hybrid species, *H. paradoxus*, were transplanted into the salt marsh habitat of *H. paradoxus* (Lexer et al. 2003). QTLs underlying traits that distinguish *H. paradoxus* from its parent species, including genes for leaf succulence and mineral ion uptake, were found to respond to strong directional selection among the BC2 plants and QTL alleles from both parental species were significantly positively selected in the *H. paradoxus* habitat. The experiments in these two studies demonstrate that the same combinations of parental alleles required to generate extreme phenotypes in artificial backcrosses also occurred in ancient hybrids. Thus, hybridization facilitated both ecological divergence *and* hybrid speciation in sunflowers.

In section 8.2.3 we learned about speciation between *Heliconius cydno* and *H. melpomene*—two butterfly species that differ in wing color patterns (Naisbit et al. 2001). *H. cydno* is black with white markings on the wings whereas *H. melpomene* is black with yellow and red markings. Hybrids between the two species have low fertility, female hybrids being sterile whereas males are fertile. Moreover, hybrid females have low mating success because males of either parent species discriminate against females with intermediate wing markings. However, in central Colombia a third *Heliconius* species, *H. heurippa*, resides that shares phenotypic similarity with both *H. cydno* and *H. melpomene*

H. cydno   *H. heurippa*   *H. melpomene*

Figure 8.18 Hybrid trait speciation in *Heliconius* butterflies. The red and white wing markings of *H. heurippa* appear to be derived from ancient hybridization between *H. cydno* and *H. melpomene*. The hybrid wing markings act as a barrier to gene exchange as all three species mate assortatively based on wing color pattern. Photo courtesy of Chris Jiggins and the MacGuire Centre. The photographs are from the Neukirchen collection.

(Mavárez et al. 2006; Jiggins et al. 2008) (Figure 8.18). *H. heurippa* has both red and white wing markings. Thus, the researchers speculated that it may have originated through hybridization between *H. cydno* and *H. melpomene*. Although female hybrids are sterile, gene flow between the two species could occur through fertile males. Mavárez et al. (2006) backcrossed male *cydno* × *melpomene* hybrids with either parent species and were able to recreate backcrosses with a similar red and white wing pattern as *H. heurippa*. Genome sequence analysis suggests that *H. heurippa* is genetically very similar to *H. cydno*, but that certain wing color traits, including the red wing band, are introgressed via hybridization with *H. melpomene* (Jiggins et al. 2008). Interestingly, however, the introgressed traits act as barriers to gene exchange between *H. heurippa* and both parent species. Mate choice experiments demonstrated strong assortative mating between all three species, and in *H. heurippa* the wing color pattern and color elements derived from the two parent species are critical for mate recognition by males.

In section 8.1.4 we introduced the Italian sparrow (*Passer italiae*) which is a homoploid hybrid species derived from past hybridization events between the house sparrow (*P. domesticus*) and Spanish sparrow (*P. hispaniolensis*)(Elgvin et al. 2017). Using genomic cline analysis, Hermansen et al. (2014) investigated barriers to gene exchange between the parent species and compared these with barriers between the Italian sparrow and either parent species identified in a parallel study (Trier et al. 2014). Interestingly, a subset of the genes that normally isolate the parent species now act as barriers to gene exchange between Italian and Spanish sparrows, whereas a different subset of parental incompatibilities acts

as barriers between Italian and house sparrows. Apparently, genes that are incompatible between the parent species have been purged in the hybrid lineage. For instance, the Italian sparrow has inherited mtDNA and several nuclear genes with mitochondrial function from the house sparrow. A comparative population genomics study also showed genome regions containing nuclear genes with mitochondrial function were repeatedly fixed for house sparrow ancestry in independent Italian sparrow populations on different Mediterranean islands (Runemark et al. 2018). Spanish sparrow alleles at these genes have evidently been selected against and are now almost absent among Italian sparrows. These mitochondria associated genes are incompatible between the parent species and also between Italian and Spanish sparrows. Yet, the Italian sparrow has a functional set of mitochondrial genes as it has inherited the functional gene complex from the house sparrow and gotten rid of the incompatible Spanish sparrow alleles through selection. Similarly, a functional set of sex-linked genes inherited from the Spanish sparrow act as a barrier to gene exchange both between the parent species and between Italian and house sparrows (Hermansen et al. 2014; Trier et al. 2014).

8.4.3 Hybridization and adaptive radiation

The examples of hybrid speciation discussed in the previous two sections clearly demonstrate that hybridization is not always a dead end. Hybridization followed by backcrossing and introgression can be an important source of genetic variation and novel combinations of genetic varieties that natural selection can shape further for adaptive purposes (Abbot et al. 2013; Seehausen et al. 2014). Rather than impeding diversification, as many zoologists believed at the time of the modern synthesis (e.g. Dobzhansky 1937; Mayr 1942), current evidence actually suggests the opposite. Seehausen (2004) argued that hybridization is likely to be common when populations invade new environments and will potentially elevate rates of response to selection by increasing standing genetic variation. Therefore, hybridization may predispose colonizing populations to rapid adaptive diversification under disruptive and divergent selection. Extensive hybridization appears to characterize several conspicuous adaptive radiations, including Darwin's finches (Grant & Grant 2011) and cichlid fishes in the Great African Lakes (e.g. Seehausen et al. 2008; Joyce et al. 2011).

African cichlid fishes are a particularly striking example of this and are arguably one of the most breathtaking and extensive adaptive radiations on Earth. The haplochromine cichlids of Lake Victoria in Africa encompass more than 700 diverse species that all evolved during the last 150 000 years. Using genomic data on cichlid fishes from Lake Victoria as well as from the major African river systems that drain into the lake, Meier et al. (2017b) demonstrated that the radiation can be traced back to ancient hybridization between two divergent lineages that entered Lake Victoria from different river systems— the Upper Nile and the Congo rivers. Many phenotypic traits that contribute to ecological adaptation and barriers to gene exchange have diverged in multiple speciation events in the Lake Victoria radiation, including tooth shapes, male nuptial color, and opsin alleles related to color vision. Indeed, some of these traits are also divergent between the Upper Nile and Congolese species outside the lake. Interestingly, this includes the different alleles on the long-wavelength sensitive opsin gene that we discussed in section 8.3.6 regarding speciation in *Pundamilia* cichlids. The red-shifted alleles are derived from the ancestral lineage from the murky water of the Upper Nile, whereas the blue-shifted alleles are derived from the lineage from the clear water of the Congo. Thus, hybridization between the ancestral lineages facilitated the Lake Victoria radiation by providing genetic variation that later became recombined and sorted into a range of new species.

Study questions

1. Discuss relative merits and problems with the genetic cluster species concept, the biological species concept, and the phylogenetic species concept.
2. Review the four main classes of barriers to gene exchange discussed in this chapter (ecological incompatibility, assortative mating, sexual incompatibility, and developmental incompatibility).
   a. Discuss how these barriers relate to the distinction between prezygotic and postzygotic barriers to gene exchange.
   b. Explain what kind of evolutionary mechanisms can cause each of the four types of barriers.
   c. Discuss which of these barriers are more and less likely to evolve among populations that continue to exchange genes.
3. Explain why sexual selection sometimes can make speciation less likely.
4. Review the Bateson–Dobzhansky–Muller model for the evolution of developmental incompatibility.
   a. What is a Bateson–Dobzhansky–Muller incompatibility (BDMI or DMI)?
   b. Discuss alternative ways that developmental incompatibilities could arise.
   c. Discuss whether an ecological incompatibility can constitute a BDMI.
5. Explain Haldane's rule and the mechanism that can explain it.
6. Why are sex chromosomes hot spots for accumulation of barriers to gene exchange?
7. Explain why gene flow is antagonistic to adaptive divergence and progress towards speciation. What factors may facilitate divergence in the face of gene flow?
8. Explain the following concepts:
   a. Allopatric speciation.
   b. Peripatric speciation.
   c. Parapatric speciation.
   d. Sympatric speciation.
9. Review Feder and co-workers' conceptual four-phase model for speciation with gene flow. Discuss how the model could be tested.
10. Explain how genome scans may be used to identify candidate loci that act as barriers to gene exchange.

11. Explain how cline analysis across a hybrid zone can be used to identify candidate loci that act as barriers to gene exchange.
12. Review the long-term fate of hybridizing populations.
13. What is reinforcement and which role might it play in speciation?
14. Discuss other mechanisms that may result in coupling of barriers to gene exchange.
15. Explain the process of polyploid hybrid speciation. Which factors may facilitate the process and which factor may constrain it?
16. Explain the process of homoploid hybrid speciation. Which factors may facilitate the process and which factor may constrain it?
17. Explain the role hybridization can play in adaptive radiations.

CHAPTER 9

Reconstructing the past

How can we use genetics and genomics to understand the evolutionary history of organisms? In the preceding chapters, we learned about how processes such as selection and genetic drift can shape the genetics of populations and, in some cases, can build to the evolution of different species. Although evolutionary change can be rapid, it mainly occurs on a timescale that extends far beyond the average human lifespan. For this reason, we must turn to other means in order to reconstruct a picture of the evolutionary history of life on our planet. In this chapter, we will focus on these methods. We begin by introducing the field of phylogenetics as a way to visualize and quantify the evolutionary relationships among species. We explain how we go from aligning DNA sequence data to building gene trees. We will demonstrate that "tree-thinking" is fundamentally important for understanding evolution. We will also go beyond phylogenetic trees to focus on phylogeography, i.e. the understanding of evolutionary relationships in a spatial context. More recently, the explosion of genomic data from ancient and modern human populations has made this an extremely exciting field which is transforming our understanding of our own evolutionary history. Before that though, we should start with learning about how modern phylogenetics has arisen from historical efforts to classify life on Earth.

9.1 Phylogenetics and phylogenomics

9.1.1 The origins of tree thinking

Humans have long pondered the question of how species have come to be and how they are interconnected and related. As we saw in the last chapter, the process of how species have evolved underlies the relationships that exist among them. Prior to Darwin's seminal work and the foundation of evolutionary biology in the nineteenth century, most people thought species were different kinds of animals and plants created by a divine entity. Creation myths from many religions reflect this view, as did some of the thinking of early Western philosophers such as Plato. The belief in a divine order among living organisms was an important motivation for Carolus Linnaeus when he began to develop a biological classification system at the time of the Enlightenment in the eighteenth century. Linnaeus, a Swedish botanist, wanted to categorize living things in order to understand the logic behind God's creation. In 1735, Linnaeus published the *Systema Naturae*, which laid the groundwork for the **binomial nomenclature** we use today. You should already be familiar with this system since we have used it throughout the book so far! In short, all species have a scientific name with two parts, the genus and the species; for example, we are *Homo sapiens* and our close relatives the chimpanzee are *Pan troglodytes*. Linnaeus was not the first to try and classify life in this way but he was the first to develop a hierarchical system, where species occur within genera, genera within families, families within orders, and so on (Table 9.1). In other words, he established the system of **taxonomy**. Linnaeus' system quickly gained popularity during his lifetime and soon he and many of his students began to catalogue and classify a huge number of different animals and plants. However, since Linnaeus and his contemporaries lived before the field of evolutionary

Evolutionary Genetics: Concepts, Analysis, and Practice. Glenn-Peter Sætre & Mark Ravinet, Oxford University Press (2019).
© Glenn-Peter Sætre & Mark Ravinet 2019. DOI: 10.1093/oso/9780198830917.001.0001

Table 9.1 Some examples of ranked hierarchical classification using Linnean taxonomy. Note that additional ranks occur and that they also differ between botany and zoology. For simplicity, we use only the seven major ranks here.

	Kingdom	Phylum	Class	Order	Family	Genus	Species
Human	Animalia	Chordata	Mammalia	Primates	Hominidae	*Homo*	*H. sapiens*
Domestic dog	Animalia	Chordata	Mammalia	Carnivora	Canidae	*Canis*	*Canis lupus familiaris*
Chicken	Animalia	Chordata	Aves	Galliformes	Phasianidae	*Gallus*	*G. gallus*
Alaskan blueberry	Plantae	Angiosperms	Eudicots	Ericales	Ericaceae	*Vaccinium*	*V. ovalifolium*
Salmonella	Eubacteria	Proteobacteria	Gammaproteobacteria	Entereobacteriales	Enteroobacteriaceae	*Salmonella*	*S. bongori*

biology was established, their attempts at grouping species were based only on arbitrary shared morphological characteristics without an understanding of how species were truly related.

In 1859, Charles Darwin published the first edition of *On the Origin of Species*, and with it he fundamentally changed how we think about life on Earth. Although Darwin lacked our present-day knowledge about the principles of heredity (chapter 1), he saw relationships among living organisms as genealogical. Crucially, Darwin introduced the concept of **common ancestry**. Inspired by the diversity of animals and plants he witnessed during his voyage around the world on the Beagle, Darwin argued that any two living species are descended from a common ancestor. For two different bird species, living on different but geographically close islands, this common ancestor might have occurred quite recently, whereas for two bird species on different continents, a common ancestor could have lived much longer ago. Going back far in time, Darwin suggested that all life on Earth is descended from a single common ancestor. In a famous illustration from one of his notebooks, drawn not long after he returned from his global voyage, Darwin sketched a branching tree and annotated it with a single phrase "I think" (Figure 9.1). He spent a great deal of time developing his "tree thinking" and by 1859, when he published the *Origin*, he included only a single diagram of an evolutionary tree, showing how species branch and diverge from one another over time and he referred to it as a "tree of life."

Darwin suggested common ancestry as the basis for relationships among species. By doing this, he provided taxonomists with a more rigorous way to classify species; it was now possible to group species by characters shared by descent. Darwin therefore paved the way for **evolutionary taxonomy** and

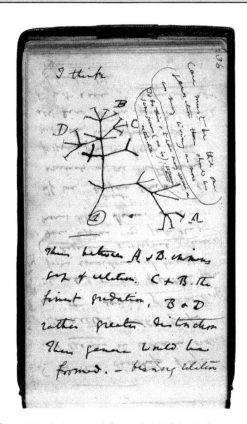

Figure 9.1 Charles Darwin's famous "I think" sketch of an evolutionary tree, written in his personal notebook from 1837. Source: Wikicommons.

the origins of **phylogenetics**, which we will turn to in more detail in the next section. Evolutionary taxonomy differs from the approach used by Linnaeus and his followers as it is a classification system which also reflects the historical process by which organisms have evolved. Drawing on Darwin's concept of shared descent, species within a genus all share a common ancestor more recently with one another than any of them does with a species from a separate genus. In other words, organisms are classified on a genealogical basis. You can reflect on your own genealogy for some insight into how this works. For example, imagine you have two siblings and three cousins. You and your two siblings are more closely related to one another than any of you are to your three cousins and vice versa. This is because you and your siblings share recent ancestry with your parents. However, you, your siblings, and your cousins all share more recent ancestry via your grandparents than with another randomly chosen person. You can already probably see that simply explaining the evolutionary relationships among organisms using words alone is sub-optimal—a much better way is to use a graphical representation similar to Darwin's tree concept. In fact, that is exactly what modern evolutionary biologists do and it is why we will now turn our attention to phylogenetics and phylogenetic trees.

9.1.2 Trees, phylogenies, and evolutionary relationships

The field of phylogenetics grew from Darwin's observations of common ancestry. Rather than simply grouping organisms because they exhibit similar traits or features, phylogenetics attempts to recreate the evolutionary history and relationships among species. Modern evolutionary biologists build **phylogenetic trees** to visualize these relationships and to help disentangle difficult taxonomic groupings. Trees or **phylogenies** in this sense are models of these relationships. This is a crucial point since all the trees we will examine here are best estimates of the evolutionary history of a set of taxa. That is not to say they are wrong; instead, by summarizing complex relationships in a graphical model, we are likely to make oversimplifications. After we have gotten a better grip on how phylogenetics works, we will turn our attention to the limits of phylogenetic inference.

In Figure 9.2A, we can see a basic phylogenetic tree with four taxa. At the tips of each tree are the species we are interested in—humans, chimpanzees, gorilla (*Gorilla gorilla*), and orangutan (*Pongo pygmaeus*)—in other words, this is a phylogeny of the great apes or Hominidae. The species are connected to one another at internal **nodes** which in our tree model represent a shared ancestor between groups.

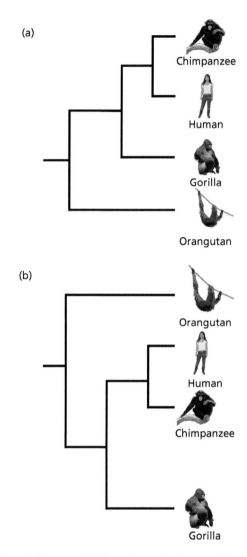

Figure 9.2 Two rooted phylogenetic trees showing the relationships between the Great Apes or Hominidae. Although A and B are drawn differently, they show the same evolutionary relationship among the four taxa—i.e. they have the same topology. For example, in both cases, humans and chimpanzees have a more recent common ancestor than either does with gorillas.

For example, the node between humans and chimpanzees on our basic phylogeny represents the common ancestor for these species. Since humans and chimpanzees are descended from a common ancestor, they are known as **sister species**. The next node in our phylogeny is the common ancestor of humans, chimps, and gorillas. Since all three of these species are descended from a single common ancestor, they form a **monophyletic group**. This also corresponds to a taxonomic division within the family Hominidae—the subfamily Homininae. Both nodes are connected to one another via **branches**; branches that extend from nodes to the tips are known as **lineages**. At the base of our tree is the **root**—the common ancestor of all the taxa the tree contains and the point in the tree marking the deepest evolutionary split among the species we are examining. Therefore, from Figure 9.2A, we can conclude that the deepest split is between the orangutans and all other great apes. Note that it is also possible to summarize the tree in Figure 9.2A using a simple notation known as the **Newick format**—e.g. (((H, C), G), O). We will generally avoid Newick format in this chapter, but readers should be aware that it is commonly used to represent phylogenetic trees in programming environments. Returning to the tree, lineage or branch length typically represents some sort of information about time or the amount of evolution that has occurred between two divergent taxa. The majority of modern phylogenies are inferred from DNA sequence data and so branch length might be proportional to the amount of sequence divergence between two taxa (i.e. the number of nucleotide substitutions). For simplicity at this stage, we will assume that the branch lengths of our tree represent time. So, from Figure 9.2A we can see that the **time to most recent common ancestor** (TMRCA) between humans and chimpanzees is lower than the TMRCA between humans and gorillas. In short, the common ancestor of humans and chimpanzees lived more recently than that of either species and gorillas. It is important to note that in our example here, the tree shows relative but not absolute time; our branch lengths do not reflect exactly when the ancestors might have lived. It is crucial to remember that it is the branching order of the tree that represents the evolutionary relationships among species and not the order of the tips. This can be an unfortunate source of confusion for some when looking at a phylogeny for the first time. We could quite easily alter the order of the species at the tip, but provided that the **tree topology** (i.e. the shape of the relationship represented by the tree) remains the same, our phylogenies are equivalent. The tree in Figure 9.2B has our tip taxa in a different order but otherwise shows exactly the same relationship or topology.

Trees can actually be drawn in a number of different ways, each conveying different types of information. It is not our aim here to give an overview of all of them (however see Page & Holmes 1998 for a good introduction), although we will encounter several types as we delve further into the topic of phylogenetics. Nonetheless, the presence or absence of a root is fundamentally important for determining how a tree is visualized and also how it can be interpreted. The two trees we have seen so far in Figures 9.2 are both **rooted**. As we saw previously, the root of a tree is the common ancestor of all taxa in our phylogeny. For simplicity, with our example we used the split between orangutans and all other great apes. However, if we were less sure about the evolutionary relationships among the taxa we were inferring a phylogeny for, we might instead use an outgroup—a species we are certain shares a much more ancient common ancestor with all our taxa of interest than any of them do with one another. If we had no outside information about relationships within the Hominidae, we might instead use a gibbon (Hylobatidae) or an Old World monkey such as the rhesus monkey (*Macaca mulatta*). There are other methods to root trees too, such as choosing the midpoint between the two most divergent taxa on a tree. Why do we care so much about the root? Well, it determines how we interpret a phylogeny. Without it, the tree has no representation of direction or relative divergence in evolutionary time. Figure 9.3 shows our Hominidae tree, but this time it is unrooted. In fact, there are multiple ways to draw this tree, which is why there are three versions of it in the figure.

You might wonder at this point what is the purpose of an unrooted tree? They are however, important for many of the phylogenetic inference methods we will learn about shortly. They are also useful for demonstrating that any phylogenetic tree is a hypothesis about evolutionary relationship and that there are limits and challenges of phylogenetic inference. How many unrooted trees can we construct from *n* taxa? We can use the following equation from Yang (2014):

$$U_n = \frac{(2n-5)!}{2^{n-3}(n-3)!} \qquad (9.1)$$

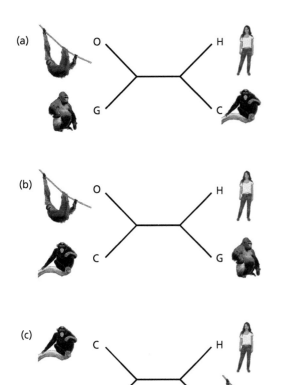

(a)

(b)

(c)

Figure 9.3 The three possible unrooted topologies for our Hominidae tree. Note that in each of them there are four branches for the tips and an additional branch joining the two halves of the tree—so five in total. We can place the root on any of these five.

Don't be put off by the exclamation mark; it is just signifying that we need to use factorials here. A factorial of n is simply the product of all numbers less than or equal to it. In other words, $4! = 4 \times 3 \times 2 \times 1 = 24$. In our higher primate example, so far we have focused on only four species; so, using equation 9.1, this means there are 3 possible unrooted trees—all of which you can see in Figure 9.3. On any unrooted tree, there are $2n-3$ branches, so for our four taxa all unrooted trees will have 5 different branches, one for each tip and also an additional branch joining the two segments of the tree. Since the root is a hypothesis, we can place it on any of these branches, meaning that to calculate all possible rooted trees from a set of n taxa, we should use the following equation:

$$R_n = U_n \cdot (2n - 3) \qquad (9.2)$$

So, for our simple four species tree, we have 3 unrooted trees (Figure 9.3), each with 5 branches on which we can place the root, meaning that we have $R_n = 3 \times 5 = 15$ possible rooted trees. We can use equations 9.1 and 9.2 to demonstrate just how difficult it might be to identify evolutionary relationships among even just a relatively small group of taxa. For example, let's say we wanted to create a phylogeny for 20 species; that means we can create approximately 2.22×10^{20} unrooted trees and 8.20×10^{21} rooted trees—that's 8.2 sextillion possible rooted trees. To put that in perspective, that's more than the approximate 100–400 billion ($2.5 \times 10^{11} \pm 1.5 \times 10^{11}$) stars in our galaxy, the Milky Way! Such huge numbers pose an obvious problem if our aim is to identify which tree represents the true evolutionary relationship among species.

9.1.3 The basis of phylogenetic inference

After Darwin laid the groundwork for evolutionary taxonomy, early systematists built upon his ideas of common descent to infer relationships among taxa. Many of the principles set out during those early years of evolutionary taxonomy are still fundamental to phylogenetic inference today, although the methods have changed. The first important principle recognized by Darwin and those who came after him is that **homologous** traits are the key to identifying relationships among taxa. Homology arises as a result of common descent and so closely related species will show a greater similarity in such traits. In phylogenetics, homologous traits are typically called **characters** and they can exhibit different **character states**. On a phylogenetic tree, a taxon or monophyletic group that has a character state that is different from the root and the outgroup has a **derived state**. In contrast, the shared outgroup and root state is **ancestral.** The second important principle established by early taxonomists is that it is crucial to distinguish between homology and **homoplasy**— i.e. when similar looking characters have arisen independently of one another. There are many examples of this in evolution—i.e. whales and snakes are both vertebrates which have experienced hindlimb loss but are evolutionary highly divergent and likely shared a common ancestor some 353 million years ago (Hedges et al. 2015). Homoplasy is caused by the repeated evolution of a character state either in **parallel** (i.e. from the same ancestral background) or by **convergence** (i.e. from different ancestral backgrounds). We have already seen some good examples of parallelism in previous chapters—for example the parallel loss of armor plates in threespine stickleback (Colosimo et al. 2004). Convergent evolution is typically considered to have occurred across deep evolutionary time, much like our whale and snake example or the repeated evolution of eyes in invertebrates and vertebrates. It is important to note, however, that there is some debate over how to draw the line between parallelism and convergence and whether the distinction is even worth making. Indeed, homology is also somewhat dependent on what sort of character we are examining—the phenotypic traits themselves might not be homologous but the genetic basis or developmental basis for them might be. Indeed, there is evidence that this might be the case for hindlimb loss (Chan et al. 2010).

The distinction between homology and homoplasy actually provides a nice demonstration of a third important principle of phylogenetics, **parsimony**. Parsimony, also known as Occam's razor, means that when choosing between alternative hypotheses we should choose the one that requires the fewest assumptions. This is actually an important principle for science and hypothesis testing in general, but in the case of phylogenetics it means that when determining relationships among taxa, we should pursue those that that require fewer leaps of logic. The evolution of wings from vertebrate forelimbs provides a good demonstration of how parsimonious thinking works. Birds and bats are both vertebrates and by rough approximation it could be argued they share homologous winged forelimbs. Under this hypothesis, we would group bats on a phylogenetic tree and show them as sharing a more recent common ancestor with birds than any other form of mammal. However, in order for this hypothesis to be a fitting explanation, it means that bats have independently acquired all the traits we use to define them as mammals *after* they diverged from the common ancestor they shared with birds. This of course is extremely unlikely; a more parsimonious explanation is that wings in bats and birds are

homoplasmic and that instead winged flight has evolved at least twice in vertebrates.

Up until this point, you will notice that there has been little mention of the role of genetic or genomic data in phylogenetics. This is intentional, to demonstrate that the principles that underlie traditional phylogenetic inference can also be applied to contemporary data. Modern phylogenetics almost exclusively uses DNA or protein sequence data in order to determine evolutionary relationships. Thus, characters are variable positions in a DNA sequence and the character states used to determine relationships are the nucleotide bases—A, C, T, and G. Similarly, the principles of homology and homoplasy are also applicable to DNA sequence data. For example, at a given variable position for three sequences, two of them might share an A whereas the third has a T. Are these character states at this position homologous? This is only true if the two A alleles we observe are identical by descent—i.e. they are derived from the same common ancestor. Alternatively, they might be homoplasious in the sense that the A allele has arisen twice, independently, via a recurrent mutation in the same position of the sequence. For a set of sequences that share a very recent common ancestor, the probability of recurrent mutation is quite low but, as we shall see later, the probability increases for sequence data from taxa that have been divergent for a long time.

At a step up from the level of the DNA sequence, we can also consider homology in terms of genes. Evolutionary relationships among genes can follow a branching, bifurcating pattern much like organisms. Two genes that occur in two species and share a common ancestor are **orthologous**. However, a gene can be duplicated leading to multiple copies of similar genes segregating in a species. We learned about gene duplication in chapter 2. When this occurs, the duplicate is not orthologous but is instead **paralogous**. Genes that are paralogs of one another experience divergent evolutionary histories; for example, a duplication event might mean that one paralog experiences a loss of function mutation and becomes a pseudogene and is then free to evolve neutrally as further mutations have no fitness consequence. Alternatively, paralogs may diverge and take up slightly different functions, as with the gene duplication of red opsin pigment genes in Old World monkeys and hominids. In this case, one of the paralogs has diverged to become sensitive to wavelengths in the green light spectrum, enabling trichromatic vision (see chapter 2, section 2.3.3). Confusing orthologs and paralogs can therefore cause problems with phylogenetic inference, as they will no longer accurately reflect the evolutionary relationship among taxa.

9.1.4 Alignment and models of sequence evolution

As we have seen, homology is fundamentally important for constructing a phylogeny. How do we determine homology with molecular data? To achieve this, we perform an **alignment** of sequences. In principle alignment is straightforward; in a simple case with four homologous sequences (Figure 9.4), we find which bases in sequence 1 match the bases in sequence 2 and then line the two sequences up so they correspond (Figure 9.4). Now we add sequence 3, which is three base pairs shorter than 1 and 2. There are several possible ways we could align this sequence but it is clear that it is missing three nucleotides relative to sequence 1 and 2, most probably due to one or more insertion–deletion (indel) polymorphisms (see chapter 2). We define this in our alignment as **gaps**. When placing gaps in the alignment we assume as few mutations as possible. In the example in Figure 9.4 the sequences align nicely if we assume that sequence 3 has three consecutive gaps and is missing the sequence CTC. Adding sequence 4, we see that it is the same length as 1 and 2, and it lacks the indel carried by sequence 3. However, sequence 4 has a C at the forty-third position whereas all other sequences carry a T. This is a polymorphism and it

is a **substitution** in our alignment. With a combination of allowing both gaps and substitutions, we can therefore align any set of sequences.

Although we could allow any number of gaps or substitutions, our aim when aligning is to produce the alignment that best preserves homology among the sequences. In other words, we want our alignment to be as true as possible to the evolutionary relationships among the sequences. Therefore, we need to calculate a statistic that gives some kind of cost or weight to different alignments.

$$D = s + wg \qquad (9.3)$$

Where s = the number of substitutions, g is the total length of gaps, and w is a **gap penalty**. This last parameter is based on our assumption that the evolution of gaps between sequences is less common than substitutions. If we set w to 1, we impose no penalty on including gaps in our alignments—i.e. they are just as permissible as substitutions. However, if we increase w to 3, a gap costs three times as much as substitution does. Irrespective of how we set the gap penalty, the simple equation in 9.3 allows us to calculate a cost for each of the alternative alignments for a set of sequences. We then choose the alignment with the lowest possible cost. Things become a little more complicated when we are aligning more than two sequences—i.e. a **multiple alignment**. However, despite this the overarching principle is the same—we seek to choose the alignment with the minimum cost across all different alignments.

Once we have aligned a set of sequences and we are confident in the homology among base pair positions, we next seek to calculate some form of **genetic distance** among them. It is this genetic distance that will allow us to construct our phylogeny.

```
1 AGTCCCTGAACCTTTTAGCAGCTCGTGTACTGTGCAACCAGGTTGGAAAA
2 AGTCCCTGGACCTTTTAGCAGCTCGTGTACTGTGCAACCAGGTTGGAAAA
3 AGTCCCTGAACCTTTTAGCAGGTGTACTGTGCAACCAGGTTGGAAAA
4 AGTCCCTGAACCTTTTAGCAGCTCGTGTACTGTGCAACCAGGCTGGAAAA


1 AGTCCCTGAACCTTTTAGCAGCTCGTGTACTGTGCAACCAGGTTGGAAAA
2 AGTCCCTGGACCTTTTAGCAGCTCGTGTACTGTGCAACCAGGTTGGAAAA
3 AGTCCCTGAACCTTTTAGCAG---GTGTACTGTGCAACCAGGTTGGAAAA
4 AGTCCCTGAACCTTTTAGCAGCTCGTGTACTGTGCAACCAGGCTGGAAAA
```

Figure 9.4 A simple example of a sequence alignment. In the upper panel, four sequences of different lengths are unaligned with one another. In the lower panel, the sequences are fully aligned to each other based on nucleotide homology. Note that the different length of sequence 3 means that a gap is introduced (denoted by the dashes). Also note that there are two polymorphic positions.

When sequences are fairly closely related to one another, this is a straightforward task. We simply count the number of substitution differences between a pair of sequences, assuming that any difference between them is as a result of a substitution on the branch extending from their common ancestor. So, if we align two 500 bp sequences which differ due to substitutions at eight positions, the genetic distance between them is 1.6 percent or $d = 0.016$ where d is the proportion of nucleotides that are different between sequences. In principle, this is similar to the statistics for sequence diversity we learned about in chapter 7—i.e. nucleotide diversity and the number of segregating sites. However, researchers working on phylogenetics are often faced with a set of sequences from species separated by large evolutionary distances. In this case, what we observe as a single substitution difference can actually mask a large number of possible histories and different substitution types. For example, multiple substitutions might occur at a single nucleotide but we are only able to observe a single difference. Similarly, even invariant sites might have undergone repeated substitutions resulting in no apparent difference between sequences, despite the fact that substitution has actually occurred. This has the clear implication that the sum of differences between two sequences is not necessarily a reflection of the true level of divergence between them. We can see this more clearly when we look at the number of differences between genes and the estimated divergence time between them; genetic distance between sequences is linear for a relatively short period of time but eventually reaches a saturation point where multiple substitutions at the same site occur and so we are unable to detect further divergence (Figure 9.5).

How then can we more accurately measure genetic distances between sequences? The answer to this is to use a model of sequence evolution in order to correct the observed genetic distance. There are a large number of different sequence evolution models; Yang (2014) lists eight commonly used models but many others exist too. It is beyond our scope to introduce all of them here. We will instead investigate some of the common features of the models and also introduce the simplest models in detail. In doing so, we hope it will become apparent that

many sequence evolution models are actually closely related, differing mainly in the number of different parameters used to define them. For more detailed overviews of sequence models, we refer readers to Page & Holmes (1998) and also Yang (2014). The most basic parameter a sequence model requires is Pij—i.e. the probability that base i will undergo a substitution and become base j at the end of a given period of time. Since there are four nucleotides, there are 16 different possible substitution probabilities, including the probability that a nucleotide will remain the same over time. A typical model will summarize all 16 probabilities in a substitution probability matrix—P^t (see Figure 9.6). Another key component of a sequence model is the frequency of the different bases f_i where i is A, C, G, or T. These summarize in a vector f and must equal 1.

At the moment, sequence models will likely seem a bit abstract to you but by exploring the simplest sequence model, the Jukes–Cantor (JC), it should hopefully become clearer. The JC model simplifies all its parameters by first assuming equal frequencies for all bases (i.e. $f_A = f_T = f_C = f_G = 0.25$) and then assuming that all substitutions have an equal probability too. Therefore, P_{AT} is the same as P_{AC} and

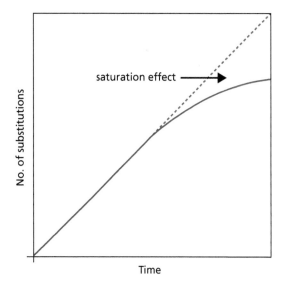

Figure 9.5 The issue of saturation with DNA sequence evolution. The number of substitutions in a sequence does not scale linearly with time because of recurrent mutations at some positions. This leads to a saturation of substitutions and thus inaccuracy in phylogenetic inference based solely on uncorrected genetic distance.

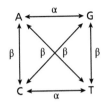

Figure 9.6 An example of two commonly used sequence models. On the left is the Jukes–Cantor model and on the right is the Kimura 1980 model. In the Jukes–Cantor model, all transitions between bases are equally probable and thus the model requires only a single substitution probability parameter, α. In the Kimura 1980 model, there are two substitution probabilities, α for transitions and β for transversions. Note that both models also include an additional parameter not shown here, f, a vector of nucleotide frequencies.

is denoted by the probability α (see left panel, Figure 9.6). With these assumptions in place, we calculate the corrected genetic distance, dc as follows:

$$d_c = -\frac{3}{4} ln\left(1 - \frac{4}{3}d\right) \qquad (9.4)$$

Where d is the same as we defined previously—the proportion of bases that are different between our chosen two sequences. Unfortunately, actually deriving equation 9.4 from the assumptions we made for the JC model is quite a complicated process. Again, we refer interested readers to Page & Holmes (1998) for an introductory overview of sequence models and also Yang (2006, 2014) for a more in-depth assessment of this topic. Nonetheless, the correction in 9.4 is made using the assumed substitution rates and nucleotide frequencies we described above.

One of the key assumptions of the JC model is that all substitutions occur with equal probability. However, as we learned in chapter 2, nucleotides are either purine (A and G) or pyrimidine (C and T) and so transitions (i.e. purine to purine or pyrimidine to pyrimidine) are more common than transversions (i.e. purine to pyrimidine or vice versa). To account for this in models of sequence evolution, Motoo Kimura extended the JC model such that there are two probabilities for substitutions—α for transitions and β for transversions (right panel, Figure 9.6). For example, in Kimura's model (known as Kimura 1980 or K80), $P_{CT} = α$ but $P_{CG} = β$ (right panel, Figure 9.6). The JC model also assumes that the frequencies of the different nucleotides in our

sequence are equal. In 1981, Joseph Felsenstein altered the JC again to allow different frequencies for each nucleotide—this is known as the F81 model. Hasegawa et al. (1985) then combined both the F81 and the K80 models into the HKY85 model—which allows nucleotide frequencies to vary and also for different substitution probabilities for transitions and transversions. Finally, a more complicated model, the GTR or general reversible model was developed (Yang et al. 1994), allowing variation in nucleotide frequencies but also giving each different type of substitution its own frequency.

All the models we have described here are variations on the basic features of the JC model. The GTR has 16 substitution parameters and four nucleotide frequency parameters. However, if all four nucleotides are equal in frequency and we set all substitutions to equal, we reduce our model to just two parameters and we once again have the JC model. Therefore, in this case, we can refer to these models as **nested**—constraining parameters in the more complex model reduces to a simpler model. We note once again that there are in fact many more models than we have touched upon here. Faced with such a multitude of models to use, how can we choose the correct one for our data? A common way to do this is to use likelihood calculations to determine how well a model fits the data. We will learn more about likelihood in section 9.1.6.

9.1.5 Building trees using distance and character data

Genetic distance among sequences provides one means of summarizing the evolutionary relationship among them too. **Distance methods** in phylogeny are therefore based on the assumption that genetic distance reflects evolutionary relationship among taxa. We can leverage the genetic distances that we learned to estimate in the previous section to construct and visualize a phylogeny for our sequences. Figure 9.7 shows two simple distance trees derived for four sequences. These are an example of **additive distance**—i.e. the sum of all the branch lengths between two sequences is equal to their genetic distance in the distance matrix. With just four taxa, it is possible to easily construct distance trees like this by hand. However, with ten taxa, we have 17 branches and 45 pairwise differences—you can see it quickly becomes far too difficult, time consuming, and generally boring to create a tree by hand! Thankfully, modern evolutionary biologists use computer-based algorithms. Two of the most well-known are the **unweighted pair group method with arithmetic means (UPGMA)** and the **Neighbor Joining** algorithm (Figure 9.7).

Despite its unwieldy name UPGMA is a very simple method. From a distance matrix (Table 9.2), we first choose the pair of sequences with the smallest distance between them and combine them; we then recalculate the distance between this pair and all others and choose the next pair. This is repeated until all sequences or taxa in our matrix are joined

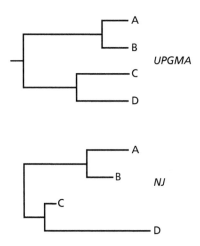

Figure 9.7 Examples of additive tree construction. The upper tree is an UPGMA, which assumes a constant rate of evolution across the tree under a molecular clock mode. Therefore, the UPGMA tree is ultrametric, i.e. all taxa on it equally distant from the root. The lower tree is a neighbor-joining tree, which is unrooted and first joins the pair of taxa with the shortest genetic distance between them. The rest of the tree is then constructed using an algorithm that joins subsequent nodes together based on distance too.

Table 9.2 A pairwise genetic distance matrix for the four taxa in Figure 9.7.

	A	B	C	D
A				
B	0.2			
C	0.8	0.8		
D	0.8	0.6	0.3	

to one another. UPGMA assumes that there is a constant rate of evolution across the tree—i.e. it conforms to the idea of a constant molecular clock (see chapter 7). This means the tree produced by UPGMA is rooted, with all taxa equally distant on the tree from the most divergent. UPGMA is fast and efficient but because of its assumption of a constant clock rate among sequences, it does not work well when applied to datasets spanning large distances of evolutionary time.

Neighbor Joining (NJ) uses a cluster approach to construct a tree that makes it very fast and straight-forward to use. The algorithm begins with a completely unresolved network among taxa, which initially resembles a star. Based on the shortest distance between two taxa in the network, they are joined at a new node and their distance to the node is recalculated. Then the algorithm finds the next closest taxa and combines these into another new node until all taxa in the network are resolved as a tree. At face value, UPGMA and NJ may seem equivalent, however it is important to stress they are not! Crucially, NJ recalculates the distance from the paired taxa and the new node it creates for them whereas UPGMA defines distances from the original distance matrix. This means that NJ produces an unrooted, additive tree while UPGMA produces a rooted, **ultrametric** tree—i.e. all taxa are equally distant from the root.

Distance based methods for constructing trees are only ever as good as the estimates of distance between the taxa or sequences included in the analysis. As we saw previously (section 9.1.4), it is not easy to estimate distance over long evolutionary timescales. For this reason, distance methods are rarely used to resolve modern, large-scale phylogenetic datasets that include many taxa. Distance measures also lack information; if we summarize relationships between taxa using the proportion of nucleotide differences between them, we overlook information on which substitutions have occurred and where in the evolutionary history they arose. Nonetheless, distance measures are often employed for analyses of population data, i.e. when only SNP data is available, or to reconstruct phylogenies among closely related species. As we will see shortly, more sophisticated methods exist to exploit exactly the sort of information distance methods cannot incorporate.

Unlike distance methods for inferring a phylogeny, discrete or character-based methods use sequence data to reconstruct where in the phylogeny substitutions have arisen. This allows us to identify which substitutions contribute to the length of the branches on a tree. The most basic discrete method is **maximum parsimony (MP)**. We already learned about how the principle of parsimony is applied to phylogenetic thinking in section 9.1.3; here we apply it directly to sequence data. So, when faced with a set of sequences, a maximum parsimony approach is concerned with estimating a tree that requires the smallest amount of changes to explain the observed data. In Figure 9.8, we can see that there are multiple ways to construct a phylogeny for our sequences. We can imagine multiple substitutions and back mutations or we can go with a much simpler, more parsimonious explanation where a single mutation can explain the tree topology.

Clearly, we need a metric that will allow us to evaluate how many changes have occurred on a tree—this is known as the tree length or **parsimony score**. It is simply the sum of all the site substitutions required for the data to fit a proposed tree. Not all sites from sequence data are useful for an MP approach. Most obviously, invariant sites are uninformative, but this is also true for singleton sites—i.e. those occurring in just one taxon—as they will always require just one substitution to be placed on the tree. For sites to be informative, they must have at least two states (i.e. be polymorphic) and occur in at least two taxa, like in the left panel in Figure 9.8. However, there are also derivations of

MP which include the weighting of sites, to boost the importance of certain informative sites; for example, we might add a weight to transversions as opposed to transitions, reflecting the rarity of the former. Finally, an important point is that the character/discrete approach used by MP also allows us to reconstruct the character states at ancestral nodes—this is known as **ancestral state reconstruction**.

One thing that we have not dealt with so far is a method to test for how sampling error might affect our tree construction. For example, what if the sequences we use are incomplete and so we miss out on including some phylogenetically informative sites? How can we get a sense of how precise our phylogenetic estimates are from our data in this case? To do so, we can use **bootstrapping**. This is a straightforward statistical approach where we resample from our data with replication. In the case of tree construction, we might randomly resample the number of informative sites from our data multiple times and then redraw the tree each time. If we did this say 1000 times and in 980 of the trees we see the same support for a node between two of our taxa of interest, we can say that node has 98 percent bootstrap support. Bootstrap support is a common metric in phylogenetic studies, and as a general rule of thumb any node with a value of 95 percent or greater can be considered well-supported by the data (Page & Holmes 1998).

9.1.6 Maximum likelihood tree estimation

Likelihood provides us with a statistical framework to evaluate whether a model fits the data we observe; it can also be used to choose between alternative models. Although in this chapter we will focus on the application of likelihood in a phylogenetic context (i.e. to assess tree models), it actually has a wide array of uses in evolutionary biology and statistics beyond this. Likelihood is essentially the probability that we would observe our data given that the assumptions for how the data came to be (i.e. our model) are true. To demonstrate the principle of how likelihood is applied, we will briefly step away from the world of trees and phylogenetics to imagine a more illustrative, real world example. You are on a hike through the forest and you come across a large pawprint in the mud ahead of you. It is large,

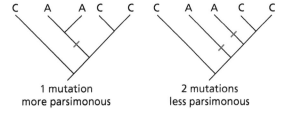

Figure 9.8 Constructing a tree with maximum parsimony. The two panels show different tree topologies. On the left, a single mutation can explain the relationship between the two A taxa. On the right, the alternative topology requires us to assume that two independent substitutions have occurred. Since the left-hand tree requires only a single substitution, we choose this topology as it is the most parsimonious.

bigger than your hand, with a wide pad and clearly five, very long claw marks. For a moment, you wonder whether a dog could have made it—but that would have to be an extremely large dog! Perhaps it was a tiger? The size would certainly fit that model but tigers only leave four claw marks and besides you are in a country with no wild tigers, so the likelihood of this is low too. Then you remember there are grizzly bears in this region, they are certainly large enough to make a paw print like this and thus the likelihood this is the correct model for the data is high.[1] We can more formally characterize likelihood (*L*) as follows:

$$L = P(D \mid M) \qquad (9.5)$$

Where $P(D \mid M)$ is the probability of the data conditional (i.e. dependent) on the model we assume. In our forest-based example, D is the paw print and M is the animal we suspect as having made it. We also saw from this example that when evaluating likelihood, we want to find the model that maximizes the probability of our data—i.e. we search for the **maximum likelihood**. Back to phylogenetics, we can apply the same approach—just that this time D is our set of sequences or taxa and M is our tree. The likelihood for any given tree is a function of the tree topology, the branch lengths, and the substitution rate, which is dependent on the model of sequence evolution used. Likelihood calculations for a tree first find **the maximum likelihood estimate** of the branch lengths and then the same for the topology. If the method works correctly we will have an estimate of our tree that best fits our sequence data, under our assumptions, such as a given sequence model.

A particularly useful feature of the ML framework is that it is possible to compare different models. One way to do this is to simply compare likelihoods among different models, looking for the highest likelihood among them. This, however, runs into problems when the magnitude of difference between different likelihood estimates is small. It is not a strict test of the difference in likelihood and should be avoided when formal tests are possible. The most classical formal test is the likelihood ratio test. The likelihood ratio statistic D is defined as:

$$D = 2\big(ln(L_1) - ln(L_0)\big) \qquad (9.6)$$

Note that since likelihood values are often very small, we transform them with the log function (ln) to produce the **log likelihood**. D is therefore the difference between the log likelihood of the two models, L_0 and L_1. Strictly the log-likelihood test is best applied when models are nested—i.e. one model is a simplified case of the other, achieved by constraining a parameter in the more complex model. We demonstrated this principle in 9.1.4 where setting all nucleotide frequencies as the same and substitution probabilities as equal will reduce the GTR to the Jukes–Cantor model of sequence evolution. When this is the case, we can use the χ^2 distribution to test the significance of D.

9.1.7 Bayesian phylogenetic inference

Probability theory is generally something that humans find hard to grasp[2] but the beauty of **Bayesian probability** is that it is closely modeled on how we actually observe and update our understanding of the world. Imagine, for example, you enter a 100 m race with 9 other people. With no prior knowledge, you might assume you all have a 1/10 (or 0.1) probability of winning. Then you learn that the person running next to you recently returned from the Summer Olympics with a gold medal for the 100 m sprint—would you give everyone an equal probability of winning now? This simple example shows that prior knowledge can alter our estimates of probability and this is the essence of a Bayesian approach. The Reverend Thomas Bayes, an English statistician and church minister, formalized this way of thinking in the eighteenth century with Bayes theorem, the basis for Bayesian probability:

$$P(A \mid B) = \frac{P(B \mid A)P(A)}{P(B)} \qquad (9.7)$$

This equation may look daunting at first but it is actually fairly simple and we will break it down with a simple example before turning to its applications to phylogenetics. We have a box in front of us filled with two types of cookies—light and dark chocolate. We're quite hungry, so we take 10 cookies

---

[1] The likelihood you will tread carefully after this discovery depends on how fresh the paw print is…

[2] We certainly include ourselves in this category!

from the box at random. We end up with 2 light cookies and 8 dark cookies. In a standard, forward probability framework, we might know the proportion of light chocolate cookies and we then would estimate the probability of getting our sample. However, if we don't know the proportion of light cookies in the box, we could use the information from our sample to estimate it; this is the Bayesian approach and is an example of reverse probability. To see how this maps to equation 9.7, we will use A to represent the proportion of light cookies in the box and B to represent the light cookies in the sample. We can see then that the first part of 9.7, $P(A|B)$ is the reverse probability problem—i.e. what is the probability that A of the cookies in the box are light given that we sampled B light cookies? In other words, what is the probability of our parameter given the data? In Bayesian inference, this is referred to as the **posterior probability.** However, we cannot actually estimate this without specifying some prior knowledge of what we think the value of A is. We do this using a **prior**, $P(A|B)$ in equation 9.7. Typically, we draw our value for a prior from a prior distribution that encompasses all possible values we might want to assign to A. With no knowledge at all, we might allow any value between zero and one but here we already know it cannot be zero, as we sampled two light cookies. The other properties in equation 9.7, $P(A)$ and $P(B)$ are the

probabilities of A and B independently; these are the marginal probabilities but they can only be calculated if we integrate over all possible values of both. In a simple example like this, it is possible to do this by hand but it is not straightforward and certainly not something we have space for here! For more complicated applications of Bayesian probability, such as to a phylogeny, integration is impossible without computational assistance. Since the method requires high levels of computational brute force, it has only become popular in the last few decades, but as we will see it is a very powerful and important methodology.

To demonstrate how Bayesian inference works in a phylogenetic context, we will again use a simple example of a tree with just three taxa, the domestic dog (*Canis lupus familiaris*), the gray wolf (*Canis lupus*) and the red fox (*Vulpes vulpes*). This is a simplification of the true Canidae tree, which contains many more species (Lindblad-Toh et al. 2005). Just as we saw earlier with the primates, there are three different possible tree topologies with these three taxa (Figure 9.9). Like with all other tree construction methods, to run a Bayesian phylogenetic analysis we require a sequence alignment, a model of sequence evolution. However, we also require a prior. As we saw in our demonstration of how general Bayesian inference works, we use a prior to give an indication of what we expect might be the true parameter or

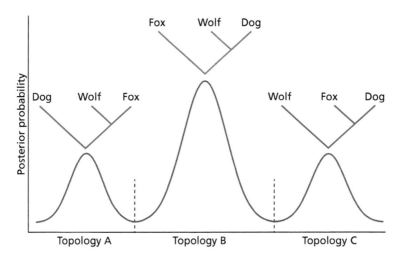

Figure 9.9 Simplified Bayesian estimation of a phylogeny between domestic dogs (*Canis lupus familiaris*), wolves (*Canis lupus*), and foxes (*Vulpes vulpes*). Here we test the three possible topologies and use posterior probability to determine which of these trees is best supported by the data.

model. If we have no information, we use an **uninformative prior**, where all three possible trees have the same probability (0.33 in this case). If we had some suspicion of the relationships among our taxa already, perhaps from fossil or archaeological data, we would use an informative prior, putting a greater weighting on a given tree topology—perhaps one where domestic dog is a sister species to the gray wolf (Lindblad-Toh et al. 2005).

Our aim with our Bayesian analysis is to find the posterior probability distribution of our different tree models. Recall from our earlier explanation of Bayesian cookies that the posterior probability is the probability of the model given the data. In our example here, we are estimating the probability of the tree given the data we have (and also given the sequence model we specified), so once we have run the analysis, we have our posterior distribution. From Figure 9.9, we see that our posterior indicates that the most likely tree is indeed one where dogs and wolves are sister taxa and the red fox is the outgroup. It might be helpful to think of Bayesian inference as updating our prior knowledge, like we did with our race example at the start of this section (Ronquist et al. 2009). However, for our simple example, we have so far ignored the fact that actually our model contains at least two parameters, the tree topology and the branch lengths, and may contain even more depending on what model of sequence evolution we choose. For this reason, depicting the posterior probability in two dimensions is an oversimplification; nonetheless it is a useful demonstration of the fact that there are different peaks of probability for phylogenetic inference and we are using our method to try and identify the highest one.

For a long time, it was simply impossible to apply Bayesian inference to phylogenetics because estimating the posterior probability distribution is incredibly difficult. To solve the Bayesian equation properly, we need to integrate over possible values of our parameters (i.e. the denominator in equation 9.7). For three topologies this is fairly straightforward, but then when we add branch lengths, which are continuous, we quickly see that the posterior probability distribution is huge. The problem becomes worse with more taxa in our phylogeny; we learnt from equation 9.2 that with 20 species, we end up with

8.20×10^{21} rooted trees. We cannot simply randomly sample from the posterior as it is just too big. To provide an analogy to help clarify this, imagine we are standing on a landscape, looking for the highest peak. Now imagine that landscape is huge—the size of North America for example. We could try walking randomly around to find it, but since the peak is so small relative to the rest of the landscape, it is very likely we would have to walk for an extremely long time before we could find it. The solution is to use a method that increases our probability of moving towards a peak; this is known as **Markov chain Monte Carlo sampling (MCMC).** MCMC essentially makes small changes to the parameters provided to the Bayesian equation in 9.7 and then recalculates the probability of the model given the data. If a suggested parameter value increases the probability, i.e. takes a step toward the peak on the probability surface, we accept it. Alternatively, a decrease in the probability, i.e. a downhill step, will only be accepted if it does not prove too great a misstep (Ronquist et al. 2009). In this way, MCMC allows us to explore the probability space more effectively and using multiple MCMC algorithms that independently converge on the same peak is a good way of having confidence in our estimation.

9.1.8 Gene tree and species tree discordance

For most of the examples and methods so far, we have treated the phylogenies estimated as the **species tree**, representing the evolutionary relationships among the species studied. However, if we were to construct a phylogeny from only a single DNA sequence from a single gene of each species, our phylogeny would not be a true phylogeny and would instead be a **gene tree**—i.e. the evolutionary relationship of that gene among species. In some cases, gene trees are consistent with the species tree and properly reflect the evolutionary history of the species studied. However, there are also many examples of **gene tree/species tree discordance**, where the two classes of trees do not agree. There are multiple reasons for why discordance might occur. We have already dealt with some; homoplasy at the sequence level, for example, can easily confound phylogenetic analysis and infer a gene tree that does not reflect the true species history. Furthermore, orthologs and paralogs, if we are not aware of them, might lead to an erroneous gene tree (see section 9.1.3). Another factor generating discordance is horizontal gene transfer, although this is less of an issue for eukaryotes (see chapter 2). We also learned about the effects of gene flow in chapter 8; phylogenetic discordance can be used to detect this and we will learn more about this in section 9.2.4. Although problematic, for accurate species tree inference these processes are likely to be rare enough in the genome to not cause widespread discordance (although in phylogenies with young species, gene flow has the potential to be quite significant).

The same cannot be said for **incomplete lineage sorting** (ILS) which is a consequence of population history and likely to occur in many taxa (Maddison 1997). To properly understand ILS, we should cast minds back to chapter 7 where we first encountered coalescent theory. Under the coalescent, any two genes or alleles will coalesce into a common ancestor backwards in time. Imagine we have three species, with the phylogeny ((A, B), C). If we pick two alleles within species A, some will coalesce before the divergence between A and B, whereas others will coalesce in the ancestral population, prior to the split between these species. This is because they were polymorphic in the ancestral population; this is referred to as **ancestral polymorphism**. As we learned in chapter 7, variation in coalescent timing like this is expected. However, problems arise for phylogenetic inference when we introduce our third species C. We sample three alleles from A, B, and C. If lineage sorting is complete, the alleles from A and B will coalesce in the AB ancestral population and the alleles from C will coalesce with the AB ancestor in the ABC ancestral population. However, what if there is a lot of ancestral polymorphism and all three alleles coalesce in the ABC ancestor? In this case, the order of coalescence is essentially random. So, if we were to examine gene trees from 90 sets of alleles that coalesce in the ancestor, 30 of them (i.e. one third) would follow our expected gene tree but the other two thirds would not, they would reflect alternative, discordant gene trees where A and C appear to show a more recent common ancestor than A and B. Another term for ILS is **deep coalescence** which better reflects the fact that the pattern arises in the ancestral population of a set of species (Maddison 1997).

There are many real examples of ILS, but one of the most striking comes from our own taxonomic family, the Hominidae. The species tree for the higher apes (Figure 9.9A) clearly shows that humans and chimpanzees share a more recent common ancestor with one another than either does with gorillas or orangutans. Yet there are three possible gene trees for this phylogeny (Figure 9.10 B, C, & D). However, based on coalescent theory, Hobolth et al. (2011) estimated genealogies for 0.9 percent of the human genome should share a more recent common ancestor with orangutans compared to chimpanzees, despite the fact that the human–orangutan divergence date is estimated to be at least 10 million years old, twice as large as the human–chimpanzee divergence time of around 5 million years (Scally et al. 2012). Using thousands of trees estimated from whole

(a)    (b)    (c) 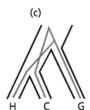   (d)

H   C   G   H   C   G   H   C   G   H   C   G

Figure 9.10 Examples of how gene trees for humans, chimpanzees, and gorillas can vary in node-depth and topology. Tree A) shows a gene tree consistent with the species tree. Tree B) shows a similarly concordant gene tree but this time with a deeper split between chimpanzees and humans which occurred in the ancestral population of all three species. Tree C) is an example of a gene tree that arises due to incomplete lineage sorting, where humans are more closely related to gorillas. Tree D) also occurs due to incomplete lineage sorting, but this time shows that chimpanzees and gorillas are more closely related than either is to humans. Trees B, C, and D all arise because coalescence occurs in the ancestral population of the three species.

genome alignment for humans, chimps, and orangutans, Hobolth et al. (2011) indeed found that 1 percent of the genome showed this human–orangutan relationship. The proportion of ILS is even higher when we consider the gorilla, a much closer relative than the orangutan; Scally et al. (2012) reported 30 percent of bases in a human, chimp, gorilla alignment showed either a MRCA for chimp–gorilla or human–gorilla (i.e. Figure 9.9 C & D).

Some authors have argued that viewing discordance generated by ILS as a problem for phylogenetics is the wrong perspective (Maddison 1997). Instead ILS should be viewed as an emergent property of how divergence unfolds between species; a species tree can therefore be considered a composite or summary of all the different gene histories. Indeed, we can actually infer a great deal from the proportion of ILS we observe in a phylogeny. ILS is predicted to be highest when ancestral effective population sizes are large and the divergence times between species are close to one another in time. Using similar predictions derived from the coalescent, Hobolth et al. (2011) used their estimates of ILS to infer ancestral effective population size for hominid lineages. Although 30 percent phylogenetic discordance between humans, chimps, and gorillas is considerable, speciation among the taxa clearly occurred separately in time and also with reasonable population sizes. In other examples, such as the recent split between three house mice subspecies, divergence has occurred much more recently (around 0.5 Myr) with effective population sizes in excess of 100 000, resulting in a much higher rate of discordance (White et al. 2009). Splitting the genome into 14 081 loci of an average size of approximately 100 Kb, White et al. (2009) estimated 39 percent of loci were consistent with the majority tree, whereas the two other possible phylogenies were supported by 36.3 percent and 24.7 percent of the loci respectively. The authors concluded that the imbalance among these phylogenies might also reflect the influence of gene flow, as well as incomplete lineage sorting.

In the early days of molecular phylogenetics, when only a single or few loci were used to generate trees, gene tree/species tree discordance posed a much more serious risk to the accuracy of inference. Phenomena such as ILS or confusion between orthologs and paralogs could easily distort phylogenetic inference and lead us to completely the wrong tree model. Now that thousands of loci are available due to high-throughput sequencing, this is less of a concern. Nonetheless, as phylogenetics has transitioned to the genomic era, this has led to new challenges which we discuss in the next section.

9.1.9 Phylogenetics at the genomic scale

In our overview of the roots of tree thinking (see sections 9.1.2 and 9.1.3), we saw how phylogenetic inference developed from phenotypic traits to molecular sequence data. Like most of the different fields of evolutionary biology we have covered in this book, phylogenetics has begun to adapt to the increasing availability, affordability, and ease of high-throughput sequencing. Species trees are no longer estimated using a handful of selectively amplified loci but instead from hundreds of genes and thousands of genome-wide sequences—phylogenetics has become the field of **phylogenomics** (Eisen & Fraser 2003; Delsuc et al. 2005). With the rise of phylogenomics new and sophisticated inference methods have been developed, specially tailored for some of the challenges that datasets of thousands of homologous characters entail (i.e. SNPs and genome-wide sequence data). However, the core principles of phylogenetics also lie at the heart of phylogenomic inference. Accurate reconstruction of a species tree relies on proper sequence alignment to find homology, a suitable model of sequence evolution to properly estimate divergence, and an appropriate tree estimation method, whether we are using a single gene or 250 (Rannala & Yang 2008).

Since phylogenomics is defined by large datasets, the approach easily overcomes the issue of **stochastic error**. Stochastic error occurs when we use an inadequate data source to infer a phylogeny—i.e. relying on only a single gene that might be discordant because of incomplete lineage sorting or horizontal gene transfer. As we learned in the last section, failing to take this into account can lead to acceptance of an incorrect tree model. Stochastic error is a case of distinguishing the phylogenetic signal from the noise and can only be solved with the inclusion of more data (Sanderson 2008). However, we still need to account for **systematic error**, when we mis-specify our model of evolution and therefore infer the wrong phylogeny (Lemmon & Lemmon 2013). A simple demonstration of systematic error would be to apply the wrong model of sequence evolution to properly calculate divergence among our aligned sequences, which would lead to inaccurate estimates of topology and branch length, no matter how much data we have. In this sense, the large size of phylogenomic datasets can actually exacerbate systematic error if handled poorly. If our dataset is of several hundred genes, do we apply a single model of evolution for all of them, or should we fit one for each? Since we know rates of mutation vary in the genome and recombination occurs among genes, this is not a valid assumption (Rannala & Yang 2008). However, fitting several hundred models is also undesirable, statistically unreliable, and computationally difficult (Rannala & Yang 2008).

Within closely related taxonomic groups, alignment of sequences or even whole genomes is possible. For example, the phylogenomic relationships among humans, chimps, and gorillas were reconstructed in this way (Scally et al. 2012). However, identifying homologous and orthologous data among extremely distant taxa (i.e. across different kingdoms of the tree of life) is much more challenging (Dunn et al. 2008; Burki 2014). Efforts to recreate the entire tree of life with phylogenomic data demonstrate the difficulty of this, as they typically rely on sequence data from highly conserved genes such as ribosomal proteins (Dunn et al. 2008; Hug et al. 2016). Often DNA from such divergent species is targeted with a capture sequencing array harboring probes specially designed to anneal to highly conserved genes or genome regions (Lemmon & Lemmon 2013). More recently, conserved genes have been assembled in divergent taxa using whole-genome sequencing. Sequencing and rapid assembly of genomes, known as metagenomics (see chapter 2), has revealed huge levels of microbial diversity (Hug et al. 2016). Completed and published genome sequences also provide a wealth of resources for authors to access and utilize in their phylogenomic studies. Indeed, some of the examples in this section made use of already published data to increase the number of taxa in their studies (e.g. Misof et al. 2014). At the time of writing, over 256 517 genome sequencing projects

are registered in the Genomes Online Database (GOLD) and this will almost certainly have increased by the time you have read this book! Phylogenomics therefore overcomes the need to design specific primers and perform amplification with PCR but instead introduces the need to perform locus selection for orthologous, informative datasets (Trautwein et al. 2012; Lemmon & Lemmon 2013; Dunn & Munro 2016).

Regardless of the method used to obtain a set of sequences or genes from which to estimate a phylogeny, large evolutionary distances mean that loss of genes in some lineages is much more likely (Burki 2014). An early, standard phylogenomics approach has been to arrange genes or sequences in a **supermatrix**; all genes are essentially concatenated together and treated as a single sequence for each species with missing data clearly marked (Delsuc et al. 2005) and a general, simplified sequence evolution model assumed (Rannala & Yang 2008). Using a supermatrix is straightforward and can be effective, especially with closely related taxa. For example, Wagner et al. (2013) used genome-wide sequence loci arranged in supermatrices to reconstruct the phylogeny of 16 Lake Victoria cichlid fish species. The authors tested a range of supermatrices with different levels of missing data. By only including loci present in 90 percent of individuals, they were unable to properly resolve the phylogeny, but by decreasing this threshold to 64 percent they found clear structuring among species and high bootstrap support (>80 percent) for all branches. The Lake Victoria adaptive radiation is estimated to have occurred over the last 15 000 years (Meier et al. 2017c); the assumption of a simple, genome-wide sequence evolution model implicit in the supermatrix may hold, even if it is unrealistic. However, such an assumption is unlikely to hold for more divergent taxa and can introduce serious bias into a phylogenomic analysis when incomplete lineage sorting is present (Kubatko & Degnan 2007; Rannala & Yang 2008).

An alternative approach to the supermatrix is the **supertree** or consensus tree. As with a standard phylogenetic analysis, a phylogeny is estimated for each gene or sequence using an individual sequence model and then a supertree is constructed from the consensus among all of the individually drawn trees. This approach is particularly popular in speciation genomics to examine phylogenetic variation across the genome; for example, White et al. (2009) used a consensus approach to examine patterns of incomplete lineage sorting among house mouse subspecies. Similarly, Martin et al. (2013) reconstructed phylogenies in 100 Kb sliding windows across the genome in order to identify genome windows where *Heliconius* butterfly species clustered by geography and not species topology, i.e. indicating genome regions shared by gene flow. By summarizing across all these trees, the authors were also able to visualize a species tree. However, supertrees approaches also have particular setbacks, particularly statistical issues from fitting too many sequence models—i.e. one for each gene, window, or sequence analyzed (Rannala & Yang 2008). More sophisticated phylogenomic methods using maximum likelihood and Bayesian inference are now preferable as these typically incorporate the distribution of gene trees to estimate a species phylogeny (Lemmon & Lemmon 2013).

Combining phylogenetics and genomics has led to an explosion of new findings that are changing our way of understanding taxonomy. Phylogenomic inference particularly provides insight into how and when diversity and evolutionary innovations have evolved on Earth (Dunn & Munro 2016). Insects provide an excellent example of this; they are a hugely important and diverse taxonomic class inhabiting many different niches, from parasites to pollinators. Indeed, recent estimates suggest there may be as many as 5.5 million insect species (Stork 2018). Well-studied insect fossils have given us some idea of when insects likely arose and diversified but the insect fossil record has been insufficient to resolve some evolutionary relationships and clearly date the first occurrence of important adaptations in the insect lineage. For example, when did winged insects appear? Misof et al. (2014) performed a phylogenomic analysis on insects, using 1478 highly conserved genes from over 144 species that best represent insect orders alive today. In order to ensure that phylogenetic error would not bias their results, the authors took extra care to only use genes found in all species and also used fossil data to calibrate the divergence times in their phylogeny. The divergence between non-insect

Hexapoda species (i.e. springtails) and all insect species suggests an origin for insects around approximately 480 Mya, which is 68 million years earlier than the oldest insect fossil. Some controversy also exists over when winged insects arose, with unambiguously winged insect fossils dating from around 324 Mya, although some suggest that even the oldest insect fossil has evidence of traits associated with flying (Engel et al. 2004). Divergence time estimates from Misof et al.'s (2014) phylogeny indicates that winged insects did split during the Devonian period (around 406 Mya), suggesting that insect flight, the first flight on Earth, may have evolved much earlier than first thought.

Like insects, birds are extremely diverse. The exact number of bird species is unclear; the IOC World Bird List has 10 771 entries, whereas the Clements Checklist has 10 550. Nonetheless, there are more species of birds than any other terrestrial vertebrate. Birds underwent a large radiation approximately 65 million years ago, following the Cretaceous–Cenozoic (K-C) extinction event (Jarvis et al. 2014) with much of the diversity of avian species having evolved in the last 50 million years (Jetz et al. 2012). Rapid bird species diversification has hindered the accurate reconstruction of a bird phylogeny. Earlier attempts to classify birds have also been made difficult by morphological similarities between groups and repeated adaptation to similar ecological niches. Furthermore, there are a number of bird lineages that contain multiple morphologically similar species but no obvious close relatives, such as owls, parrots, and the hoatzin (Figure 9.11). The hoatzin, also unfortunately known as the stinkbird, is the only member of its family and has a very unclear evolutionary relationship with other bird species (McCormack et al. 2013). A number of studies over the past decade have set out to use phylogenomics to resolve these issues. Hackett et al. (2008) published one of the first avian phylogenomics studies, setting out a foundation for further studies to build on. The authors used a 32 Kb alignment of 19 conserved nuclear loci from 169 species and used a maximum likelihood estimation method which allowed different parameters for each of the loci. Their phylogeny identified the deepest split among all extant birds between Palaeognathae, which includes ostriches,

232 EVOLUTIONARY GENETICS

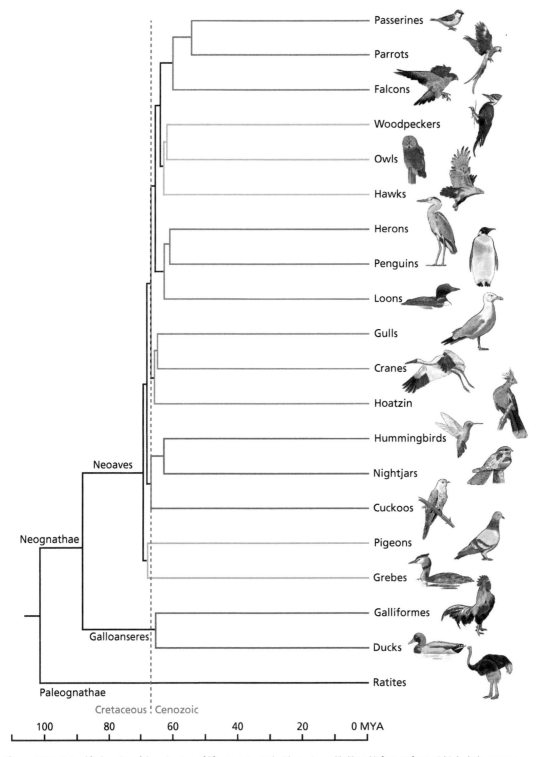

Figure 9.11 A simplified version of the avian tree of life, reconstructed with maximum likelihood inference from 48 bird whole genome sequences. The red-dashed line shows the Cretaceous–Cenozoic boundary which is the time of K-C extinction of the dinosaurs. Many lineages in the Neoaves clade diversified after this extinction event. Redrawn and simplified from Jarvis et al. (2014).

emus, and cassowaries, and all other birds, the Neognathae. They also identified a second split within the Neognathae between Galloanserae, which includes chickens and ducks, and the Neoaves. Both of these splits have been well resolved across multiple phylogenomic and morphological analyses (McCormack et al. 2013; Jarvis et al. 2014; Prum et al. 2015); as we will see, controversy in bird phylogenetics lies mainly in the Neoavian order which incorporates 95 percent, the vast majority, of extant bird diversity. Hackett et al.'s (2008) analysis also produced some unexpected results. They identified two Neoavian clades, the landbirds, including songbirds, hawks, and owls, and waterbirds, including storks and penguins. Within the landbirds, they showed that Passeriformes—the songbirds—were closely related to parrots (dark green lineages at the top of Figure 9.11). Furthermore, they found a lack of support for monophyly in some groups where it was expected. Most strikingly, falcons and hawks were not monophyletic but occur separately within the landbird clade (dark and light green in Figure 9.11). Later phylogenomic analyses focusing on the Neoaves also supported this (McCormack et al. 2013; Jarvis et al. 2014). Another major contribution to avian phylogenomics came from Jarvis et al. (2014) who assembled 45 bird genomes and incorporated three published genomes to produce a dataset which included 8251 orthologous genes and 3769 short, highly conserved genome regions to produce a 41.8 Mb alignment. From this, they constructed a maximum likelihood phylogeny which broadly supported many of the previous findings by Hackett et al. (2008) but also provided good estimates of the major events in bird diversification, including confirming that the diversification of Neoaves occurred after the K-C extinction event (see dashed red line, Figure 9.11). Importantly, Jarvis et al. (2014) also identified two new Neoavian clades—the Passerea and the Columbea, which separated out groups such as doves, flamingos, and grebes (orange clade in Figure 9.11). These clades were later challenged by Prum et al. (2015), who suggested they were artefacts of using too few species. Using target capture sequencing of 394 loci in 198 species and both Bayesian and ML methods, Prum et al. (2015) found no evidence of either clade. Prum et al. (2015) also found the hoatzin

diverged from all other landbirds around 65 million years ago.

Despite some level of disagreement, all of the avian phylogenomic studies we highlighted here have contributed significantly to our knowledge of how birds have diversified. For example, the surprising and consistent finding that hawks and falcons are not monophyletic suggests that the ancestor to all modern landbirds was likely predatory (see Figure 9.11; Hackett et al. 2008; Jarvis et al. 2014). However, the controversies of avian phylogenetics are likely to remain unresolved without more data from a much broader taxonomic perspective. Efforts are underway to achieve this—the Avian Genome project is aiming to sequence the genomes of over 10 000 species (O'Brien et al. 2014). In the meantime, however, phylogenomics studies in birds are an illustrative example of the fact that sequencing genomes alone is not a simple solution to complex challenges such as reconstructing diversification during an adaptive radiation. For example, Jarvis et al. (2014) clearly noted that none of the gene trees from their orthologous gene set matched their final species tree. Furthermore, we have not just experienced advances in sequencing technology; advances in methods for quantifying the morphology of fossils and extant species, such as microCT scans, have meant that phylogenetic inference based on morphology is still of value. This has been particularly well used to reconstruct the morphology of poorly preserved insect fossils (Trautwein et al. 2012) and to map the evolution of bird beak diversity onto the avian phylogeny (Cooney et al. 2017).

9.2 Phylogeography—genes in time and space

9.2.1 Phylogeography and its link to modern population genetics

Tree thinking and reconstructing relationships among species using phylogenetics is a powerful means of understanding evolution. However, while a phylogenetic perspective might give us an excellent viewpoint of the evolution of diversity at higher taxonomic levels, we learned from our examination of speciation in chapter 8 that a great deal of very interesting diversity evolves within species and populations. The diversity we see within species or between populations is also rarely randomly distributed—there is often a pattern of spatial structure. This is evident from phenotypic diversity within our own species—human skin color, body mass, and disease resistance all vary geographically (Novembre & Di Rienzo 2009). We also see spatial variation in human dietary traits such as the higher proportion of lactose tolerant individuals occurring in Europe and Africa, regions where pastoral agriculture has been established for long periods of time (see chapter 7). Spatial distribution is also clearly apparent when looking at regions of the genome that are not under selection too. For example, neutral population genetic structure in Europe clearly reflects geography with clustering apparent even between adjacent countries such as Ireland and the United Kingdom (Novembre et al. 2008). In order to understand the evolutionary history of such geographical patterns, we need to examine the evolutionary relationships of the genes involved in both time and space. This is the aim of **phylogeography**.

Phylogeography has often been described as bridging the gap between the macroevolutionary approach of phylogenetics and the microevolutionary focus of population genetics (Avise 2000; Edwards et al. 2016). This was certainly true in the early stages of the field, as researchers applied methods such as the phylogenetic analysis we learned about in the previous part of this chapter to intraspecific variation. As a field, phylogeography properly developed in the late 1980s and early 1990s. This allowed it to take advantage of two important developments in molecular and theoretical genetics—mitochondrial DNA sequencing and coalescent theory (Avise 2000, Hickerson et al. 2010). Phylogeography also pioneered several lines of thought that are now commonplace in modern population genetics. Most important of all was the concept that populations and lineages within species could be defined genealogically and that they could be considered as separate entities—often referred to as **operational taxonomic units** in the literature (Avise 2009; Hickerson et al. 2010). We will deal with this in more detail over the next two sections (9.2.2 and 9.2.3).

It is sometimes easy to imagine a scientific field as diverse as evolutionary biology being split into many distinct sub-disciplines that do not interact; i.e. phylogenetics, phylogeography, and population genetics. We hope that by outlining some of the origins of both phylogenetics and phylogeography in this chapter, we have not given this impression. We will spend some time on phylogeography here to illustrate the important contributions it has made to the modern approaches we take as evolutionary biologists and geneticists. Although we agree that phylogeography has in the past been a separate sub-discipline of evolutionary genetics, we argue that now it is fully integrated and its approaches form a core part of how researchers approach evolutionary questions. So, while a study might not explicitly state it is phylogeographical, it often draws deeply from this line of thinking. This is partly due to the fact that high-throughput sequencing technologies and advances in analysis methods (e.g. see sections 9.2.3 and 9.2.4) have made it much easier to tackle the sort of questions phylogeography was conceived to test (Hickerson et al. 2010). Before turning to phylogeography and the analysis of genetic structure and population history in more detail, we would argue that phylogeographical thinking has actually gone truly mainstream, beyond scientific research. The interest in human population structure and history is reflected in the popularity of personal DNA analysis kits that give estimates of ancestry, relatedness, and family history. Indeed, many of the most exciting and important recent advances in our understanding of human evolution have arisen from a phylogeographical perspective.

9.2.2 Phylogeography and biogeography

In one of the earliest phylogeography studies, Avise et al. (1979) examined mitochondrial diversity in pocket gophers (*Geomys pinetis*), a small burrowing rodent, from across Florida, Alabama, and Georgia in the Southeastern United States. They discovered mitochondrial DNA haplotypes varied among the gophers and that there was a strong geographical component to this, with apparently two lineages occurring in the eastern and western parts of the species distribution, respectively. They speculated that the geographical split might represent colonization of the eastern and western regions by different groups that had been spatially separated long enough to evolve around 3 percent divergence in mitochondrial sequence. Importantly, the authors visualized the relationships among the haplotypes using a **haplotype network** where the size of nodes are proportional to the frequency of the haplotypes and the links between them give an indication of the number of nucleotide differences separating them (Figure 9.12). In a single study, Avise and colleagues identified a connection between genealogical and spatial variation and also introduced a method to visualize intraspecific genetic variation that overcomes many of the issues of using a strictly interspecific method such as a phylogenetic tree.

Haplotype networks make it possible to visualize the fact that there are extant ancestors, to capture low divergence among haplotypes (i.e. one or two nucleotides) and to show reticulation as a result of recombination or gene flow (Posada & Crandall 2001). They can also greatly aid interpretation of phylogeographic data. Sætre et al. (2012) sequenced the mitochondrial DNA control region of 181 house sparrows from 17 different countries in order to study the genetic variation of the species across its native range (Figure 9.12). Crucially they included seven subspecies of house sparrow. They found that 60 percent of the individuals sequenced shared a single mitochondrial haplotype and the majority of the remaining 36 haplotypes differed from it by only a single nucleotide substitution. A lack of any mitochondrial

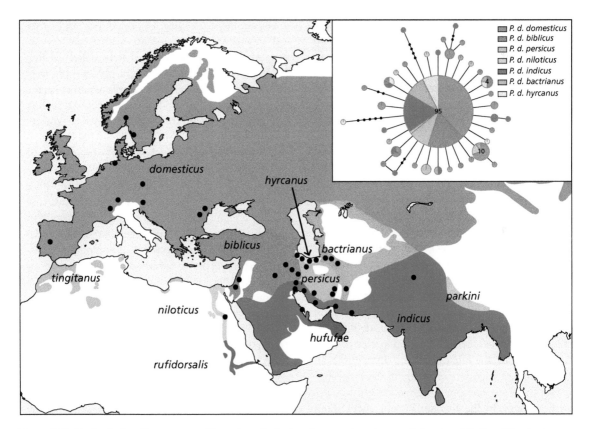

Figure 9.12 The distribution of house sparrow (*Passer domesticus*) subspecies across its native range in Eurasia and Northern Africa. Inset is a haplotype network constructed from sequences of the mitochondrial control region for 181 house sparrow individuals. The diameter of the nodes in the network is proportional to the frequency of the haplotype it represents. Therefore, the major node in the center is the most frequent haplotype and contains 60% of individuals. Haplotypes are connected to one another by branches and a single branch represents a single nucleotide difference. When more than one difference occurs, black dots on the branch indicate the number of substitutions. We can see most other haplotypes differ from the main node by only a single nucleotide difference. Redrawn from Sætre et al. (2012).

structure among house sparrow subspecies argued against a previous hypothesis that the species might be defined by ancient eastern and western lineages and nuclear loci also failed to support this pattern (Sætre et al. 2012). Since all subspecies of house sparrow except for the bactrianus sparrow (see chapter 7) are human-commensals, the authors concluded adaptation to an anthropogenic niche had likely occurred just once.

A great deal of phylogeographic research has focused on understanding how biogeographical events have shaped the genetic diversity observable in contemporary populations. Over the last 2.5 million years, during the Pleistocene,[3] the global climate underwent repeated oscillations from glacial to interglacial cycles. In Europe for example, glacial periods resulted in large-scale ice advance and then subsequent ice-free interglacial periods. At the height of the last glacial period, known as the **Last Glacial Maximum,** some 25 000 years ago, much of Northern Europe and North America were covered in colossal ice sheets, with some parts of the ice over a kilometer thick. As we saw briefly in chapter 8, this had a profound impact on the distributions of temperate species, with many surviving in more southerly **glacial refugia** before recolonizing temperate habitats following ice retreat (Hewitt 2001; Provan & Bennett 2008). Taberlet et al. (1998) used a comparative phylogeographical approach to examine the effect of Pleistocene glacial cycles on 10 diverse taxa in Europe; their study focused on mammal, amphibian, insect, and plant species in order to try and find generalities among these species. Although all species had recolonized temperate habitats following ice retreat, there was very little concordance among all 10 taxa—they apparently used different recolonization routes at different times. Nonetheless, some common patterns were apparent; firstly the highest genetic diversity in all taxa examined occurred in southern Europe, suggesting these species spread from southern refugia, most probably in the Iberian, Italian, and Balkan peninsulas. Furthermore, in Northern Europe diversity is shaped by which lineages recolonized which

areas; for example, in *Quercus* white oaks, populations in Britain and Scandinavia appear to have been derived from expansions from Iberian and Italian refugia (Taberlet et al. 1998).

9.2.3 Detecting spatial population structure with genomic data

In Chapter 3, we learned about the importance and prevalence of population structure (sections 3.3.2 and 3.3.3). Since phylogeography seeks to understand and explain the spatial distribution of genetic diversity within a species, identifying structure among populations is an obvious first question to address. Simply defined, population structure reflects the fact that when arranged in space, individuals in close proximity are likely to be more closely related to one another than those separated by a large physical distance. Thus, population structure can be generated by isolation by distance (see section 3.3.4). However, structure might reflect more than just physical distance, for example when it occurs between habitats. A good example of this is seen in multiple lake-stream threespine stickleback species pairs from Vancouver Island, British Columbia; a clear transition between lake and stream genetic clusters is seen along streams, consistent with a change in habitat conditions (Berner et al. 2009; Roesti et al. 2012). Population structure in this case is likely a result of divergent selection between lake and stream habitats. Structuring among populations might also arise from recent demographic history and the order in which different populations recolonized certain areas (e.g. Taberlet et al. 1998).

How can we detect population structure? Typically, researchers apply several methods and we will introduce two of the most well-known here. The first of these is **principal components analysis (PCA)** and requires no model of how individuals or populations might be structured. PCA has a long history of application to genetic data and was first used by Menozzi et al. (1978) to explore population structure in Europeans over 40 years ago. PCA is a multivariate statistical method with the aim of reducing many variables into a smaller, more manageable set—i.e. the principal components (Quinn & Keough 2002). Essentially, PCA aims to identify the main axes of variation in a dataset with each

---

[3] The Pleistocene ended around 11 700 years ago and the Earth entered the Holocene. More recently, the Anthropocene (i.e. human epoch) has been proposed but has not been officially recognized as a geological unit.

axis being independent of the next (i.e. there should be no correlation between them). The first component summarizes the major axis variation and the second the next largest and so on, until cumulatively all the available variation is explained. In the context of genetic data, PCA summarizes the major axes of variation in allele frequencies and then produces the coordinates of individuals along these axes (Patterson et al. 2006). PCA has become popular with large-scale genomic datasets because it is computationally fast and is typically interpreted visually, making it more straightforward to understand than model-based methods.

The huge numbers of markers made available for analysis due to high-throughput sequencing has offered some remarkable insights into relatedness and structuring in human populations. Using 500 568 SNPs from a sample of 1387 Europeans, Novembre et al. (2008) summarized allele frequency differences using PCA (Figure 9.13). The authors noted their analysis resembled a European map with clear clustering among geographically proximate populations. Indeed, they were able to detect such fine-scale structure in their PCA that it was possible to separate out German, French, and Italian speaking

inhabitants of Switzerland. Such inference is not possible without a very large numbers of variants—the average geographical differentiation measured by F_{ST} at a single SNP is 0.004 (Novembre et al. 2008). Detecting this kind of structure has important ramifications for biomedical research—population structure can influence the ability of methods such as genome-wide association studies (GWAS—see chapter 6) to detect correlations with disease variants. PCA is clearly a powerful tool but care should be taken when using it to interpret historical reasons for why structure might occur—spatial patterns of variation can arise as a result of neutral processes, not just large-scale migrations or population movements (Novembre & Stephens 2008).

A second-class of methods for detecting population structure in genetic data are **model-based clustering methods**. The most well-known of these is STRUCTURE (Pritchard et al. 2000), which we briefly mentioned in chapter 3, but other, similar methods also exist (Alexander et al. 2009; Lawson et al. 2012). As the name suggests, a method like STRUCTURE uses a model of population structure. We start with a sample of individuals of which we have no a priori knowledge of population of origin;

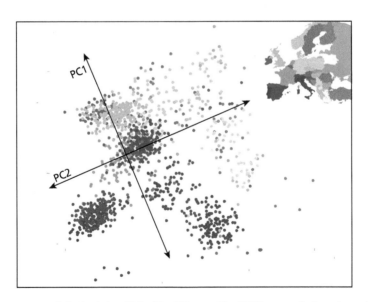

Figure 9.13 Principal component analysis of variation of 0.5 million SNPs scored for 1387 Europeans by Novembre et al. (2008). Principal components orientations are rotated to visualize the similarity between the geographic clustering among individuals and the geography of Europe, shown in the inset map. Colors of points correspond to the countries of origin shown in the map. Figure redrawn from data available on Github: https://github.com/NovembreLab/Novembre_etal_2008_misc.

we then assume that these individuals can be divided into K clusters—each of which have no structure within them. In other words, this means within the K clusters, the Hardy–Weinberg expectation is met and there is no linkage disequilibrium other than that produced by physical linkage (Pritchard et al. 2000). The algorithm in STRUCTURE works to fit individuals to these K clusters based on the probability of their genotypes given this simple model of population structure—thus STRUCTURE is a Bayesian approach. Since the algorithm also assigns cluster membership on a locus-by-locus basis, it is also possible to estimate Q—i.e. the proportion of an individual's genome derived from each K clusters (Pritchard et al. 2000). Although extremely powerful and widely used, STRUCTURE also has limitations. Chief amongst these is the fact that the algorithm does not choose between different values of K—thus several ad hoc methods based on the likelihood of the different models are used (Pritchard et al. 2000; Evanno et al. 2005). For this reason, cluster-based model methods are often paired with other analyses such as PCA (e.g. Ravinet et al. 2018b).

Clustering methods such as STRUCTURE can be extremely informative in terms of understanding recent human history. Leslie et al. (2015) also used 500 000 SNPs to investigate population clustering among 2091 people in the United Kingdom (see Figure 9.14). Since much of the UK has experienced recent migration and population movement, the study focused only on people from rural areas who had four grandparents born within 80 km of one another.[4] Since the average birth year of grandparents was 1885, the study effectively surveyed pre-twentieth century genetic diversity in the UK. Measures of F_{ST} among geographical regions was extremely low—on average just 0.0007; therefore, the authors applied fineSTRUCTURE, a variant of STRUCTURE that uses SNPs in linkage to estimate haplotypes (Lawson et al. 2012). The strongest clustering was between people in the Orkney Islands versus all others, then between people from Wales and all other samples from outside Orkney (Figure 9.14). Much of England was homogenous

for a single cluster with the exception of two clusters representing Devon and Cornwall respectively in the South West and a further one in the North representing Yorkshire (Figure 9.14). The authors also used fineSTRUCTURE to cluster 6209 individuals from the rest of Europe into 51 population clusters; this allowed them to then estimate the ancestry contributions these populations made to clusters within the UK. Ancestry profiles for British populations clearly reflect historical knowledge; for example, the Orkney Islands all show a significant contribution from Norwegian clusters, consistent with the fact that Vikings established Orkney—it was controlled by Norway until the fifteenth century. In contrast, the largest cluster in Central and Southern England has considerable contributions from western Germany, likely reflecting Saxon migrations.

Figure 9.14 Map of the United Kingdom showing the clustering of 2039 people sampled from across the UK, redrawn from Leslie et al. (2015). Note the large red cluster covers much of England with the exception of two distinct clusters (pink and blue) in Devon and Cornwall to the southwest and an orange cluster focused on Yorkshire near the border with Scotland. Contains OS data © Crown copyright and database right 2012. © EuroGeographics for some administrative boundaries. With acknowledgement to Stephen Leslie for his assistance with this figure.

---

[4] This is actually quite difficult to achieve in many parts of the UK—it would certainly exclude one of the authors (Ravinet)!

9.2.4 Testing demographic models and evolutionary scenarios

We learned of the dangers of using a single locus to reconstruct a phylogeny in section 9.1.8. These also present an issue for studying phylogeography too—a single locus can provide a misleading picture of how evolutionary history has unfolded among populations. This was also a major criticism of many early phylogeographical studies which focused mainly on using mitochondrial DNA. Once again, stochasticity and incomplete lineage sorting might result in a gene tree discordant from the true evolutionary relationship among species or populations. Furthermore, with sex-linked loci such as mitochondrial DNA or Y-chromosome markers, sex differences in dispersal might also obscure the evolutionary history of a species. For example, in a species where males are highly mobile and migrate long distances but females have a relatively restricted range, a mitochondrial-based phylogeographical study will be unlikely to accurately reflect patterns of recolonization and population movement. A good example of this is seen in domestic horses; global horse mitochondrial DNA is highly diverse whereas Y-chromosome diversity is extremely low, which is likely due to the breeding practice of mating valuable stallions with multiple mares (Librado et al. 2016).

Even from the early days of phylogeographical research, there was a clear push towards a more rigorous, hypothesis-based framework for understanding the spatial distribution of genetic diversity. Some attempts were made to draw predictions from population genetic theory and to test them. A classic example comes from work on the recolonization of new habitats following glacial retreat after the Last Glacial Maximum (LGM). In Europe, most refugia are largely thought to have been located to the south, in regions such as the Mediterranean and the Iberian Peninsula, beyond the maximum extent of Pleistocene ice sheets (Provan & Bennett 2008). A simple prediction is that as populations spread northwards from these refugia, they experienced a series of repeated population bottlenecks, such that new populations to the north were founded by only a few individuals. If this had occurred, we might expect a negative correlation between latitude and

measures of genetic diversity—i.e. lower diversity in more Northern populations (Taberlet et al. 1998).

Nonetheless, despite these efforts to introduce statistical rigor to phylogeographical analyses, the favored approach was still largely descriptive with a qualitative interpretation of the data. One of the biggest criticisms of phylogeographical hypothesis testing at this time was that it provided no effective way for evaluating the error associated with a particular hypothesis. In short, how likely is it that a hypothesized evolutionary history might explain the patterns of genetic diversity we observe today? Simulation of population genetic data under the coalescent also provided a stark demonstration of this—a simple demographic model might produce hundreds of different neutral genealogies. However, rather than striking a deathblow to phylogeography, coalescent theory and coalescent simulations have in fact opened up newer, more rigorous means of explicitly testing different evolutionary scenarios.

Approximate Bayesian computation (ABC) has been an important component of this (Figure 9.15). Although on the face of it, ABC might seem complicated, a general outline of the method is actually quite simple and it draws on the elements of Bayesian theory we learned in section 9.1.7. Let us imagine we have two recently diverged populations and we wish to know when they diverged. We can achieve this using ABC by building a demographic model in which two species diverged t generations ago—thus we have only a single parameter. As you will recall from our outline of Bayesian phylogenetics in section 9.1.7, a Bayesian approach allows us to set a prior. In this case, we can set a limit on the lower and upper bounds we believe t to be. Perhaps we know that a major biogeographical event, like the formation of an ice sheet, occurred 30 000 generations ago and we suspect that might have initiated the divergence between our two species; using ABC, we could set a prior that reflects this hypothesis. This highlights an important principle of testing demographic hypotheses in a statistical framework—we are able to incorporate external sources of evidence (i.e. biogeographical events or the presence/absence of fossils) to shape our inference (Knowles 2009). With our model defined, we then simulate data, typically using the coalescent, and then summarize that data using summary

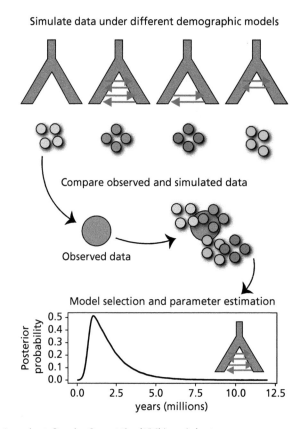

Figure 9.15 Schematic of how Approximate Bayesian Computation (ABC) is carried out.

statistics—i.e. population genetic statistics such as nucleotide diversity and F_{ST}. If we calculate the same statistics for our observed, real data we can then calculate the difference between the real and the simulated data, throwing away all simulations that are further than a specified threshold of difference. One way to visualize this is to imagine the real and simulated data distributed in space—we are essentially only keeping simulated data that is closer to our real data (Figure 9.15). By doing this, we are able to *approximate* the posterior distribution of our model (Bertorelle et al. 2010). In this simplified case, we are estimating only the t parameter and by examining the distribution of t from our retained simulations, we have a posterior for t. An approach such as ABC therefore allows us to estimate demographic parameters and also to get some measure of our certainty in them. From our posterior, we can define a **high-density posterior region**, akin to confidence intervals on our estimates.

In addition to parameter estimation, ABC can also be used to compare different demographic scenarios that might explain the observed data. Returning to our simplified example, we might build four models instead of one—the first is our original model but the other models include a new parameter, m—the fraction of migrants the two populations exchange each generation, starting at different times. For example, we might have constant migration, or only recent migration on secondary contact between lineages. We again simulate data under our models and again compare it to our observed data. This time, when we retain simulated datasets close to our observed data we are able to approximate the posterior probability of the different models. We do this by quantifying the proportion of retained simulations of each model; for example, if we retain 1000 simulations and 850 of them are from our simple model, while only 50 each are from our three models with gene flow, the posterior probabilities for the four models are 0.85, 0.05, 0.05, and 0.05 respectively. Using ABC for **model selection** in this way, we would argue our analysis provides greater support for divergence in strict isolation between these two species. A final useful aspect of ABC is that we can run a straightforward test of the reliability of our hypothesis testing framework. By rerunning our analyses with datasets simulated under a known model in place of the observed we can estimate the rate of false positives (i.e. incorrectly choosing a model when the data comes from another) and also the rate of false negatives (i.e. incorrectly rejecting the true model). Therefore, ABC provides a way to evaluate how much power we have to distinguish different evolutionary hypotheses.

ABC is extremely flexible and, in some cases, very powerful. However, it is important to keep in mind that it is not without drawbacks. Firstly, since ABC requires a very large number of simulations to properly approximate the posterior distribution it is computationally expensive. Although advancements are continually being made with fast and effective simulations (Kelleher et al. 2016), this is still a bottleneck on the application of ABC—especially to datasets derived from high-throughput sequencing with thousands of loci. Secondly, great care is needed to ensure realistic models are tested. For any given observed dataset, there is a universe of possible scenarios that could have resulted in it (Knowles 2009) and it is simply impossible to test all of them. If we fail to include the most realistic model, we might support an incorrect scenario that can easily lead to incorrect estimates of demographic parameters. This leads to a third, more critical point about using model-based approaches to testing demographic scenarios—all models are wrong (Hickerson 2014). Models are by definition a simplified approximation of reality and so will never totally reflect how two populations or species might have diverged. This need not be fatal for model-based inference in evolutionary genetics, but it is an important point that we strongly encourage you to keep in mind.

ABC is not the only method for testing among demographic scenarios. Thanks to the large numbers of loci available from high-throughput sequencing—including sequencing more individuals at lower cost—it is much easier to get an accurate estimate of the **allele frequency spectrum (AFS)**. For two populations, the joint allele frequency spectrum is a matrix of the counts of alleles shared between the two populations (Sousa & Hey 2013). The expected joint AFS can be derived rapidly for different demographic scenarios (i.e. isolation vs isolation-with-gene-flow) using coalescent simulations or analytical methods (Gutenkunst et al. 2009; Excoffier et al. 2013). This makes it possible to calculate the maximum likelihood for the observed dataset under a set of models to perform model selection and parameter estimation. Likelihood-based AFS methods are usually much faster and generally easier to implement than ABC for large genomic datasets (Excoffier et al. 2013). Furthermore, although early instances of the approach were limited to a joint spectrum between two populations, extensions mean it is now possible to include multiple populations in models (Kamm et al. 2017).

Demographic inference has proved an important tool for investigating the evolutionary history of a wide variety of taxa. Indeed, this is seen as an important first step in studies of speciation in order to place any further research on barriers to gene flow in perspective. Demographic inference is crucial to determine the geographic context in which divergence might have taken place (i.e. allopatry vs sympatry) and whether species pairs might have exchanged genes. For this reason, Nadachowska-Brzyska et al. (2013) used a suite of coalescent methods on whole genome resequencing data from 10 pied flycatchers (*Ficedula hypoleuca*) and 10 collared flycatchers (*Ficedula albicollis*). The authors used ABC to test five broad demographic scenarios under which divergence might have occurred. These included divergence in total spatial isolation (i.e. allopatry), divergence with constant migration (i.e. parapatry), and various periods of migration either recently or more anciently. For each of the five scenarios, they tested three further models where population size either remained constant or varied during the process of divergence—meaning they tested a total of 15 models in a hierarchical framework. The posterior probabilities of their ABC analysis best supported a model where divergence occurred in isolation 340 K years ago with recent secondary contact and asymmetric gene flow between the two flycatcher species (Nadachowska-Brzyska et al. 2013; Figure 9.16). The best-supported scenario also showed a decline in effective population size since divergence began. To further explore this, Nadachowska-Brzyska et al. (2013) used a coalescent analysis designed to infer past effective population size from the density of heterozygous sites present in a single genome sequence from the collared flycatcher. Regions of high heterozygote density suggest a large effective population size going backwards in time, whereas fewer polymorphisms point towards effective population size being smaller (Figure 9.16). The analysis shows collared flycatcher effective population declined from a peak of about 1.6 million 200 K years ago to around 500 000 in the last twenty thousand years, a decline broadly consistent with the onset of the last glacial period.

9.2.5 Ancient DNA—a direct insight to the past

One of the most important contributions to understanding phylogeography in the last decade has come from the study of **ancient DNA**. In 1984, Higuchi and co-workers were able to extract and amplify a small region of the mitochondrial genome from a quagga, a species of zebra that went extinct due to hunting pressures in the late nineteenth century (Higuchi et al. 1984). Not long after this, mitochondrial sequences were also published from Egyptian mummies (Pääbo 1985). Much of the initial focus of the field was on mitochondrial DNA because it was easier to amplify and usually present in large amounts in most samples (MacHugh et al. 2017). However, unlike work on contemporary DNA samples, there are a number of considerable challenges for ancient DNA research. One serious concern is that of contamination from microbes or even researchers working in the lab. This was such an issue during the early stages of the field that

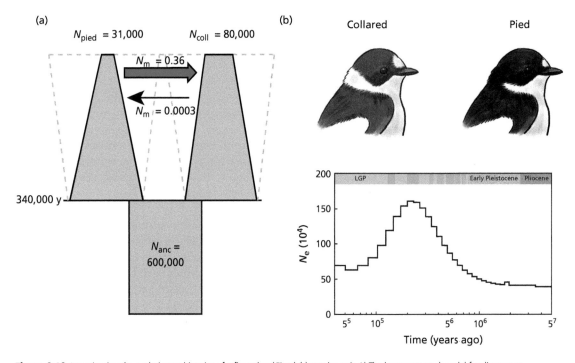

Figure 9.16 Investigating the evolutionary histories of a flycatcher (*Ficedula*) species pair. A) The best supported model for divergence between the collared and pied flycatchers suggests divergence in allopatry around 0.34 million years ago with subsequent secondary contact. B) A reconstruction of the effective population size of the collared flycatcher going backwards in time. Note peak in the effective population size in the second from last interglacial (shown by the golden color at the top of the figure). The species underwent a significant decline following the onset of the last glacial period (LGP). Redrawn from Nadachowska-Brzyska et al. (2013).

many high-profile studies were shown to be false positives (Willerslev & Cooper 2005). A particularly famous example is that of Woodward et al. (1994), who claimed to have sequenced dinosaur DNA from fossilized bones dating from the Cretaceous period, some 80 million years ago. However, later reanalysis confirmed the sequence was most likely of human origin (Young et al. 1995) not to mention the implausibility of the findings given the theoretical limit for DNA survival is between 100 Kyr and 1 million years, providing conditions for preservation are optimal (Willerslev & Cooper 2005). DNA degradation is another issue that ancient DNA researchers must contend with. In living cells, DNA is constantly being damaged by processes such as hydrolysis and oxidation but DNA repair mechanisms act to restore aberrant nucleotides. However, after an organism dies, these repair mechanisms no longer function and so DNA damage accrues over time (Willersev & Cooper 2005; Orlando et al. 2015). Damage can often modify the nucleotide sequence and a common pattern is a C to T mutation due to deamination.

Perhaps unsurprisingly, ancient DNA research has been transformed by the arrival of high-throughput sequencing. Determining the extent of DNA degradation is possible from HTS data using bioinformatic pipelines. For example, damage induced mutations tend to appear at the ends of sequence reads making it possible to construct a degradation profile from sequence data and to only utilize undamaged DNA regions for mapping and variant calling (Orlando et al. 2015). HTS has made it possible to much more easily independently verify findings to rule out contamination as an alternative explanation. Mapping to a reference genome, for example, minimizes the possibility of including sequences from sources other than the organism of interest. Damage profiles from sequence analysis can also help with contamination—modern DNA will not show evidence of damage (Orlando et al. 2015). Since ancient DNA studies often extract template DNA from samples excavated from the ground or from archaeological sites, it can be difficult to remove traces of microbes. This means that standard whole-genome resequencing approaches are wasteful for ancient DNA as a large amount of sequence will likely be derived from contaminants.

To address this, most ancient DNA studies instead use target enrichment approaches or capture arrays to target regions of interest and focus sequencing effort (Orlando et al. 2015).

When we examine range expansions or population structure in contemporary data, genetic data from individuals from the past can provide essential insight into the processes and events that might have shaped modern genetic diversity. An excellent example of this comes from recent work on the origins and domestication of the horse (*Equus caballus*). Somewhat ironically, horses and their ancestors provide a textbook example of morphological evolution (i.e. the evolution of the single toe) but we have a poor grasp on how the once diverse group of equids has been reduced to a small number of species. Today, *Equus* is represented by two groups, non caballus equids, such as zebras, asses, and donkeys, and caballus equids, which includes the domestic horse and Przewalski's horse—a wild-roaming species occurring in Central Asia that was thought to have diverged from the main horse lineage prior to domestication (Figure 9.17). The early archaeological record for domestic horses is actually quite poor compared to other livestock. However, the earliest evidence of domestication is from the Botai culture that existed 5500 years ago in present day Kazakhstan, suggesting a much more recent process than that of, say, cattle or chickens (Librado et al. 2016). Early work attempting to date the domestication and spread of horses focused on using the divergence time between the domestic horse and Przewalski's horse as a lower limit for the timing of domestication. This was difficult because of the fact that Przewalski's horse was nearly driven to extinction in the mid-twentieth century, with all modern individuals (of which there are approximately 2100) descended from just 12 horses. Such a drastic bottleneck has resulted in low genetic diversity and an excess of deleterious mutations across the genome, suggesting a high genetic load (Librado et al. 2017). Earlier genomic studies of Przewalski's horse pointed to considerable gene flow with domestic breeds (Schubert et al. 2014). Combined with a bottleneck, this gene flow might mean that much of the variation in the modern day Przewalski's horse population falls within the range seen in domestic breeds.

Figure 9.17 A group of Przewalski's horses in Hustai National Park, Mongolia. Source: Kelsey Rideout, Wikipedia/Wikicommons.

To account for this, Der Sarkissian and colleagues sequenced the genomes of all 11 lineages used to establish the modern day Przewalski's horse lineage and also sequenced five historical specimens, pre-dating the population decline (Der Sarkissian et al. 2015). They also included the ancient genomes of two horses from the late Pleistocene (dating from 14 and 46 kya). With this new dataset, they were able to show that Przewalski's horse diverged from the lineage that eventually would become the domestic horse some 45 000 years ago. However, the sequencing of another sample of ancient horses, this time from the Botai culture, the earliest evidence of domestication, revealed that actually Przewalski's horse is unlikely to be a wild species at all (Gaunitz et al. 2018). Instead Przewalski's horse groups monophyletically within the Botai horses, suggesting they are in fact a population of feral horses—i.e. once domesticated and then left to roam the Central Asian steppe. The sequencing of the Botai horses also point to another interesting

aspect of ancient DNA—no modern horse breeds are descendants of this group, suggesting the tantalizing possibility that horses were actually domesticated more than once (Gaunitz et al. 2018). The story of horse domestication clearly still has gaps to fill but it is clear that ancient DNA has added a perspective that would not be possible from sampling contemporary populations alone.

9.2.6 Human origins: insights from genomic data

What are our origins? How has our species evolved? Such questions have likely been asked by humans for as long as we have existed. Unsurprisingly then, human phylogeography and evolution continues to be a major topic of research interest. How genetics and genomic data have shaped our understanding of our own origins provides an excellent example of how the field of phylogeography has evolved in the last few decades. In a remarkable series of studies, researchers have made use of both phylogeographical and phylo-genetic approaches to disentangle the complex history of *Homo*, our own genus. A combination of fossil evidence, genomic data, and sophisticated analysis of ancient DNA have transformed our understanding of early human ancestry.

Prior to genetic data, our inference on early human origins came primarily from fossil evidence. Spectacular finds of early hominid fossils such as "Lucy," the 3.2 million-year-old *Australopithecus* specimen from Ethiopia suggested that Africa was an important region for human evolution. However, it remained unclear whether the continent played a role in the origins of anatomically modern humans (i.e. *Homo sapiens sapiens*). Around the time of the first studies of human genetic variation, two major models of modern human origins existed. The first of these was the **multi-regional model**, which suggested different groups of modern humans arose from geographically widespread populations of *Homo erectus*, an archaic human, throughout the Pleistocene. Alternatively, the **out-of-Africa model** suggested a much more recent origin for modern humans on the African continent with a later large-scale population movement into Eurasia (Nielsen et al. 2017). In 1987, Cann and co-workers analyzed mitochondrial DNA and found all current human lineages coalesced in the last 200 K years with the greatest diversity among modern Africans, suggesting Africa was a recent center for the evolution of modern humans (Cann et al. 1987). Subsequent genomic studies have provided additional support for the out-of-Africa model, shifting the focus off a single locus. Modern African populations have the highest diversity of all humans and also extensive population structure. Indeed, the deepest split between human lineages is between the click-speaking Khoe San and all other populations, dated at 110–160 K years (Veeramah & Hammer 2014). Studies using observed modern genomic data and demographic models also point towards a recent origin of modern humans and then at least one major out-of-Africa migration sometime in the last 50–100 K years (Nielsen et al. 2017).

An even more complex picture of recent human evolution has begun to emerge as a result of the successful sequencing of the ancient genomes of several archaic hominids. This began with the publication of a Neanderthal genome sequence in 2010; a composite sequence put together using DNA extracted from three separate fossils. Green et al. (2010) showed that 1.5–2 percent of the genome in non-African modern humans is of Neanderthal origin whereas Africans in contrast show almost no trace of Neanderthal genes. Evidence of Neanderthal admixture in non-Africans provides further support for the out-of-Africa model as any gene flow must have occurred after the major migration out of the African continent, some 50–60 K years ago (Figure 9.18). Ancient DNA therefore suggests a scenario where modern humans left Africa to migrate further afield where they came into contact with archaic hominid populations such as Neanderthals and exchanged genes. Further evidence of human–Neanderthal interactions comes from the ancient genome of a 37–42 K year old individual from Romania who likely had a Neanderthal ancestor in the previous 4–6 generations (Fu et al. 2015). Intriguingly, this is not the only example of gene flow between modern and archaic humans; sequencing of ancient DNA from a small finger bone found in the Denisova Cave in Siberia revealed another archaic group, distinct from Neanderthals and now known as the Denisovans (Reich et al. 2010). A comparison between Denisovan genomes and those of modern humans suggest 0.2 percent of the genome of East Asians is derived from this now extinct group of ancient hominids. However, an even greater Denisovan contribution of 3–6 percent is seen in the genomes of people from Melanesia, Papua New Guinea, and Australasia, which indicates gene flow between Denisovans and modern humans may have occurred twice (Racimo et al. 2015). In 2018, Slon and co-workers published the genome of a girl who lived approximately 90 K years ago and had a Neanderthal father and a Denisovan mother (Slon et al. 2018). This F1 hybrid demonstrates that the two archaic human lineages also likely exchanged genes with one another, despite having diverged approximately 390 K years ago. Intriguingly, Denisovans also appear to show signatures of introgression with yet another archaic hominid group, although it is not clear which. There is some speculation that this might be *Homo erectus* (Figure 9.18; Racimo et al. 2015).

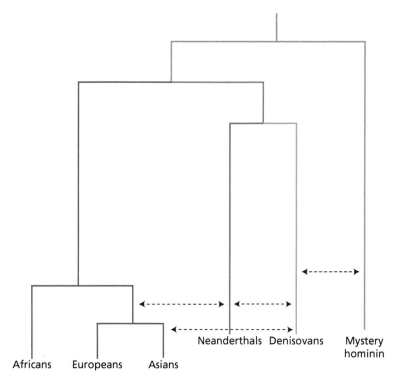

Figure 9.18 Modern and ancient DNA reveal a complex history of archaic introgression among hominid lineages. For modern humans, introgression has occurred between Neanderthals and the ancestors of all non-Africans. Some Asian populations also experienced later introgression from Denisovans too. Furthermore, there is evidence of admixture among Denisovans and Neanderthals, including the recent finding of an F1 hybrid individual (see text). Finally, the Denisovan genome appears to contain introgressed regions from another, currently unidentified "mystery hominin" which is speculated to be *Homo erectus*.

9.2.7 Using the four-population test to infer gene flow and introgression

How is it possible to detect evidence of introgression and determine the percentage of archaic ancestry in the genomes of modern humans? One way to do this is to use demographic models and the population clustering algorithms we learned about in sections 9.2.2 and 9.2.3. However, a more widely used method is the **four-population test** and its variants, often referred to as ABBA-BABA or *D* statistics.[5] The test is simple, very straightforward, and also much less computationally expensive than fitting the complicated demographic models we learned about in section 9.2.4. The method assumes a standard tree topology between four populations and then tests for an excess of shared variants between the first three populations (see Figure 9.19). To simplify our understanding here, we will first consider the test being carried out on a single nucleotide position in the genome. Using an outgroup at the fourth tip on the tree, we can determine whether the site is ancestral (A) or derived (B). We expect a number of different site patterns between the four taxa or populations on our tree but the two of interest are ABBA, where the second and third populations share a derived allele, or BABA, where the first and third populations share a derived allele. Now imagine that we traverse our genome and count the occurrences of either an ABBA or BABA pattern at each of our polymorphic nucleotide positions. We will then have an idea of the frequencies of these two different site patterns. We can therefore calculate *D* like so:

$$D\left(P_1, P_2, P_3, O\right) = \frac{\sum C_{ABBA} - C_{BABA}}{\sum C_{ABBA} + C_{BABA}} \quad (9.8)$$

[5] Note that the *D* statistic in the four-population test is completely different to the *D* we defined in equation 9.6.

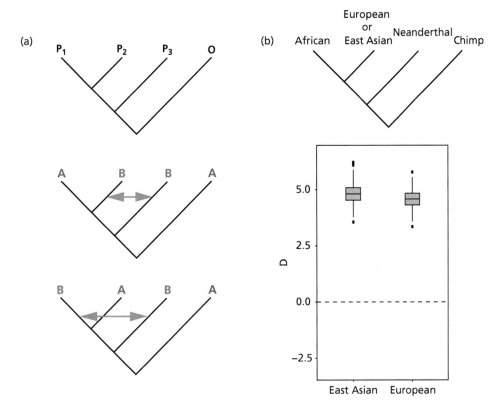

Figure 9.19 The four-population test is a straightforward way of testing for an excess of shared alleles among populations, pointing towards potential introgression. In A), the top tree represents the test topology with three ingroup populations (P_1, P_2, and P_3) and an outgroup (O). The test compares the frequencies of the ABBA and BABA site patterns, shown in the lower trees, which suggest an excess of sharing between either P_2 and P_3 (ABBA) or P_1 and P_3 (BABA). In B) we can see the test topology used by Green et al. (2010) to detect introgression between the ancestor of Europeans and East Asians and Neanderthals. The lower boxplot shows that with P_2 set to either a European or East Asian genome, D (see equation 9.4) is positive and distant from the dashed line at zero, which indicates a significant excess of ABBA sites compared to the expectation of balanced ABBA-BABA frequencies under neutrality.

Where C_{ABBA} and C_{BABA} are the counts of ABBA and BABA site patterns respectively (equation modified from Martin et al. 2014). Assuming a neutral scenario with no selection or gene flow among the populations, the frequencies of ABBA and BABA patterns should be equivalent because the only process that is able to generate them is incomplete lineage sorting. However, if gene flow occurs between any two of the three ingroup populations, this will lead to an excess in the frequency of either ABBA (i.e. second and third) or BABA (i.e. first and third) patterns.

One of the first uses of the ABBA-BABA test was to calculate the D statistic in order to determine the extent of shared sites between modern humans and Neanderthals, part of the research we learned about in the previous section. Using two topologies of (((African, European) Neanderthal), Chimpanzee)

and (((African, East Asian) Neanderthal), Chimpanzee) with alignments of single genomes, Green et al. 2010 calculated D. A positive value of D for either test would suggest a great proportion of shared sites between Neanderthals and Europeans or East Asians. Alternatively, a negative value of D in either test would suggest an excess of sharing between Africans and Neanderthals. Finally, a D of zero would suggest no excess of shared alleles in either direction. Green et al. 2010 estimated D to be 4.57 for the test with Europeans and 4.81 with East Asians. These results show the ancestor of East Asians and Europeans met and interbred with Neanderthals after their expansion out of the African continent.

In its original formulation, the four-population test was used to examine excess of shared site patterns across the entire genome. However, one criticism of the test is that the statistic D can be inflated by ancestral population structure and demographic history. For example, a population bottleneck might remove signatures of admixture in one of the test populations (Racimo et al. 2015). One way to account for this is to use the four-population test on genome windows in order to determine the location of introgression in the genome and also to provide a measure of the size of introgressed regions (Martin et al. 2014). However, D is less suited for this task as it can vary with levels of diversity in the genome (Martin et al. 2014). To overcome this limitation, Martin et al. (2014) derived a new form of the statistic, f_d, which is now widely applied in studies seeking to identify introgression between divergent taxa, and not just humans. For example, Jones et al. (2018) examined the genomes of snowshoe hares (*Lepus americanus*) in North America that are polymorphic for winter coat color. In regions with heavy snow cover during winter, most hares molt and produce a white coat. However, in less snow dense regions, hares will retain a brown coat. Jones et al. (2018) showed a signature of high f_d between brown coat snowshoe hares and the closely related black-tailed jackrabbit (*Lepus californicus*) in the genomic region surrounding the *agouti* gene, known to be involved in pigmentation. This suggests that the ability to maintain a brown winter coat in the snowshoe hare may be due to **adaptive introgression**.

Study questions

1. How did Darwin's introduction of the concept of common ancestry among living things alter the purpose of taxonomy?
2. Explain the following components of a phylogenetic tree
   a. Tip
   b. Branch
   c. Node
   d. Root.
3. You have collected sequence data for 10 species and you want to estimate their phylogeny. How many possible unrooted trees can we construct from this many species? How many rooted trees?
4. What do we mean by monophyletic group on a phylogenetic tree?

5. Discuss the differences between homology and homoplasy.
6. You plan to align the four sequences you see in Figure 9.4. You set the gap penalty to 1. Calculate the cost of the alignment. What is the cost of the alignment if you set the gap penalty to 2?
7. You have sequenced the mitochondrial genomes of a single individual each of three different species—A, B, & C. Each genome is 16 569 bp in length. Genome A has 278 nucleotide differences when aligned against B and only 15 differences when aligned against C. Genome B differs from C at 450 nucleotide positions. Calculate the pairwise genetic distance between each of them. What is the mean genetic distance?
8. Consider Hasegawa et al.'s 1985 model of sequence evolution. There are two different substitution probabilities; α (for transitions) and β (for transversions). There are also four separate nucleotide frequencies—f_A, f_T, f_C, f_G. How can we alter Hasegawa et al.'s model so that it is equivalent to Jukes–Cantor, the simplest model of sequence evolution? Since it is possible to simplify the HKY model, what special case of models do we have here?
9. What is the main difference between an additive distance tree (i.e. such as those produced by Neighbor Joining) and an ultrametric tree (such as that produced by UPGMA)?
10. Define the following concepts:
    a. Parsimony
    b. Bootstrapping
    c. Likelihood.
11. Discuss how Bayesian probability differs from a standard forward probability framework.
12. Highlight the main differences between model-based clustering analysis and principal components analysis for identifying population structure.
13. You wish to rerun the four-population tests used by Green et al. (2010) to test your hypothesis that Europeans might have met and exchanged genes with Denisovans. How would you rearrange the test topology used by Green et al. (2010) to achieve this?
14. You are studying two populations of a butterfly species which occur on either side of the Andes mountain range in South America. On the

western side of the Andes, population A is completely isolated and no other closely related butterfly species occur. On the eastern side, population B co-occurs with a second closely related species, species C. You collect 10 individuals each from population A, population B, and the third population, species C. You then sequence their genomes, plus one genome of an outgroup species from North America so that you can determine the ancestral and derived alleles. You want to conduct a four-population test to see whether population B and species C have interbred. You use the following test topology where P_1 = population A, P_2 = population b, P_3 = species C and O = outgroup. Going through your genome alignment, you count 62 955 ABBA sites and 48 887 BABA sites. What is the value of D? Is this strong evidence of gene flow?

CHAPTER 10

Sequencing the genome and beyond

It is remarkable to think how much progress has been made in genomics since the completion of the human reference genome in 2001. The data generation for a sequencing project that took over a decade, costing between $500 million and $1 billion USD, is now completely feasible in just a week or so with commercially available lab protocols and sequencing services—for around just $1000 USD. Throughout the course of the book, we have aimed to paint a picture of how this genomic revolution has completely transformed evolutionary genetics. However, science is defined by continual progress, and new technologies, approaches, and perspectives for dealing with genomes are approaching on the horizon. Some of these are already gaining a foothold in evolutionary biology, whereas others have yet to extend their reach much beyond model species. In this final chapter, we will start with introducing you to what it means to sequence and assemble a reference genome. With talk of sequences, it is easy to forget that references are drafts of a linear sequence of base pairs; the true genome is not linear but has structure and function. We will explore the genome as a 3D entity—from how it is transcribed, to how proteins interact with it, and finally to how it is actually structured. This also gives an opportunity to focus on epigenetics and how we should interpret processes such as DNA methylation in an evolutionary context. In the second part of the chapter, we move away from ways of reading the genome to focusing on ways we can interact with it. We will explore how we might test the function and role that the candidate

genes we are able to detect with population genomic data play. We will introduce you to transgenics, in particular the transformative technology of CRISPR/CAS9, and explore how this might change the face of evolutionary biology in the near future.

10.1 Assembling and building on the genome

10.1.1 Reference genomes and genome assembly

Since the beginning of the genomic revolution in biology, the introduction of new sequencing technologies and methods has been extremely fast. Yet for much of the "-omics era of biology" (Ellegren 2014), producing a high-quality **reference genome** has been a milestone or major target for many researchers working in evolutionary biology. A reference genome is more or less what it sounds like, an example of the DNA sequence and genes that make up the genome of a particular species. Many reference genomes are made up of DNA from only a single representative of a species, whereas others, such as certain versions of the human genome, are constructed from multiple individuals. However, no reference genome is definitive—as we have learned throughout this book, there is a great deal of genetic diversity within populations and species and so by definition, a single genome is unable to capture this. A useful analogy to illustrate how we might think about a reference genome is that of a **type specimen**, an individual of a species used to establish its formal

Evolutionary Genetics: Concepts, Analysis, and Practice. Glenn-Peter Sætre & Mark Ravinet, Oxford University Press (2019).
© Glenn-Peter Sætre & Mark Ravinet 2019. DOI: 10.1093/oso/9780198830917.001.0001

scientific name and recognize its phenotype. Type specimens are stored in museum collections and used as a reference for judging whether other samples from the species are similar to it. Like a reference genome, a type specimen cannot encompass all the variation within the species it defines, but it does act as a standard against which we are able to build up a wider understanding of what variation does exist.

Our comparison of a reference genome to a type specimen is however, an imperfect one. There is no clear definition for how much of the genome of an organism should actually be sequenced in order for us to have a sequence we can call a reference (Ellegren 2014). In some cases, particularly in well-studied model species, the amount of genome sequenced might be relatively close to the true size of the genome. However, the genomes of many species are filled with repetitive sequences caused by transposable elements or highly repetitive regions such as centromeres and telomeres (Ellegren 2014). These are not easy to reliably sequence, especially with short-read sequencing technologies where repeated bases are hard to distinguish from error signals. Many reference genomes also lack the hemizygous sex chromosome because of high levels of sequence divergence compared to the autosomes (Soh et al. 2014). Therefore, if type specimens were like reference genomes, this would be akin to fishing one out from a museum collection and finding that a large part of it was missing![1] So, it is important to remember that a reference genome is often likely to fall short of the true DNA sequence.

How do we generate a reference genome? Is it as simple as just sequencing DNA? Actually getting hold of the sequence is an important part of the process, but we also need to piece our sequence data together so that it is properly organized and arranged as a map of the genome of the species we are interested in. This is the aim of **genome assembly**. As we noted at the outset of the chapter, with modern sequencing technologies this is certainly much easier today than it was during the early days of genomic research—however, properly assembling a genome is still not a simple task. Firstly, we must start with sequencing a representative indi-

vidual or small group of individuals of a species. In most cases, we will want to choose an inbred individual, or one with low levels of heterozygosity, as this can make genome assembly less complicated (Ekblom & Wolf 2014). For example, when sequencing the house sparrow (*Passer domesticus*) genome, Elgvin et al. (2017) chose to use a highly inbred individual from a small isolated population in Northern Norway. We can again turn to an analogy to illustrate just what genome assembly entails. Imagine if we took 100 hard copies of this book and then shredded it all up in front of you.[2] Then we ask you to take the shredded material and put back together a single readable version. As you search through the shredded paper, you might find parts of sentences that you are able to fit together and since there were 100 copies, any given sentence will be cut up in different ways. This means you will be able to also piece together the copies and verify whether the sentences you have constructed are what you think they are. Eventually, when you have assembled enough sentences, you might be able to put together longer sequences of text—i.e. paragraphs or sections. It would take you a very long time, but eventually you would have lots of text of different lengths—but it might be hard to arrange it all properly in order. So, we decide to help you out by giving you a table of contents. Now you know which topics appear where in the book, so if you find text that mentions genome assembly for example, you know it probably belongs in chapter 10. With this guide, you can arrange your text into their appropriate chapters and reconstruct a likely semi-complete version of the book. Another commonly used analogy is that of a puzzle of a landscape, where most of the puzzle is of a blue sky (i.e. the pieces are identical) and there is no image on the box to guide us (Pop 2009; Meltz Steinberg et al. 2017).

Our book reconstruction analogy essentially summarizes a **shotgun sequencing** approach leading to a **de novo assembly** of a reference genome. Shotgun sequencing is so-called because, rather than fragments of sentences, assembly takes place using short sequence reads from random positions in the genome (Figure 10.1). Of course, none of this is conducted by hand either. With a genome assembly project, we use computational tools to assemble

---

[1] We should note that some type specimens are in fact just like this. For example, Denisovans (see chapter 9) were first identified from a single finger bone.

[2] Of course, we would never dream of doing such a thing!

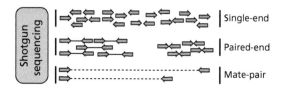

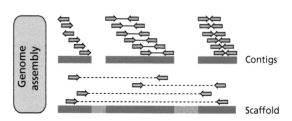

Figure 10.1 The components of genome sequencing and assembly. In the top panel, shotgun sequencing is conducted using single and paired-end sequencing strategies. Mate-pair libraries with different insert sizes are also used to span large physical distances in the genome. Assembly is then conducted using genome assembly pipelines which combine short reads (single and paired-end) into contigs and with mate-pair libraries to bring contigs together into scaffolds. Adapted from Ekblom & Wolf (2016).

overlapping short sequence reads into larger sequences known as **contigs** (Figure 10.1). Accurate assembly of contigs is made possible by sequencing a genome to high **coverage**—i.e. repeatedly sequencing the same sequence multiple times. In addition to this, **paired-end** sequencing is often used, such that short DNA fragments (i.e. < 1Kb) are sequenced at both ends. This should not be confused with the sequencing of **mate-pair libraries**; these contain large insert sequences to span considerable physical distances (i.e. 10 Kb). Mate-pair libraries are important because they are necessary to span parts of the genome that are difficult to properly sequence, such as repetitive regions. If the forward and reverse mate-pairs fall on different contigs, they can be used to join the contigs together in a **scaffold**, a much larger stretch of sequence (Figure 10.1). Furthermore, since the size of the insert between the mate-pairs is known, we are aware of the size of the gap between two contigs and this can be denoted on the scaffold as a null sequence, typically depicted as a sequence of Ns (Ekblom & Wolf 2014). Often, a range of mate-pair libraries with different insert sizes is used in order to properly span difficult to sequence regions. The next step beyond contigs and

scaffolds is assembly into either linkage groups or chromosomes (Ellegren 2014), similar to the transition from paragraphs to chapters in our book analogy. In most studies, the "table of contents" is actually a **linkage map**; this makes it possible to arrange and orientate scaffolds in relation to one another, based on an understanding of recombination and linkage in the genome. In other words, it is the fusing of a genetic map and a physical (sequence) map.

There is a multitude of different computational pipelines for genome assembly and the number is growing constantly (Nagarajan & Pop 2013; Ekblom & Wolf 2014). It is beyond our scope to explain them all fully here and, furthermore, given the speed at which the field develops it would be out-of-date before you could read it! Nonetheless, we can focus on the most commonly used approaches, in order to give you a good idea of how genome assembly is actually carried out. To assemble contigs from reads, assembly pipelines mostly use a De Bruijn graph approach (Flicek & Birney 2009). Reads are essentially broken down into smaller sub-sequences known as **kmers**—i.e. a sequence of k base pairs. The kmers form nodes in a graph network and then any kmers sharing a subsequence of k−1 are linked in the network. Assembly pipelines then traverse the network graph, assembling kmers into larger sequences in a computationally efficient manner (Pop 2009; Compeau et al. 2011). As we mentioned above, assembly pipelines then use mate-pair libraries with different insert sizes to assemble contigs into scaffolds. In some cases, this can be done using another closely related reference sequence—i.e. reference-assisted assembly. Repetitive regions also pose a major issue for genome assembly, as they can occur throughout the genome and can make it very difficult to properly construct contigs (Pop 2009). For example, imagine a 2 Kb repetitive sequence occurring across a genome, perhaps as a result of a transposable element. With short reads of around 150–200 bp, we are likely to sequence somewhere in the middle of this repetitive sequence—our reads will rarely include the unique flanking sequence that might be useful for distinguishing between repeats. Long-reads >2 Kb, however, will be able to capture such flanking sequences and therefore make it more straightforward to tell repeats apart. In other words, it is only possible to properly resolve repetitive elements if they are shorter than the read length used. This is certainly a major issue in the genomes of some organisms; for example, in wild grasses over 65 percent of the genome consists of repeats generated by transposable elements (Jia et al. 2013). For these reasons, hybrid assembly approaches combining both short- and long-read sequences are also available; for example, long-read sequences can be used to act as a backbone to arrange contigs into scaffolds (English et al. 2012). Indeed, for species with a large number of repeats in their genome such as the cod (*Gadhus morhua*) (Tørresen et al. 2017) and the rough periwinkle, *Littorina saxatilis* (Westram et al. 2018), such hybrid assemblies have been essential for improving genome assembly quality. High-density linkage maps, constructed with whole-genome resequencing data, can also further assist with properly orientating and assembling difficult to sequence genome regions (Glazer et al. 2015). As long-read sequencing technologies become cheaper and more mature, they are likely to be used more and more as longer reads make the construction of contigs much more straightforward—i.e. they are more likely to overlap one another (Meltz Steinberg et al. 2017).

Determining the quality of a reference genome isn't a simple task either—mainly because we rarely have the true genome to compare it to! One simple way to assess whether a *de novo* assembly has been effective is to compare the size of the assembled sequence with an independent estimate of genome size, such as from flow cytometry (Ekblom & Wolf 2014). However, this is only a measure of the completeness of a genome and it gives us no information on whether the genome has been well assembled and is a good representation of reality. The **N50 statistic** is also widely used for assessing assemblies; it is the smallest size of scaffold or contig on which 50 percent of bases in the assembly occur (Yandell & Ence 2012; Ekblom & Wolf 2014). As an example, imagine we have assembled a genome which we know the true length of is 280 Mb and our assembly pipeline has produced six scaffolds, which when ordered in size are 85, 67, 46, 25, 16, and 11 megabases each. Firstly, if we sum our scaffolds we can see we have actually assembled only 250 Mb, so our assembly contains approximately 90 percent of the true genome. 50 percent of our assembled 250 Mb is 125 Mb, so our scaffold N50 is actually 67 Mb as 85 + 67 Mb is greater than 50 percent. Again, N50 is a useful statistic for determining the efficiency and performance of an assembly pipeline, but it gives little or no information about how "true" an assembly might be—i.e. its level of quality (Nagarajan & Pop 2013). This is arguably only achievable with comparative approaches, including mapping to very high-quality reference sequences from closely related species that might reveal large assembly errors or misplacements (Ellegren 2014). However, it is important to remember that all reference genomes are somewhat incomplete and likely to contain errors. For example, the number of errors in an assembly of a human chromosome can vary wildly between the bioinformatic pipelines used (Salzberg et al. 2012). Obtaining a high-quality assembly of a genome is, however, only the first step to producing a useful reference for downstream analysis; **genome annotation** is also required. We will return to the challenges of genome annotation in more detail in the next section (section 10.1.2). For these reasons, reference genomes are best thought of as a working best approximation of the true genome rather than a finished product. This is why reference genomes that are publicly available are often also referred to as **draft genomes**, indicating that, while they have met the criteria for submission to a public database, there is work to be done to improve them.

Despite the caveats to genome assembly we have outlined here, the huge and increasing number of reference genomes available for study is a fundamentally important resource in modern evolutionary biology. Reference sequences are essentially a physical map of the biological organization of the genome—where genes occur, how they interact and how properties of the genome vary spatially (Ellegren 2014). For example, comparisons between genomes across evolutionarily divergent taxa can be hugely informative on gene function and how genes have been conserved by purifying selection. A reference genome is also the basis for learning more about population genomic variation using **resequencing** approaches—i.e. (typically) lower coverage sequencing of multiple individuals. Resequenced genomes are then **mapped to the reference** sequence and variation can be characterized by identifying how individuals differ in relation to the reference. Nonetheless, our working model for the genome is still very much a simplified one of a linear, haplotype sequence. In other words, it fails to incorporate the genetic variation that we know occurs in nature (Meltz Steinberg et al. 2017). This way of formulating a reference sequence can actually introduce bias—i.e. if a small genome region is deleted in the reference individual, it will be impossible to survey variation in this region occurring in other individuals through a resequencing study (Paten et al. 2017). New methods and approaches are challenging how we think about the reference genome. For example, the concept of a **pangenome**, a collection of multiple genome sequences representing variation within a species, has been suggested as an alternative approach (Meltz Steinberg et al. 2017). This could potentially be combined with **genome graphs**—i.e. linear reference structures that allow "bubbles" in the reference to identify regions where variation is occurring (Paten et al. 2017) (Figure 10.2). Recently individual *de novo* assemblies have been produced for multiple individuals to more accurately identify structural variation; these too can be combined with genome graphs (Maretty et al. 2017). Our standards for what it means to produce a reference genome are clearly in the process of changing. However, as we will explore throughout the rest of this chapter, DNA sequence provides an important foundation for our understanding of the genome as a whole.

10.1.2 Genome annotation, gene expression, and gene interactions

So far, we have only dealt with the genome in terms of DNA sequence data—but obviously the genome is much more than that, it contains a multitude of genes! As we mentioned briefly in the previous section, a genome assembly is just the first part of the puzzle. If we want our reference genome to be of use, we need to incorporate our understanding of genes and their function. This is the purpose of genome annotation. When we annotate a genome, we carry out two tasks. First we align all our available evidence of gene positions and function to the reference genome and then we use this evidence to build gene models—i.e. our best approximations of the physical position and structure of genes (Yandell & Ence 2012). In order to be able to do this effectively, the reference genome we have assembled has to be of a certain quality and, again, this is where statistics such as the N50 are useful. Recall that the scaffold N50 statistic is the minimum scaffold size of the group of largest scaffolds that contain 50 percent of the assembled genome. If our scaffold N50 is around the size of the average gene in our genome, 50 percent of all genes will be likely to occur on at least one scaffold (Yandell & Ence 2012).

However, as we saw above, proper annotation requires that we align all our available evidence of genes to the reference genome. What constitutes evidence? Usually this is coding DNA sequence

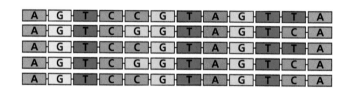

Figure 10.2 Linear vs non-linear genome models. A standard linear reference genome model ignores the variation that might occur in a population. In the top panel, we see that no single sequence can accurately represent population level variation. The lower panel shows a genome graph approach where "bubbles" in the graph allow the assembly to account for polymorphism.

or **cDNA**, although sometimes proteins are used too (Morozova et al. 2009; Yandell & Ence 2012). Early genome studies on model species such as *Drosophila melanogaster* and mice also used **expressed sequence tags** or **ESTs**—i.e. Sanger sequenced cDNA to annotate the genome (Wang et al. 2009). Such model systems benefited from a rich history of work by many different scientists on many different genes and gene functions, meaning there was a large body of available gene evidence for annotation (Yandell & Ence 2012). However, even the genome of a well-studied organism such as *Drosophila melanogaster* has around 17 702 genes[3] and not all have been studied extensively—in other words, there is a bias towards evidence from genes of interest. Furthermore, such resources are not available for new organisms; imagine you were to sequence the genome of an exotic animal with no genomes available from closely related species—what gene evidence could you use then?

This is where **transcriptomics** has been a hugely important part of the "omics" revolution in biology (Ellegren 2014). High-throughput sequencing of millions of RNA transcripts, more commonly known as **RNA-seq** has made it relatively straightforward to sequence the coding portions of the entire genome (Wang et al. 2009; Morozova et al. 2009). RNA-seq essentially has two major uses; the first is that it provides us with evidence of genes and the second is that it enables us to quantify the expression levels of these genes (Ozsolak & Milos 2010). At present, the majority of RNA-seq is conducted using short-read sequencing technology (i.e. Illumina—see chapter 1). Therefore, in order to build gene transcripts from short reads, the data can either be mapped to the reference genome or *de novo* assembled into a standalone transcriptome (Grabherr et al. 2011). Challenges for transcriptome assembly bear a lot of similarity to those we encountered in the previous section and so we will not go into detail on them here. However, empirical and simulation studies have shown that both mapping and separate assembly of RNA-seq data are robust approaches to reconstructing gene transcripts

(Hornett & Wheat 2010; Vijay et al. 2013). Properly assembled transcripts can help identify the identity and function of genes, their spatial positions in the genome, the location of intron–exon boundaries and also different splice variants that might occur (Morozova et al. 2009).

Transcriptome sequencing also provides a quantitative measure of gene expression—i.e. the level of a gene product being produced via transcription. This is because the number of reads of a transcript in RNA-seq is proportional to the abundance of the mRNA transcript in the cell, tissue, or organism assayed (Trapnell et al. 2012). Transcriptomics therefore allows us to study how the genes are regulated and switched on in different tissues, at different stages of development and in different environments. We can therefore think of the transcriptome as dynamic—it responds to different stimuli and environments with different genes and networks functioning at different stages of the life history of an organism. This does not mean that the genome is static and unchanging over a lifetime; as we will see in section 10.1.4, this is clearly not the case at all. However, understanding how and why gene expression patterns differ between taxa is an important component of understanding the link between changes at the level of the genome and the phenotypes that selection ultimately acts upon. We have actually encountered several examples of where transcriptomics has been used to quantify gene expression already through the course of the book. For example, in chapter 8 we learned about how sexual conflict has shaped the evolution of sex chromosomes in Trinidadian guppies (*Poecilia reticulata*). Wright et al. (2017, 2018) used RNA-sequencing to quantify differences in expression between males and females. They showed that the ratio of male to female expression was skewed in non-recombining regions of the sex chromosome, suggesting the guppies have evolved sex-biased expression patterns to prevent antagonistic selection between sexes. Gene expression analysis was also used to further investigate gene function in hooded and carrion crows (Poelstra et al. 2014, 2015). By identifying genes significantly differentially expressed in the heads and torsos of the crows, and strongly associated with plumage color differences, the authors were able to characterize the melanogenesis gene pathway that

---

[3] Estimated from release 6 of the *Drosophila melanogaster* genome: https://www.ncbi.nlm.nih.gov/genome/?term=drosophila+melanogaster

is involved in this striking phenotypic difference (Poelstra et al. 2015).

10.1.3 Proteomics—bridging the gap between gene expression and the phenotype

An important goal for any evolutionary biologist is to develop an understanding of how the genetic changes they might observe in the genome sequence result in different phenotypes, be it plumage color, behavior, or physiological differences. As we have seen here and throughout the book, knowledge of the genome provides us with a foundation for understanding the genotype, but how do we approach the phenotype? Transcriptomics can be thought of as a step above the genome, upwards toward an understanding of how genetic change shapes phenotypic change. But is it sufficient to link changes in the genome and the phenotype? Arguably no—transcriptomics quantifies and measures messenger RNA, but not the final protein products of genes that actually make up our phenotype of interest (Valcu & Kempanaers 2015). The field of **proteomics** involves the study of the **proteome**, i.e. the proteins that are encoded and transcribed from the genome (Diz et al. 2012). Characterizing the proteome is important because there are a number of steps between the transcription of mRNA from DNA sequence and the production of the final protein that can play a role in regulating gene expression and protein abundance (Valcu & Kempanaers 2015; Vogel & Marcotte 2012). These include the production of different protein isoforms due to alternative splicing of gene exons during transcription and also post-translational modification of polypetides after they have been translated from mRNA. Indeed, there is a surprisingly high variation in the strength of the correlation between protein abundance and the levels of expression of their corresponding mRNA transcripts (Vogel & Marcotte 2012). For example, in humans, mRNA expression explains only about 22 percent of the variance in protein abundance, whereas this is 47 percent and 58 percent in *E. coli* and yeast respectively (de Sousa Abreu et al. 2009). In other words, analyzing proteins directly captures information on post-transcriptional regulation that cannot be detected using transcriptomics approaches that measure the abundance of mRNA (see section 10.1.2). We do not want to give the impression that any one approach is superior to the other. In fact, a combination of "-omics" approaches including genomics, transcriptomics, and proteomics is important to gain a proper understanding of the processes involved in the translation of genotype to phenotype.

Advances in mass spectrometry are central to the development of high-throughput proteomics. **Mass spectrometry** is an analytical chemistry technique that measures the mass/charge ratio of ions with very high accuracy in order to identify the chemical structure of different molecules. For proteomics, the mass/charge ratio of ions can be used to identify the amino acid composition of the peptide analyzed (Diz et al. 2012; Valcu & Kempanaers 2015). There is a high diversity of different proteomic pipelines and methods; however, all of them involve mass spectrometry in some way (Diz et al. 2012; Valcu & Kempanaers 2015). Once proteins are extracted and isolated for analysis, they are usually separated; one way to do this is to use gel electrophoresis (see chapter 1). Separated proteins are then excised from the gel and are broken down into peptides using enzymatic digestion. These peptides can then be identified from the signatures they exhibit during mass spectrometry; this is known as peptide mass fingerprinting and can be used to reconstruct protein sequences (Valcu & Kempanaers 2015). This is also an example of where a combined "-omics" approach is advantageous, as the genome or transcriptome is necessary to identify peptides. Therefore, to properly link proteomic data with a genome or transcriptome assembly, knowledge of bioinformatics is necessary (Diz et al. 2012). As well as characterizing the complement of proteins expressed by the genome, proteomics can also be used to measure the abundance of proteins. This can be achieved by labeling proteins or peptides with isotopic markers or using label free methods that rely on the intensity of mass spectrometry signals to distinguish protein amounts (Valcu & Kempanaers 2015).

Proteomics has proven particularly useful for studying reproductive behavior and reproductive barriers with a focus on the role of proteins in speciation. Reproductive proteins play an important role in speciation and sexual reproduction; for example, the coevolution of protein recognition complexes are necessary for determining whether or not sperm can fertilize an egg (Swanson & Vacquier 2002). Proteomics, combined with isotopic labeling of proteins, has been used to identify the transfer of seminal fluid proteins from males into the reproductive tract of females during mating in *Drosophila melanogaster* (Findlay et al. 2008). A combination of proteomic and genomic data was also used to identify seminal fluid proteins transferred between two field cricket species *Gryllus firmus* and *G. pennsylvanicus*. Although genomic differentiation between these species is generally low, differentiation among seminal fluid proteins was high and appears to be driven by positive selection, consistent with these proteins playing a role in barriers to gene exchange (Andres et al. 2008).

10.1.4 DNA methylation, chromatin binding, and epigenomics

So, we have focused on the fact that variation occurs beyond the level of base pair differences in the genome. What about the factors that influence the level of gene expression and transcription? One mechanism altering gene expression is **DNA methylation,** which we first encountered in chapter 1. This is when a methyl group (CH_3) is added to cytosine positions in the genome. Like chromatin proteins, methylation can also alter gene expression when it occurs in the promoter region of a gene. Methylation mechanisms differ quite substantially among taxa but the canonical examples typically come from mammals (Suzuki & Bird 2008). In mammals, CpG dinucleotides (i.e. a consecutive C and G nucleotide pair) experience methylation, although there are some genome regions with a high density of CpG sites but no methylation—known as CpG islands. The extent of DNA methylation in a genome can be easily identified using **bisulfite sequencing**. Genomic DNA is first sequenced using a standard HTS method, then the DNA template is treated with sodium bisulfite, which causes unmethylated cytosine positions to change to uracil. Therefore, any remaining cytosine sites are methylated positions (Suzuki & Bird 2008; Trucchi et al. 2016).

The structure of how DNA is stored can also contribute to how gene expression is regulated. Back in chapter 1, we learned about the basics of the genome—i.e. that it consists of a sequence of nucleotides. In chapter 2, we then learned that DNA is stored in the nucleus of cells as chromosomes. However, if we were to extract the ~3235 Mb human genome from a single nucleus and stretch it out so that it was one single sequence, it would be 2 meters in length. Yet the average nucleus is just 5 μm in diameter (Buenrostro et al. 2013)! How is it possible that so much DNA is compacted so that it can be stored in the tiny nucleus of a single cell? Hierarchical folding and packing of the DNA molecule is the answer. First of all, DNA is wrapped around bundles of proteins known as **histones**, much like you might wrap a piece of string around a cylinder of some sort to keep it together (see chapter 2). Together, the DNA and the histone complex are known as **nucleosomes** and these are spaced out along the DNA like a series of beads on a string. Continued folding and compaction of nucleosomes forms 30 nm fibers and these are then folded into tighter and tighter coils, eventually resulting in chromosomes. The DNA, histones and additional proteins that together make up chromosomes are known as **chromatin**. This packaging and compression of DNA has important consequences for the machinery of DNA transcription. Very tightly packed regions that cannot be easily accessed by RNA polymerases are known as **heterochromatin** whereas the opposite, loosely packed regions are easier to access and are referred to as **euchromatin**. Histones can be modified with methylation and demethylation, which can result in an enrichment of heterochromatin and euchromatin regions and modulates gene expression. Histone modification is also known to play a role in silencing selfish genetic elements such as TEs, and likely play a role in intragenomic conflict (Presgraves 2010; Yoshida et al. 2018). Divergence in chromatin binding profiles between species might account for some genomic incompatibilities that lead to hybrid sterility (see chapter 8) as hybrids lack sufficient means to suppress selfish elements (Yoshida et al. 2018).

Understanding the extent of variation in chromatin binding can help us understand how gene expression is regulated beyond the sequence level and a wide arrange of techniques have been developed to quantify this (Meyer & Liu 2014). ChIP-seq (chromatin immunoprecipitation with sequencing) essentially enriches genome regions that bind to specifically modified histone proteins (and also other proteins involved in gene expression, such as transcription factors). The enriched regions are then sequenced and mapped back to the genome, to identify where these proteins bind (Park 2009). ATAC-seq is a method of sequencing open chromatin regions that are accessible by transposons—i.e. where chromatin is accessible for gene expression (Buenrostro et al. 2013).

Taken together, chromatin binding and DNA methylation form the **epigenome** and are examples of epigenetic changes to gene expression (see chapter 1). Epi- is a Greek prefix meaning on top of or in addition to—signaling that the role of these processes is not encoded in the DNA sequence. Epigenetics is closely linked to the concept of phenotypic plasticity where phenotypic variation occurs that cannot be explained by variation in genotypes. The role of both epigenetics and phenotypic plasticity in evolution remains controversial and there is considerable debate over how they might contribute to evolutionary change (Schlichting & Wund 2014; Verhoeven et al. 2016). There is considerable evidence that epigenetic change contributes to phenotypic variation in response to environmental stress or changing conditions (Verhoeven et al. 2016). In this sense, it may contribute to evolutionary change by modifying the strength of selection on phenotypes or exposing cryptic genetic variation to selection. However, more controversial is the concept of **genetic accommodation**, where epigenetic change and the phenotypic plasticity it results in allows populations to persist in the face of an environmental change until genetic mutations occur and lead to adaptation (Schlichting & Wund 2014). To properly assess whether epigenetics can facilitate evolution in this way, we need to know whether epigenetic change can expose cryptic variation to selection, whether it can be passed across generations, how many generations such changes can be sustained for, and whether patterns of methylation themselves are heritable, are associated with environmental variation, and are under selection (Verhoeven et al. 2016). As we learned in

chapter 1, transgenerational epigenetic effects appear to sometimes occur and to persist over a few generations, at least in laboratory conditions and in some taxa, particularly plants (Schmitz 2014). In *Arabidopsis thaliana*, there is good evidence that (experimentally) epigenetically modified genome regions can persist for some generations and can apparently occasionally contribute to very large proportions of phenotypic variation (Cortijo et al. 2014). There is also evidence of a genetic basis for different methylation patterns among *A. thaliana* populations inhabiting different environments, although it remains unclear whether selection has acted on these (Schmitz et al. 2013). At present, there is no consensus on the importance of epigenomics for shaping how evolution plays out, but it is clearly an exciting area for future research.

10.1.5 The 3D structure of the genome

There are times when it is necessary and even helpful to think of the genome as a linear 2D sequence (i.e. when mapping sequence reads to it); however, the reality is far from that—the genome is expressed and translated as the transcriptome and the proteome. As we learned in the previous section, genome structure plays an important role in regulating rates of transcription and translation of genes, as well as having functional consequences observable at the phenotypic level. However, the genome also has a 3D structure when it occurs in the nucleus, with interactions among regions which are separated by large distances in terms of nucleotides. One of the main approaches to identify 3D genome structure is chromosome conformation capture (3C). 3C approaches identify genome regions that are physically close to one another in 3D space, but not necessarily close by when laid out on the linear genome sequence (Spielmann et al. 2018). The most recent 3C method is Hi-C; this incorporates high-throughput sequencing to identify the frequency of interactions across the entire genome and can also be used to improve genome assemblies (Belton et al. 2012).

The 3D nature of the genome has somewhat been neglected in the context of evolutionary genetics, but it clearly could have important implications for evolution. For example, structural variations such as inversions have implications at the level of 3D genome organization and can disrupt genome regions that show a high density of interactions. This has been associated with human disease phenotypes such as finger malformations (Spielmann et al. 2018). A great deal of work is necessary to better characterize 3D genome structure for many species and to unravel its importance in evolution. One barrier to our understanding is the fact that any single analysis of structure is only a temporal snapshot; genome structure is dynamic and changes through time. Indeed, this led to the recent call for a large-scale project to characterize the spatial and temporal nature of the genome—i.e. in 4D (Dekker et al. 2017).

10.2 Manipulating the genome

10.2.1 From candidate genes to an understanding of function

The advances we have seen in population and evolutionary genetics since the onset of the "-omics" revolution have been breathtaking. Throughout this book, we have introduced you to a large number of examples where genome-sequencing technologies have provided us with previously unimaginable power to examine evolutionary questions. As we commented in chapter 1, it is truly remarkable to have been working at a time when we might have begun our careers working on only a handful of markers, to being able to use millions of markers in just a few years. Nonetheless, it is important to be mindful of the fact that genomic data (and other "-omics" data too for that matter) can only get us so far. It is reasonable to ask whether swathes of new data have advanced our understanding as much as we might have expected (Baird 2017). It is important to be cautious; when the human genome was sequenced there was considerable hype and expectation that it would help us find cures for a myriad of diseases and disorders. Similarly, the introduction of genomic data to evolutionary biology research brought with it an optimism that it might help us properly answer difficult questions such as whether regulatory or coding changes underlie adaptations, whether these muta-

tions are of small or large effect and which genes are involved in speciation (Hoekstra & Coyne 2007; Ellegren 2014). Although we hope that you are convinced that genomic data has given us insight into such evolutionary questions, it is fair to say that there is still a lot to be done in order to move towards a proper understanding of even basic questions, such as how a mutation at the genome level can alter a phenotype and bring about a change in fitness for an organism carrying it.

Perhaps one of the most important caveats to modern evolutionary genetics is the fact that a great deal of research focuses on identifying statistical correlations and associations rather than demonstrating a true functional link (Dean & Thornton 2007). We can think back to chapters 6 and 7 for good examples of this. Association mapping using QTL analysis or GWAS might help us identify genomic regions or even candidate genes that are associated with a trait. However, these are again only statistical correlations and often a great deal of further work is necessary to properly investigate the functional roles of the genes they identify. Similarly, genome scan studies looking for signatures of selection might point towards candidate genes, but without functional validation we cannot be certain that the evidence we see actually represents a selective sweep that might have occurred at a gene. If we accept that further evidence is necessary to demonstrate that a gene might be involved in an adaptation, how can we go about this? Functional validation of candidate genes using molecular biology techniques may often be the only way. For example, these might include transgenics or gene knockout assays—both topics we will cover in more detail in the next section. Indeed, a "functional synthesis" merging correlative inference from genomic data and experimental validation of candidate genes using molecular biology provides a powerful framework for investigating evolutionary questions (Dean & Thornton 2007).

This is not just a fanciful idea—there are good examples of such a functional synthesis in organisms not traditionally considered model species[4]

---

[4] However, we concede that with a well-annotated reference genome, a wide number of biological resources, and over two decades of intensive research, it would be fair to describe threespine stickleback as a model species now!

and we have encountered several in this book! Perhaps the most illustrative example for our purposes here is that of the threespine stickleback (*Gasterosteus aculeatus*) that we first encountered in detail back in chapter 6. Since the end of the last glaciation, threespine stickleback have repeatedly colonized freshwater environments exposed by ice retreat in the Northern Hemisphere (McKinnon & Rundle 2002). Threespine stickleback are actually a predominately marine species but they have been shown rapidly to adapt to freshwater environments (Bell et al. 2004). A suite of adaptations characterizes this remarkable example of parallel evolution, but none more so than the loss of armor plating on the lateral flank of the fish; marine fish are fully armor plated but freshwater fish have a reduced armor phenotype.

Understanding the genetic basis of armor plating has been the major focus of research for a number of researchers and it demonstrates nicely how it is possible to go from a statistical association to an understanding of the functional basis of the trait. An early stickleback linkage map constructed using microsatellite markers demonstrated it was possible to associate phenotypic trait variation with genomic regions (Peichel et al. 2001). However, as this study used a cross between two freshwater fish, they did not map any major phenotypic variation in plate number. A follow-up study used finer-resolution QTL mapping with a marine and freshwater cross to identify a QTL occurring in a 400 Kb region on chromosome IV that was able to explain 78 percent of the variance in armor plate morphology (Colosimo et al. 2004). One year later, the same group used even finer-scale association mapping using a polymorphic population to narrow this down to a 16 Kb block that contained a candidate gene—*Ectodysplasin (Eda)*—and showed high linkage disequilibrium with armor plate number (Colosimo et al. 2005). Indeed, the region acted much like a simple biallelic

major effect locus—fish homozygous for the *L* allele had 4–6 plates, fish homozygous for the *C* allele had 28–32 plates, and *LC* heterozygotes were partially plated with 15–19 plates. A number of population genetic studies used microsatellite markers found in the introns of *Eda* to demonstrate evidence of strong F_{ST} differences between completely plated marine and low-plated freshwater sticklebacks (Raeymaekers et al. 2006, 2014; Kitano et al. 2008). Genomic data has also clearly pointed to a role for *Eda* in the evolution of plate morph differences in parallel populations; whole-genome resequencing data further shows increased divergence and allele frequency differences at the 16 Kb block surrounding the *Eda* gene (Jones et al. 2012). Other population genomic studies in marine–freshwater threespine stickleback have also shown high differentiation measured by F_{ST} close to this region (Hohenlohe et al. 2010; Kusakabe et al. 2017).

However, in their seminal 2005 study, Colosimo and co-workers went beyond using only a QTL analysis to link their candidate region to function. They first created progeny from two low-plated individuals (i.e. all progeny homozygous for the *LL Eda* haplotype) and then injected early stage embryos with a construct of an *Eda* allele from mice to create **transgenic** fish. A small proportion of these transgenic fish developed a higher number of armor plates than expected given their *LL* genotype. Clearly this demonstrates the role *Eda* plays in underlying this phenotype, although a clear causative mutation remained elusive—other than it being present in the 16 Kb narrow region surrounding the gene. In a further study from the same group, O'Brown et al. (2015) compared complete marine and low freshwater haplotypes to identify a single base pair change from T in the marine to a G in freshwater. They then cloned a 3 Kb region surrounding this SNP and attached it to a green-fluorescent

Figure 10.3 (overleaf) A combination of evolutionary genetics tools have led to the identification of the causal SNP underlying armor plate number variation in threespine stickleback. Top panel shows a 50-Kb sliding window estimate of F_{ST} between freshwater and marine stickleback, with a region of high divergence in the vicinity of the *Ectodysplasin (Eda)* gene (redrawn from Kusakabe et al. 2017). The middle panel shows variation in armor plate phenotypes between a marine stickleback from Japan (JAMA) and two freshwater stickleback from Japan (NAKA) and Western Canada (PAXB). The nucleotide sequences show a single T to G nucleotide difference is fixed between the marine form and all freshwater forms just downstream from the *Eda* gene (O'Brown et al. 2015). The lowest panel shows two transgenic stickleback with a GFP reporter construct attached to the marine *Eda* haplotype. In the top stickleback, the high level of fluorescence indicates where the *Eda* gene product is expressed along the lateral flank. In the lower fish, the SNP has been altered using genome editing to reduce expression, demonstrating the functional role of *Eda* in plate development. © 2015, O'Brown et al. for middle and bottom panels.

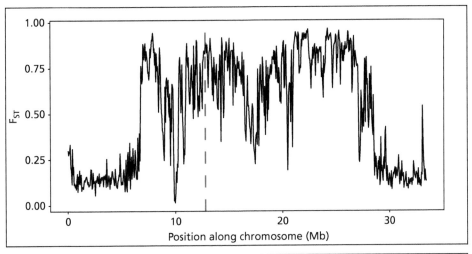

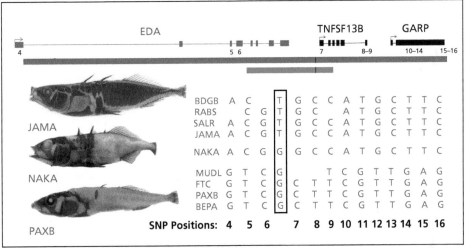

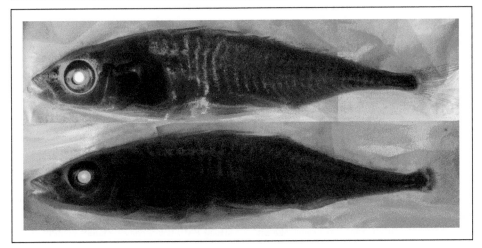

protein (GFP) **reporter construct**. Reporter assays like this attach genes that fluoresce or are easily visualized to regulatory regions of candidate genes, meaning it is easy to detect where the gene of interest is expressed during development. The GFP assay clearly showed *Eda* is expressed on the lateral flanks of young fish during early development (O'Brown et al. 2015). To confirm the single SNP altered expression of the *Eda* gene and therefore led to plate development, they then used genome editing to alter the SNP in their clone-GFP reporter construct and showed that expression was greatly reduced on the flanks of developing young fish (Figure 10.3).

The series of papers that led to the discovery of the single causative mutation underlying the loss of armor plating in threespine stickleback is an excellent example of the power of a multidisciplinary approach to evolutionary genetics. Population genetics and quantitative genetics pointed to a genome region and functional genetics approaches demonstrated causality. However, while we now know that a single SNP underlies this phenotypic trait difference, there are still many questions that remain unanswered. Most notable is the fact that the exact reason as to why a low-plated phenotype has a higher fitness in freshwater populations is not clear. A large number of different hypotheses have been suggested, including buoyancy (Myhre & Klepaker 2009), lower abundance of predators, calcium concentration, and growth rate differences (Barrett et al. 2008; Marchinko & Schluter 2007). An exciting avenue for future work in this system will be to combine functional genomic techniques such as **genome editing** with some of the experimental assays and field fitness tests that have been established for the species (Arnegard et al. 2014).

10.2.2 CRISPR/Cas9 and the future of genetic manipulation

As our previous section highlights, molecular biology tools for genetic manipulation have enormous potential for applications to evolutionary genetics research. However, despite recognition over a decade ago that molecular biology approaches should be adopted as part of a "functional synthesis" (Dean & Thornton 2007), they still remain relatively underused. This is most likely because the tools required

for genetic manipulation have typically been expensive in terms of time and money, as well as being extremely difficult to apply to non-model organisms. With the advent of **CRISPR/Cas9** and related RNA-guided genome editing tools that are cheap, efficient and relatively easy to use, this is beginning to change (Bono et al. 2015).

CRISPR stands for clustered, regularly interspaced short palindromic repeats and is a form of adaptive immune system first recognized in bacteria (Marraffini & Sontheimer 2010). In microbes such as bacteria and archaea, CRISPR acts as a defence mechanism against maladaptive horizontal gene transfer. Within the CRISPR complex, spacer sequences sit between short repeats. These spacer sequences allow the complex to produce an approximately 20 bp CRISPR RNA (crRNA) sequence that interacts with CRISPR associated proteins (**Cas**) and guides them to the foreign target (see Figure 10.4; Terns & Terns 2014). The Cas protein family, in particular Cas9, are nucleases, i.e. enzymes that cleave DNA sequence; other forms of Cas proteins are able to cleave RNA, making the CRISPR/Cas system effective against viral derived RNA too. When guided to a target sequence by the crRNA, a second crRNA sequence, the trans-activating crRNA or tracrRNA activates the Cas nuclease which then cleaves a double-stranded break in the target invasive DNA/RNA, rendering it harmless (Marraffini & Sonthemier 2010; Terns & Terns 2014) (Figure 10.4). As an adaptive immune pathway that can target or modify DNA, the CRISPR/Cas complex has enormous potential as a genome-editing tool and intensive research effort has focused on developing the CRISPR/Cas9 system for this purpose over the last few years (Sander & Joung 2014).

CRISPR/Cas9 can effectively be used to target any region of a genome. In principle, the only requirement is that the target region is upstream of a protospacer adjacent motif or **PAM**, a small DNA sequence that allows the CRISPR/Cas9 system to recognize the target (Sander & Joung 2014; Terns & Terns 2014). Since PAM sequences are extremely simple—i.e. NGG or GC—this requirement is easily met (Figure 10.4). Repurposed as a genetic manipulation tool, the CRISPR/Cas9 system has also been simplified; the two crRNA sequences are joined together to produce a single guide RNA or sgRNA

(Figure 10.4; Sander & Joung 2014; Terns & Terns 2014). This sgRNA, in particular the guide RNA target sequence, can easily be synthesized, making CRISPR/Cas9 design rapid and straightforward. In contrast, other genetic manipulation techniques such as TALEN (Transcription activator-like effector nucleases) and ZFNs (zinc finger nucleases) require artificial enzyme design and are much more expensive (Terns & Terns 2014; Bono et al. 2015).

Once CRISPR/Cas9 is used to cleave a DNA or RNA sequence, what happens next? If the target is cleaved without an additional donor sequence nearby, DNA repair mechanisms within the cell will result in **nonhomologous end joining (NHEJ)** that typically introduces a small insertion or deletion (Bono et al. 2015). Alternatively, if a donor sequence is present alongside the CRISPR/Cas9 system,

homology directed repair (HDR) can incorporate it into the sequence (Terns & Terns 2014; Bono et al. 2015). These two different repair mechanisms mean that CRISPR/Cas9 can be leveraged for different genome editing purposes (Figure 10.4). For example, NHEJ can be used to disrupt gene function by shifting the open reading frame of a gene, in other words it can produce a **gene knockout**. Gene knockouts are important for helping to determine gene function, for example by disabling a gene and then studying the phenotypic consequences. On the other hand, HDR is able to incorporate sequences, meaning it can be used to produce a **gene knockin**, i.e. to create a transgenic organism with an introduced gene sequence in the target region. However, it is worth noting that CRISPR/Cas9 protocols are under continuous development; for example, it is now possible

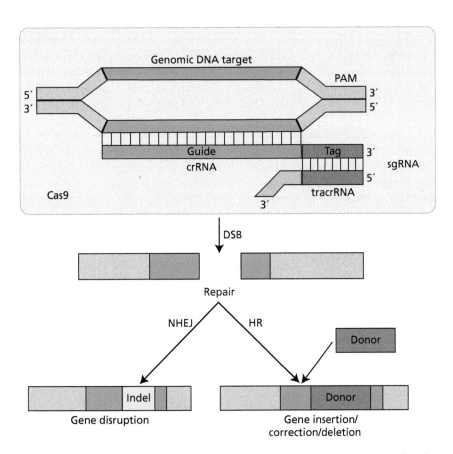

Figure 10.4 CRISPR-based genome editing. The CRISPR/Cas9 construct is guided to its genomic DNA target by a single guide RNA (sgRNA) provided a protospacer adjacent motif or PAM is present upstream. The Cas9 protein then cleaves the DNA sequence, resulting in a double-strand break (DSB). DNA repair machinery then induces either an indel as a result of nonhomologous end joining (NHEJ) or gene insertion using homology directed repair (HDR). Adapted from Terns & Terns (2014).

to efficiently perform gene knockin using the NHEJ pathway too (Watakabe et al. 2018). Nonetheless, the major strength of the CRISPR/Cas9 system is that if genetic manipulation is performed at an early enough developmental stage to be incorporated into the germline, the genome edits will be heritable.

As well as genome-level editing, CRISPR/Cas9 can also be used to regulate gene expression at the transcriptomic level. Similar molecular biology tools such as **RNA interference (RNAi)**, also based on prokaryote adaptive immunity, already exist to do this (Ketting 2011). RNAi works on the basis of gene disruption by intercepting and interrupting the translation of mRNA (Ketting 2011). Disrupting and suppressing gene expression is known as **gene knockdown**. While effective in some cases, RNAi also has **off-target effects**, meaning it can often disrupt the function of other genes. This is particularly problematic if the method is being used to study the functional role of a specific gene. **CRISPRi** or CRISPR-based interference can be used to suppress gene expression whereas **CRISPRa** can also be used to increase it (Bono et al. 2015). The approach is very similar to CRISPR based editing but uses altered Cas9 proteins, meaning it is again easy to design and potentially highly specific. Furthermore, CRISPRi/CRISPRa can again be used to make heritable changes to gene expression, an important feature for establishing experimental lines of study organisms.

10.2.3 Prospects for genome editing in evolutionary genetics

Unlike other genome-editing methods, CRISPR/Cas9 is easy and cost effective to design and use. Since the method was first outlined in 2013, it has exploded in popularity and it is now possible to buy CRISPR/Cas9 kits and protocols directly from biotechnology companies. Several databases also now exist to help identify targets for the system in published genomes.[5] Furthermore, CRISPR/Cas9 demonstrations or protocols have been published for a number of species widely used in evolutionary research. For example, there are protocols for threes-

[5] For example, the brilliantly named CC-Top: https://crispr.cos.uni-heidelberg.de/

pine stickleback (Erickson et al. 2016), *Arabidopsis thaliana* (Jiang et al. 2013), wild mice (Hirose et al. 2017) and several butterfly species, including *Heliconius erato* (Zhang et al. 2017). Clearly then, the pieces are in place for CRISPR/Cas9 to be an effective tool for evolutionary geneticists, even if they are not working on an extremely well-established model system. This leads to an interesting avenue for speculation—what might widespread gene editing enable us to do?

Gene disruption and the production of gene knockdowns/knockouts using the CRISPR/Cas9 NHEJ pathway is the most straightforward use of the technology. Therefore, this seems the most likely avenue for immediate future application. Knockdowns and knockouts are important tools for understanding gene function; disrupting a gene to see how it plays a role in development of a phenotype is a classic reverse genetics approach. There are already some exciting examples of CRISPR/Cas9 based gene knockouts being used to identify the functional roles gene play. For example, Zhang et al. (2014) used CRISPR to investigate the function of the gene *dmrt6* in Nile tilapia (*Oreochromis niloticus*)—a fish species, they hypothesized *dmrt6* was involved in spermatogenesis and indeed knockout individuals lacked spermatocytes in their testes compared to control fish. CRISPR knockdowns of pigmentation genes in threespine stickleback have been used to create a fish model for studying albinism. Using genome-sequencing, Hart & Miller (2017) identified a single SNP difference in the *Hsp5* gene associated with albinism in a wild-caught individual, using CRISPR/Cas9, they then deleted the same region in developing young and established an experimental line with the albinism phenotype that they named *casper*. As well as demonstrating the function that this gene plays in pigmentation loss, *casper* mutants can be used to effectively visualize GFP reporter assays within the body (Figure 10.3, lower panel)—in other words, it is possible to see fluorescence signals through their translucent skin! Knockouts have also been used in a comparative context. Focusing on four butterfly species, Zhang et al. (2017) used CRISPR/Cas9 to knockout the *optix* gene, known to play a role in wing pattern development. This resulted in a stark replacement of color pigments with melanins in the wings of each of the species, producing a dark-gray phenotype. Furthermore, for the buckeye butterfly (*Junonia coenia*), *optix* disruption resulted in blue iridescence in the wing scales. Therefore, gene manipulation in this example confirmed that *optix* is highly conserved across the species and also has an unexpected functional role in some species (Zhang et al. 2017).

Medical genetics also provides some interesting examples of how genome editing can be used to create study models and potentially rescue phenotypes. Duchenne muscular dystrophy (DMD) is a muscular disorder that causes muscle wasting and affects around one in every 5000 boys, severely limiting life expectancy and often resulting in death due to respiratory failure (Amoasii et al. 2017). DMD is X-linked in humans and is typically caused by mutations that cause a deletion around exon 50 in the *dystrophin* gene, disrupting the production of the dystrophin protein, which is an important protein for proper muscle function. Thus, introducing mutations that can cause the neighboring exon (51) to be skipped can actually rescue the production of dystrophin and is potentially an effective gene-based therapy for DMD (Amoasii et al. 2017, 2018). Of course, like most medical research, such a treatment must first be tested in animal models before it can be applied to humans. To achieve this, Amoasii et al. (2017) used CRISPR/Cas9 to delete the corresponding exon 50 in the *dystrophin* ortholog in mice, resulting in individuals with severely reduced dystrophin levels and muscle wasting. The researchers then designed a CRISPR assay to target a small 9-bp region in exon 51 that causes it to be skipped during transcription. Applying this new edit to mice with the deleted exon 50 resulted in a restoration of dystrophin expression in muscle tissue (Amoasii et al. 2017). In a follow-up study, the same authors also demonstrated that *dystrophin* protein expression could be rescued in dogs that experienced a similar condition to human DMD (Amoasii et al. 2018).

Knockouts using CRISPR/Cas9 are clearly already helping us understand gene function. It is also possible to easily multiplex CRISPR assays, meaning they can be used to study the function of multiple genes simultaneously. There is already a move towards high-throughput assays of gene function using genome-editing arrays on cell cultures for cancer biology (Shalem et al. 2015). Gene knockins, using the HDR pathway, by contrast are less straightforward and more difficult to achieve than knockouts (Terns & Terns 2014; Bono et al. 2015). At present, examples of how this is being used in evolutionary genetics are lacking. Nonetheless, efficient knockins would open up a new range of potential experimental avenues. For example, a classic method in ecological speciation research has been the reciprocal transplant experiment; i.e. taking two species or ecotypes adapted to different environments and switching them and measuring their growth and fitness (Schluter 2000b). With a CRISPR knockin, we could perform the same type of experiment but this time swapping genes between the populations, as a demonstration of the role of the gene in maintaining fitness (Bono et al. 2015). Additionally, "restoring" phenotypes in experimental lines (as we saw in the stickleback *Eda* example) or across species boundaries are now much more feasible in a wider array of taxa thanks to CRISPR technology.

As with all new methods and approaches in any scientific field, it is important to approach CRISPR with a good understanding of its limitations. One of the biggest difficulties with using the technology is effective delivery of the CRISPR/Cas9 complex to early stage embryos (Bono et al. 2015). Many of the studies we have highlighted here use either plasmids (Amoasii et al. 2017) or micro-injection (Zhang et al. 2017; Hart & Miller 2017) of CRISPR constructs directly into embryos. This is obviously relatively feasible for taxa such as stickleback but clearly much more challenging for organisms such as birds or reptiles. Similarly, not all species can be easily kept in a laboratory and establishing experimental lines is potentially impossible in some cases (Bono et al. 2015). There are also obvious concerns about the level of biosecurity surrounding the creation of transgenic organisms; strict protocols and procedures are necessary to ensure that lab-based experimental lines remain in the lab and are not able to interbreed with wild populations.

Concerns over the biosecurity of CRISPR-based experiments are an ethical issue that needs to be addressed at a societal level, not a scientific one. One of the most well-known and potentially controversial speculative applications of the technology is

CRISPR-based **gene drive** (Figure 10.5). This is the idea that CRISPR could be used to introduce gene constructs alongside homing endonuclease genes that break the strands of homologous chromosomes and then copy themselves (Esvelt et al. 2014; Webber et al. 2015). Essentially a gene drive construct would act in a similar way to a selfish genetic element by distorting segregation and promoting its growth in the population it is introduced to (Figure 10.5). Suggested uses of gene drive include the eradication of pest species by introducing genes that reduce fertility or skew sex ratios; i.e., gene-based mosquito control (Esvelt et al. 2014; Webber et al. 2015). Indeed, gene drive has already been demonstrated in a laboratory population of *Anopheles gambiae*, the mosquito that spreads malaria. Kyrou et al. (2018) used CRISPR/Cas9 to disrupt the *doublesex* gene and cause female sterility. The gene drive construct swept through the test populations to full fixation within 7–11 generations and resulted in population collapse because of the disruption of egg production. There are, however, some real potential issues with this kind of technology. Imagine we introduce a gene that reduces fertility in order to eradicate a population of an invasive species. Our gene drive construct works well and it spreads rapidly, reducing the population size in a few generations. However, we did not properly account for the fact that our invasive species is able to disperse very efficiently, therefore our introduced gene reaches its native range and begins to spread there too, decimating the species entirely. Use of gene drive to control pests might therefore only be possible in isolated populations, where we can be certain that it will not spread further (Esvelt et al. 2014; Webber et al. 2015).

10.3 The future of evolutionary genetics?

Niels Bohr, the Nobel Prize winning Danish physicist, once quipped "Prediction is very difficult, especially about the future." He was right; grand predictions about the future nearly always turn out to be wrong (Gardner 2010). In this final section of the book, our aim is not to be speculative or to make predictions; instead we hope to provide a short perspective. Evolutionary genetics has clearly come a long way since the modern synthesis merged population genetic theory and evolutionary biology in the early twentieth Century. It is likely that in the not too distant future, most evolutionary biologists will look back and agree that the "-omics" age was another major synthesis event, bringing about substantial change in how we view evolutionary processes (Ellegren 2014).

The technological advances of the last decade are clearly indisputable; if you start an evolutionary genomics project today you can easily have millions of loci at your disposal within a few months. As we have explored extensively in this chapter, new kinds of data and analyses are also on the horizon. Datasets and sample sizes are growing all the time, so much so that data storage is much more of a hurdle for large-scale studies than data generation. Perhaps an even more pressing question is, "How can we make the most of this data?" Analysis of genomic data has arguably lagged behind the

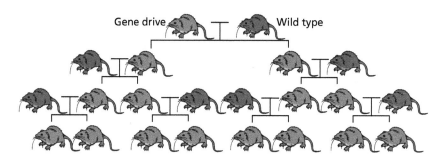

Figure 10.5 Gene drive as a method for biocontrol. Genome editing is used to create a gene drive construct that distorts segregation ratios in the germline and increases the number of copies of itself each generation. By introducing this into the population, it can quickly spread to fixation more rapidly than it would if under selection and certainly by drift. By linking such a gene drive element to a deleterious gene that distorts sex ratio or reduces fertility, this could be used to drive a population to extinction. Adapted from Esvelt et al. (2014).

advances in sequencing platforms that allow it to be generated with ease. This is changing—there are huge numbers of R packages, GitHub repositories, and software pipelines out there dedicated to genomic inference. An exciting new avenue already on the way is the application of machine learning to large-scale evolutionary genomic datasets (Schrider & Kern 2018), an approach that has proven extremely powerful in identifying selective sweeps for example (Schrider & Kern 2016) and also gene flow (Schrider & Kern 2018).

We shouldn't, however, get swept away with the idea that technological revolutions will solve all problems and answer all questions. We can take two examples to illustrate why this is. Over a decade ago, prior to the onset of large-scale genomic datasets in many different species, Hoekstra & Coyne (2007) addressed whether coding or regulatory mutations played a more important role in adaptation. They concluded at the time that there was not enough evidence to support either type of mutation as being more important, only that both likely played a role in evolution. Arguably, this question remains unresolved—we have excellent examples of both types of mutations occurring in evolution, but not enough to clearly come down on either side of the debate. Speciation research provides our other example. The onset of genomic data brought about optimism that genome-wide patterns of genetic differentiation would point towards genes involved in reproductive isolation (Turner et al. 2005). Yet very few such genes have been successfully identified, despite a great deal of sequencing effort (Baird 2017). On the other hand, genomic data has been hugely important for developing our understanding of the processes involved in speciation over the past decade (Seehausen et al. 2014; Westram & Ravinet 2017).

The most important point is that to be a modern evolutionary geneticist, you must be prepared to be flexible. As technological advances develop, you might need to alter how you conduct your research or spend time learning a new method. The fact that bioinformatics and statistics have become invaluable, essential tools that any evolutionary biologist should have on hand was our primary motivation for writing this book! Although it can be difficult at times to grasp new methods which feel beyond your area of expertise, the rewards are rich. There are many exciting questions and problems that remain to be resolved in evolutionary biology. New data and methods will likely open up even more avenues to investigate. The field is continually evolving and we should too.

Study questions

1. How is a *de novo* genome assembly performed using shotgun sequencing with short reads?
2. What is the difference between paired-end sequencing and mate-pair libraries?
3. You have sequenced the genome of your favorite organism and also assembled it using a standard pipeline. This organism has a true genome size of 2.5 Gb. You have assembled eight scaffolds; 452 Mb, 426 Mb, 407 Mb, 359 Mb, 274 Mb, 196 Mb, 91 Mb, and 47 Mb. Calculate:
   a. The percentage of the true genome you have assembled.
   b. The scaffold N50 statistic for your assembly.
4. Explain how a genome graph is a different philosophy on genome assembly than standard approaches.
5. Discuss the two main uses for transcriptomic data.
6. How can euchromatin and heterochromatin influence gene expression?
7. What are the two pathways that CRISPR/Cas9 can use to perform genome editing?
8. Describe the differences between the following:
   a. gene knockin
   b. gene knockout
   c. gene knockdown.
9. Discuss the potential advantages and disadvantages with using CRISPR/Cas9 based gene drive technology for biocontrol. Do you think it is a technology we should use or not?

Glossary

Absolute fitness Usually, the expected number of fertile, surviving offspring of an individual with a certain genotype or trait value.

Adaptive introgression Gene flow from one species into another that has an adaptive role in the recipient.

Adaptive landscape The graphical relationship between mean population fitness and the frequency of an allele (or a combination of alleles in cases of multiple loci) in a population. Adaptive landscapes can also depict the relationship between fitness and variation in one or more phenotypic traits.

Additive distance A tree drawing method where the sum of branch lengths between two taxa or sequences is equal to their genetic distance in a distance matrix.

Additive fitness When total fitness equals the sum of fitness contributions from alleles at different loci fitness is said to be additive rather than epistatic.

Additive variance The component of the genetic variance in a trait that is due to additive effects of alleles.

Alignment The final product of lining up the homologous character states among a set of sequences (DNA, RNA, amino acid) in order to represent an evolutionary relationship between them.

Allele One of two or more forms of a gene or a locus.

Allele frequency spectrum The distribution of allele frequencies of the polymorphic sites along a DNA sequence within a population. In a single population this can be visualized using a histogram. The joint frequency spectrum between two populations is often visualized using a heat map.

Allopatric speciation Speciation in geographic isolation— i.e. without gene flow.

Allopatry The literal meaning in Greek is "other fatherland." A population or species lives in allopatry when it is geographically isolated from others.

Allopolyploidy A form of polyploidy (having more than the usual diploid chromosome set) that involves hybridization between species.

Alternative splicing The controlled process in which different sets of the exons encoded by a gene are included or excluded in the resulting processed mRNAs. This enables a gene to code for more than one protein.

Ancestral A character state inherited from a common ancestor. For example, hair is an ancestral trait that is common to all mammals.

Ancestral polymorphism A polymorphic site or locus that arose in the ancestral population, prior to a speciation event.

Ancestral state reconstruction A method for reconstructing a character state at an ancestral, internal node within a phylogeny using the distribution of character states among the descendant taxa.

Ancient DNA DNA isolated and sequenced from an ancient or historical specimen. There is no strict definition on how old a specimen should be before it is considered ancient, but typically it is hundreds or thousands of years.

Aneuploidy An imbalance in the number of chromosomes in a cell or an organism. Down syndrome in humans is one example, in which the individual carries an extra (complete or partial) copy of chromosome 21.

Antagonistic selection Opposing selective forces affecting a genotype or a phenotypic trait.

Approximate Bayesian Computation (ABC) A method that uses simulation to approximate the likelihood equation and to estimate the posterior probability of complex evolutionary and demographic models.

Arms race A process in which coevolving species (natural enemies or competitors) affect each other's evolution in such a way that an adaptation in one species induces a counter-adaptation in the other species, etc. Arms races can also occur within a species, for instance between the sexes when evolutionary interests of the sexes conflict – see also **sexual conflict**.

Assortative mating A tendency for individuals with similar phenotypes to interbreed more frequently than expected from random mating.

Autopolyploidy A form of polyploidy (having more than the usual diploid chromosome set) that involves interbreeding within a species.

Autosomal chromosomes The chromosomes that are not sex chromosomes, which occur in homologous pairs in diploid organisms.

Balancing selection A form of natural selection that maintains genetic polymorphism at a locus. Heterozygote advantage and negative frequency-dependent selection are the most important mechanisms.

Baldwin effect The situation in which a learned trait affects the organism's subsequent evolution.

Barrier to gene exchange A phenotypic trait that reduces gene flow between differentiated populations or species. It is sometimes, somewhat imprecisely, referred to as a reproductive barrier or an isolating mechanism.

Bateson–Dobzhansky–Muller incompatibility (BDMI) Developmental incompatibility between a pair of species caused by epistasis between alleles at two or more loci, which results in hybrids with low fertility and/or viability.

Bayesian probability A type of thinking that uses prior expectations as a means to estimate the probability of an outcome.

Binomial nomenclature The binary naming system introduced by Linnaeus in which all species are given two scientific names in Latin (or Greek). The first name represents the genus the species belongs to and the second name represents the species.

Bioinformatics The use of computer science, mathematics, and statistics to analyze and interpret biological data. A new and increasingly important interdisciplinary field.

Biological species concept A definition of a species as a group of actually or potentially interbreeding populations that are reproductively isolated from other such groups.

Bisulfite sequencing A method for identifying DNA methylation patterns in the genome. Template DNA is treated with sodium bisulfite that alters all unmethylated cytosine positions to uracil. Any remaining cytosine sites are therefore methylated.

Bootstrapping Resampling of a dataset with replacement in order to get a measurement of sample error. For phylogenetic trees, this means repeatedly resampling informative loci and estimating trees in order to get an idea of the support for branching patterns in the tree.

Bottleneck A sharp reduction in the size of a population.

Branch A connection between nodes or tips on a phylogenetic tree, often representing time or genetic distance.

Breeder's equation An equation in quantitative genetics describing the change in the mean value of a trait as a function of the strength of selection multiplied with the heritability of the trait.

Candidate gene A gene that we have reason to believe affects a particular phenotypic trait in some way, for example, encoding a protein, changing gene expression, or altering a genetic pathway.

cDNA Coding DNA sequence—i.e. the part of the genome that is used to code the amino acid sequences that make up proteins.

Central dogma of molecular biology A theory for the main pathways of genetic information flow in living organisms. The three most important processes are: DNA replication, the process where a DNA molecule copies itself; transcription, the process where a DNA molecule dictates the nucleotide sequence of an RNA molecule; and translation, the process where the genetic code of DNA is translated (via an mRNA intermediate) into a protein product. Other pathways include reverse transcription, the process in which an RNA molecule is reverse transcribed into a complementary DNA copy.

Centromere The part of a chromosome that links sister chromatids (the two identical copies of a replicated chromosome) and that acts as attachment site for the kinetochore. During cell division spindle fibers attach to the centromere via the kinetochore, which then drags the two sister chromatids to either side of the cell.

Character Homologous phenotypic or genetic states shared among taxa. For example, a nucleotide position is a character.

Character displacement Evolutionary divergence in heritable traits in a pair of species as a consequence of their ecological interaction.

Character state The different states a homologous character can take. Returning to the example of a nucleotide position, the four possible character states are A, C, T, and G.

Chromatin The material made up by the tight packaging of DNA, histones, and other proteins. Together, chromatin forms the chromosomes.

Chromosome Cellular bodies that contain DNA. In prokaryotes, the chromosome is a single, circular DNA molecule. In eukaryotes, the cell nucleus contains a variable number of paired, linear chromosomes, consisting of single DNA molecules packed by special histone proteins. Most eukaryote chromosome pairs consist of homologs—autosomal chromosomes containing the same genes. In addition, many species have a pair of sex chromosomes in which one of the sexes has a homologous pair (e.g. XX females in mammals) and the other sex a pair of different sex chromosomes (e.g. XY males in mammals).

Class I TE Replicative transposable element that transposes by means of being transcribed to RNA, reverse transcribed back to DNA, and then the DNA copy is reinserted at a new location in the genome. Also called a retrotransposon.

Class II TE Conservative or replicative transposable element that transposes either by excising it itself and reinserting at a different location in the genome, or by the element being replicated into a DNA copy which is then reinserted elsewhere in the genome. Also called a DNA transposon.

Cline A measurable gradient in a single character or in the frequency of an allele along the geographical range of a species or across a hybrid zone between a pair of hybridizing species.

Coalescent theory A collection of retrospective population genetic models seeking to infer how past evolutionary events have shaped the current pattern of genetic variation in natural populations.

Codon A triplet of nucleotides that code for a specific amino acid in protein synthesis. In addition, the start codon is a signal for initiation of translation and the stop codons signal for ceasing translation.

Coefficient of relatedness—*r* The probability that two individual organisms share an identical copy of an allele inherited from a recent common ancestor.

Coevolution Evolutionary change in one species triggered by evolutionary change in another species. Coevolution occurs between species that are ecologically connected, such as between parasites and hosts, predators and prey, between mutualists, and between competitors.

Command line A text-based interface for interacting with a computer. Unlike a graphical user interface (which you experience when using apps on your phone or a software package on your computer), it relies on text commands provided over separate lines.

Common ancestry The principle that all life on Earth can be traced backwards on a phylogeny to a period of common ancestry. Recently diverged species share a more recent common ancestry.

Common garden experiment A colloquial term for a transplant experiment aimed at disentangling the effect of genetic and environmental components on a phenotypic trait. In a classical common garden experiment phenotypically (and genetically) different individuals are transplanted from their native environment into a common environment.

Competitive release Niche expansion in the absence of one or more competing species.

Concerted evolution The process that causes the copies of a tandem repeated DNA segment within a species to be more closely related to each other, in terms of sequence similarity, than to members of the same repeated DNA segments in a different species. Recombination events such as gene conversion and unequal crossing over prevent the copies from diverging within a species.

Conjugation Transfer of DNA between bacteria by means of direct contact between cells.

Conservative transposition Transposition caused by the transposable element itself being excised and reinserted elsewhere in the genome (cut-and-paste transposition).

Conspecific sperm precedence The phenomenon that sperm from the same species as the female are more likely to fertilize her eggs when she engages in both con- and heterospecific mating.

Contig A contiguous DNA sequence constructed from overlapping reads, typically using genome assembly pipelines.

Convergence The evolution of similar phenotypic or genotypic states among phylogenetically distant organisms. Wings in bats, moths, and birds are an example of convergence.

Coupling of barriers to gene exchange Evolutionary processes that cause different barriers to gene exchange to coincide and cause stronger genetic isolation. Reinforcement is one such process.

Coverage The repeated sequencing of the same region of a genome multiple times in order to distinguish polymorphic sites from sequencing error. For example, if we sequence a single 1GB genome on a high-throughput sequencing machine that has an output of 10 GB, we are able to sequence the genome ten times. This is often denoted as 10X coverage.

CRISPR Stands for clustered, regularly interspaced short palindromic repeats. A form of adaptive immune system that was first recognized in bacteria as a defense mechanism against deleterious horizontal gene transfer. Now repurposed as a genome editing tool as CRISPR/Cas9.

CRISPR/Cas9 A method that repurposes CRISPR and the associated Cas9 protein as a genome editing tool. The system is cheap, easy to customize, and is more specific than other genome editing methods.

CRISPRi/CRISPRa CRISPR based interference and activation of gene expression. Used to alter gene expression in a similar way to RNAi but has the added advantage of being heritable.

Crossing over The exchange of genetic material between homologous chromosomes during meiosis, resulting in recombinant daughter chromosomes in the gametes.

Cross-population extended haplotype homozygosity A statistic for detecting selective sweeps that compares the lengths of haplotypes between two populations (see **extended haplotype homozygosity**).

Cryptic choice A form of mate choice in which a female (or the eggs) use physical or chemical mechanisms to control which sperm fertilizes the egg.

Cryptic species A pair of species that are phenotypically very similar.

Deep coalescence When two genes coalesce in the ancestral population prior to the split between species. See also: ancestral polymorphism.

De novo assembly A novel assembly of a reference genome without using any other closely related genomes to guide the process.

Derived state A character state that distinguishes an organism or group of organisms on a phylogeny from all others. For example, primates possess an opposable thumb.

Directional epistasis A non-additive interaction between alleles at a pair of loci that is biased in the same direction with respect to fitness or phenotypic value.

Directional selection Selection for a trait value that is higher or lower than the current population mean, and selection for the allele or haplotype with highest marginal fitness.

Disassortative mating A tendency for individuals with different phenotypes to interbreed more frequently than expected from random mating.

Disruptive selection Selection against intermediate trait values and selection against heterozygotes.

Distance methods A method for drawing phylogenetic trees based on pairwise genetic distance (i.e. the number of nucleotide substitutions) between taxa.

Divergent selection Natural selection in different directions in different populations.

DMI See **Bateson–Dobzhansky–Muller incompatibility (BDMI)**. Because Bateson's contribution to the theory was rediscovered as late as 1996, the shorter "Dobzhansky–Muller incompatibility (DMI)" was already well established and is still frequently used in the literature.

DNA methylation When a methyl group (CH_3) is added to a cytosine position in the genome. When this occurs in the promoter region of a gene, it can alter gene expression.

DNA repair and complementation hypothesis A hypothesis for the origin and maintenance of sexual reproduction stating that the main benefits of sex are more efficient DNA repair during recombination and masking of deleterious alleles by dominance in outcrossing populations.

DNA replication The process in which a DNA molecule copies itself. A mother cell passes on genetic information to its daughter cells during cell division. DNA replication can also be performed *in vitro* in the laboratory to amplify targeted DNA sequences (e.g. polymerase chain reaction) or as a step in DNA sequencing.

DNA sequence The order of nucleotides that carry adenine (A), thymine (T), cytosine (C), or guanine (G) bases in a DNA molecule.

Dominance variance The component of the genetic variance in a trait that is due to interactions between alternative alleles within each locus.

Dominant A gene variant (allele) is said to be dominant when its phenotypic effect overrides that of another (recessive) allele in heterozygotes. An allele A_1 is fully dominant over A_2 if the genotypes A_1A_1 and A_1A_2 yield the same phenotype, which is different from the phenotype of A_2A_2. Allelic dominance forms a continuum from fully dominant via additive inheritance, in which the phenotype of the heterozygous is intermediate between the two, to fully recessive.

Draft genome A reference genome that has met the standards required for publication and storage on publicly available databases. However, as the name implies, they are drafts that require improvement and should not be considered an exact representation of the true genome.

Effective population size (N_e) One can think of effective population size as a statistical measure of how many copies of a gene there actually are in circulation in a population (contributing to the gene pool), expressed in terms of number of genetically contributing individuals. More technically it can be defined as the number of individuals a population would have had if it had been an idealized population of size N, measured in terms of rate of genetic drift or some other population genetic quantity. The idealized population would have a constant population size, non-overlapping generations, a balanced sex ratio, each parent would have the same number of offspring, there would be no selection, and there would be random mating among the individuals. In real populations individuals contribute differently to the gene pool for a number of reasons. Accordingly, the effective population size is typically smaller than the census size.

Enhancer Regulatory sequence that binds transcription factors (proteins) that enhance gene expression.

Epigenetics The study of cellular and physiological characteristics that are inherited in cell division, but that are not caused by changes in the DNA sequence. Examples include DNA methylation, which alters how genes are expressed without altering the DNA sequence of the genes.

Epigenome The genome-wide modification of changes to gene expression—i.e. changes to expression that are not encoded in the genome itself.

Epistasis An interaction between two or more loci on the phenotype or fitness in which their joint effect differs from the sum of the loci taken separately.

Epistatic variance The component of the genetic variance in a trait that is due to interactions between alleles at different loci.

Euchromatin Less densely packed chromatin regions that can be easily accessed and transcribed by RNA polymerases.

Eusociality The highest level of social organization in animals, in which the individuals engage in cooperative care of offspring and a division of labor among reproducing and non-reproducing individuals.

Evolutionary stasis A prolonged period in evolutionary history in which a species does not experience net morphological change.

Evolutionary taxonomy The classification of organisms based on their evolutionary relationships, rather than just morphological similarity (i.e. as was practiced by Linnaeus).

Evolvability The capacity of a population to respond phenotypically to selection.

Exon Continuous stretches of protein coding sequence within a gene. The coding sequence of a eukaryote gene is typically compartmented into exons interspersed with segments of noncoding DNA (introns).

Exon shuffling Exon duplications within genes and exchange of exons between different genes.

Expressed sequence tag (EST) A method for Sanger sequencing coding DNA. Now largely replaced by direct sequencing of RNA.

Extended haplotype homozygosity A haplotype statistic for detecting selective sweeps from sequence data. The statistic starts with a focal allele—e.g. one putatively under selection—and then measures the probability that the region surrounding this SNP is identical by descent in any two randomly chosen haplotypes from a population. Identity by descent is calculated using the homozygosity of SNPs surrounding the focal allele. Therefore, EHH measures the length of the haplotype surrounding the allele under selection at the population level.

Fertility selection Selection on a trait that affects fertilization success or fecundity of a mating pair.

Fitness Expected *per capita* growth rate (reproductive success) of a biological entity. A biological entity can for instance be a group of individuals that share a genotype that affects a phenotypic trait.

Fitness component A factor that contributes to total fitness, including survival probability and fecundity of different age classes.

Fitness function The mathematical relationship between a genotype or phenotype and its fitness.

Five prime untranslated region (5′ UTR) The section of mRNA immediately upstream of the start codon of the open reading frame. It contains various regulatory sequences for the translation of the open reading frame into protein.

Founder effect Loss of genetic variation by increased rate of genetic drift in a small population that establishes itself in a previously unoccupied habitat.

Four-population test A simple test for an excess of shared sites based on a tree topology with four populations. Specifically tests for a deviation in the frequency of ABBA or BABA patterns (where A is ancestral and B derived) from an average of zero under neutrality.

Frame-shift mutation A mutation that distorts the reading frame of a gene. Insertions or deletions that are not divisible by three would distort the reading frame consisting of codons of nucleotide triplets.

Frequency-dependent selection The fitness of a phenotypic trait or a genotype depends on its frequency in the population.

Gap A space in a sequence alignment used to represent an insertion/deletion polymorphism among sequences.

Gap penalty A penalty applied to gaps when constructing a sequence alignment. Can be tuned to make the algorithm more or less permissive of the presence of gaps.

Gene In molecular genetics a gene is defined as a segment of DNA that specifies a particular protein or RNA product. In evolutionary genetics, it is sometimes loosely defined as a heritable unit that affects a phenotypic trait.

Gene conversion The process during meiosis in which one DNA sequence converts the homologous sequence to become similar to itself by means of DNA repair enzymes. Gene conversion can also occur between non-homologous (paralogous) sequences in cases of misalignment of repeated DNA segments.

Gene drive Genome editing that results in the introduction of a selfish construct that copies itself and distorts segregation ratios. Can be used in biocontrol in order to spread a deleterious gene through a population of pests.

Gene duplication The duplication of any region of DNA that contains a gene. Genes can be duplicated by means of unequal crossing over, transposition, and through whole genome duplications. Gene duplication is the main mechanism for the origin of new genes, as the resulting copies can diverge and take up progressively more different functions.

Gene expression The variation in amount and timing (during development) of gene product (mRNA) being produced by a gene. Thus, gene expression profile varies between tissues and during the life of an organism.

Gene family A set of similar genes that have similar, but not necessarily identical functions. Gene families originate through (often repeated) gene duplication events.

Gene flow The exchange of genes between populations through migration and interbreeding.

Gene knockdown Experimental manipulation or interference with a gene to prevent gene expression.

Gene knockin Experimental introduction of a gene or allele to a target region.

Gene knockout Complete disruption of gene function through the experimental introduction of insertions or deletions to shift the open reading frame.

Gene pool The set of alleles at all loci that are segregating across generations in a population.

Genetic accommodation A conceptual model where epigenetic change and the phenotypic plasticity it causes allows a population to persist in challenging environments until a mutation in the genome can occur and drive adaptation to these conditions.

Genetic architecture The characteristics of genetic variation responsible for heritable variation in a phenotypic trait.

Genetic cluster species concept A definition of a species as groups of individuals that form a genetic (and/or morphologic) cluster with few or no intermediates to other such clusters.

Genetic constraint Adaptive evolution hindered or slowed down because of the way the trait is inherited.

Genetic drift Random changes in allele frequencies within a population.

Genetic distance The number of nucleotide substitutions between two taxa.

Genetic linkage Genes or other loci that do not segregate independently of each other during meiosis because they are located close together on the same chromosome.

Genetic load A reduction in mean population fitness caused by the presence of genotypes with low fitness.

Genetic map The order of a set of genetic markers and the recombinational distance between them on a chromosome estimated from the recombination rates between the loci in a crossing experiment or from a pedigree. Recombinational distance is typically measured in units of centimorgans. One centimorgan equals an expected number of 0.01 intervening crossing over events between a pair of loci per generation.

Genetic marker Short, polymorphic DNA sequences in which the allelic variants can be distinguished by a genotyping technique.

Genetic revolution A hypothesis suggesting that adaptive evolution mainly occurs in small founder populations because gene interaction networks would break down by genetic drift in small populations.

Gene tree A phylogenetic relationship among genes and alleles, well described by the coalescent.

Gene tree/Species tree discordance When the topologies of gene trees and species trees conflict; caused by a number of factors including gene flow and incomplete lineage sorting.

Genome assembly A reconstruction of the true genome sequence of an organism, put together from sequence reads. Genome assemblies are best thought of as models of the true genome as they are not without error and may also be incomplete.

Genome annotation A method to identify the position, distribution, and structure of genes in the genome. Achieved by first aligning evidence of genes to the genome (i.e. expressed sequences, transcripts, proteins) and then creating gene models with intron and exon boundaries.

Genome editing Experimental genomic techniques that make it possible to delete or insert gene regions.

Genome graphs A way of representing the reference genome and the variation that occurs at the population level in the form of SNPs and structural variants. For example, these can be shown as "bubbles" in the assembly, whereas invariant regions can be adequately represented by a linear sequence.

Genome scan A colloquial term for genetic research methods in which the entire genome (or a significant proportion of it) of an organism is searched systematically using statistics and then interpreted. Applications of genome scans include identifying alleles that associate with a phenotypic trait (such as genome wide association study, GWAS), identifying alleles or genomic regions that have been subjected to recent selection, and identifying signatures of demographic changes such as changes in population size and admixture between populations.

Genotype The set of alleles an individual carries at a specific gene or locus.

Germline mutation A mutation that occurs during gamete production and that therefore can be passed on to future generations.

Glacial refugia Geographic regions that allowed temperate species to persist during glacial periods during the Pleistocene and recolonize previously inaccessible or uninhabitable regions during interglacials.

Group selection The controversial hypothesis that differential survival of groups of individuals can lead to adaptations at the group level.

Haldane's rule The observed pattern that when there are differences in hybrid fitness between the sexes it is usually hybrids of the heterogametic sex (XY males and ZW females) that have the lowest fitness.

Haplotype The allele combination at two or more loci an individual has inherited from one of its parents, or in other words, the allele combination at two or more loci present in one of the gametes upon fertilization.

Haplotype network A method for visualizing intraspecific variation in haplotypes (mitochondrial and nuclear). Node diameter is relative to the frequency of a haplotype, links in the network represent the number of nucleotide substitutions between them.

Haplotype statistics A set of statistical methods that take advantage of the fact that positive selection at a focal allele results in a selective sweep due to strong linkage disequilibrium between the focal allele and alleles in its physical vicinity. A sweep will therefore produce a region around the selected allele exhibiting increased homozygosity (see **extended haplotype homozygosity**).

Hardy–Weinberg expectation (HWE) The genotype frequencies expected from the Hardy–Weinberg null model given the observed allele frequencies.

Hemizygous Having only a single copy of a gene instead of the customary two copies. In humans and other mammals, males are hemizygous for genes on their single X-chromosome because these genes have no homolog on the reduced Y chromosome.

Heritability The proportion of the phenotypic variance in a trait that is due to additive variance (heritability in the narrow sense).

Heterochromatin Tightly packed chromatin regions which cannot be easily accessed by RNA polymerase and therefore are typically not expressed.

Heterozygous A genotype consisting of different alleles (e.g. A_1A_2).

High-density posterior region The 95 percent interval of a posterior probability distribution that is expected to contain the true value of a parameter of interest. Analogous to a confidence interval for Bayesian estimation.

Hill–Robertson effect A reduction in the efficiency of natural selection due to random linkage disequilibrium between loci.

Histones Structural proteins that DNA wraps itself around, much like a spool for a piece of string.

HKA test (Hudson–Kreitman–Aguadé test) A neutrality test using the prediction from neutral theory that genetic variation within populations should be positively correlated with divergence between populations across loci that differ in (neutral) mutation rate.

Homologous A trait or state—i.e. phenotypic or genetic—which is shared among organisms and derived from the same origin through inheritance. The wings of all birds are homologous and this is an example of homology.

Homology directed repair A form of DNA repair mechanism that is able to incorporate a donor sequence into a genome. Less common and more difficult to achieve than nonhomologous end joining.

Homoplasy The alternative to homology; similar traits, structures, phenotypes, or nucleotide substitutions but derived independently. For example, the wings of birds and bats are an example of homoplasy. These traits can also be described as **homoplasmic**.

Homoploid hybrid speciation (HHS) Speciation resulting from hybridization between two divergent species without change in chromosome number.

Homozygous A genotype consisting of similar alleles (e.g. A_1A_1).

Horizontal gene transfer The movement of DNA from one organism to another by other means than by the vertical transmission from parents to offspring.

Hybrid breakdown Developmental incompatibility between a pair of species that results in reduced fertility and/or viability beyond the F1 hybrid generations (e.g. in backcrossed individuals and F2 hybrids).

Hybrid zone A geographical range where the distribution of two divergent populations or species overlap and where hybridization takes place.

Identical by descent Two individuals share an allele that is identical by descent if they have inherited the same copy from a common relative.

Inbreeding A tendency for genetic relatives to interbreed more often than expected from random mating.

Inbreeding depression A reduction in mean population fitness as a consequence of exposure of rare recessive deleterious alleles among individuals in an inbred population.

Inclusive fitness The number of offspring equivalents an individual is expected to rear through its behavior towards own offspring and those of genetic kin.

Incomplete lineage sorting An issue where ancestral polymorphism can lead to gene tree/species tree discordance. Best illustrated by a gene tree for at least three taxa that coalesces in the ancestral population. See also **ancestral polymorphism** and **deep coalescence**.

Infinite site model A statistical null model in evolutionary genetics applicable to inferring demographic and evolutionary processes that may have affected genetic variation at the level of DNA sequences.

Inheritance of acquired characteristics A hypothesis suggesting that physiological changes acquired during an individual's lifetime can be transmitted to its offspring. The hypothesis has been rejected.

Integrated haplotype score (iHS) A haplotype statistic measuring the area under the curve defined by the extended haplotype homozygosity statistic (EHH), that is, from 1 at the focal allele until the point it decays to zero. Application of the iHS normally uses an outgroup species to identify the ancestral and derived allele. The iHS is then the log of the division of these two areas. A negative iHS suggests selection has favored the derived allele, while a positive iHS points to the ancestral allele (see **extended haplotype homozygosity**).

Interspersed repeats Similar DNA sequences found at different, apparently random, locations throughout the genome. Replicative transposition is the main mechanism for their occurrence.

Introgression Incorporation of DNA from one species into that of a different species by means of hybridization and backcrossing.

Intron Stretches of noncoding DNA within a gene that intersperse coding sequences (exons).

Introns-early hypothesis A hypothesis for the origin of introns that suggests that introns evolved early in the history of life to facilitate recombination between functional domains of proteins and that introns have been lost secondarily in many organisms because of selection for rapid replication.

Introns-late hypothesis A hypothesis for the origin of introns suggesting that they first evolved in early eukaryote genes.

Inversion A 180° flip of a DNA segment on a chromosome.

Isolation by distance The phenomenon that individuals of a species that are in close geographic proximity tend to be genetically more similar to each other than individuals that live far apart as a consequence of limited dispersal. Individuals tend to reproduce with individuals in their vicinity and will therefore diverge genetically from individuals that live further away, even in the absence of geographic barriers or selection, simply by genetic drift.

Kinetochore A protein complex attached to the centromere that controls the segregation of sister chromatids during cell division. Spindle fibers attach to the kinetochore of each sister chromatid and drags them to each side of the cell before cell division is completed.

Kin selection Selection that favors the productivity of family groups even at the expense of individual survival or reproductive success.

Kmer Small sequences of k base pairs, where k is a number shorter than the length of the read. Used to match reads together into a contig, although also often used in alignment to a reference genome.

Large X-effect The phenomenon that loci that cause low fertility in hybrids are disproportionally abundant on the X-chromosome.

Last Glacial Maximum The maximum extent of the continental ice sheets during the last glacial period.

Likelihood A statistical framework for estimating how well a model we have constructed fits the data we observe. Often used for parameter estimation from models.

Lineage A grouping or subset of a phylogenetic tree that all descend from a common ancestor.

Linkage disequilibrium (LD) A non-random association between alleles at different loci in a population in which certain allele combinations (haplotypes) are statistically overrepresented relative to others.

Linkage equilibrium (LE) Statistical independence between alleles segregating at different loci. Linkage equilibrium can also be seen as the multilocus equivalent of the Hardy–Weinberg model describing the expected haplotype and genotype frequencies in a population for two or more loci under a set of ideal assumptions.

Linkage map A genetic map of the genome that orientates markers based on their pairwise recombination rate. Can be used as a means of improving a genome assembly and orientating scaffolds.

Locus A genomic location of interest, such as a gene or a genetic marker.

Log likelihood The log transformed likelihood. Often calculated because likelihood estimates are very small.

Lottery hypothesis A hypothesis for the maintenance of sexual reproduction suggesting that sexually reproducing females would produce more surviving offspring than asexual ones, due to the difference in genetic diversity of the offspring.

Mapping to a reference Alignment of resequenced reads to the reference genome. Typically used to identify population level polymorphism.

Marginal fitness The average fitness of an allele in a population taking into account the fitnesses of the genotypes the allele is confined to and the frequencies of those genotypes.

Markov chain Monte Carlo sampling (MCMC) A method that allows us to explore probability space by proposing changes to parameters in a model and then accepting based on whether they increase or decrease the likelihood of the model.

Mass spectrometry A chemical analysis method that sorts ions based on their mass-to-charge ratio. It has a wide array of uses but in proteomics it is used to identify proteins.

Mate-pair libraries A method for sequencing either end of very large insert sequences—i.e. >10 Kb. These are then able to span large physical distances in the genome, especially those that contain repetitive sequences. They are also used to join contigs together on a scaffold.

Maximum likelihood The model or model parameters that maximizes the probability of our data given the model. We use a maximum likelihood framework to return the **maximum likelihood estimate** of parameters such as branch lengths and topology in the case of phylogenetics.

Maximum parsimony (MP) A phylogenetic tree construction method that uses parsimony. It attempts to build a tree that requires the smallest amount of evolutionary changes in order to explain the observed data.

McDonald–Kreitman (MK) test A neutrality test using the prediction from neutral theory that the ratio of replacement to silent polymorphisms within populations or species should equal the ratio of replacement to silent fixations between the populations or species. Several extensions of the MK test have been proposed and are in use, including the neutrality index N_I and the direction of selection DoS test.

Mean population fitness The average fitness of the individuals in a population.

Mendel's first law The observation that the two alleles at a locus in an individual (usually) segregate independently during meiosis so that each haploid gamete is equally likely to carry either allele.

Mendel's second law The observation that alleles at different loci (often) segregate independently during the formation of gametes. That is, multilocus alleles segregate independently when they are in **linkage equilibrium**.

Metagenomics The study of genetic material derived directly from environmental samples, such as water and soil.

Metapopulation A group of spatially separated populations of the same species; a population of populations. The concept of metapopulation is useful for studying the effects of habitat fragmentation and the resulting structuring of populations.

Minority cytotype exclusion principle The constraint on establishing a viable population of a polyploid species because a new polyploid will be rare compared to its diploid parent species and will therefore hybridize with the parent species rather than mate with other polyploids.

Missense mutation A point mutation that causes a codon to specify a different amino acid.

Missing heritability problem A discrepancy between the estimated heritability of many phenotypic traits, such as heritable human diseases, and the proportion of that heritability that can be accounted for by loci that associate with the trait according to GWAS analysis.

Model-based clustering methods A class of methods used to detect population structure based on a simple model of K clusters, each meeting the Hardy-Weinberg and linkage equilibrium expectation.

Model selection A method for distinguishing between alternative models. In the context of phylogeography and phylogenetics, this is typically done based on the posterior probability or likelihood of the models.

Modern synthesis of evolution The general consensus that Darwinian evolution is consistent with modern genetic principles that developed in 1936–1950. The synthesis of different branches of biology resulted in a refined evolutionary theory that emphasized the role of mutation in generating genetic variation within populations, and natural selection and genetic drift causing evolutionary change by altering allele frequencies.

Molecular clock A hypothesis derived from neutral theory stating that genetic divergence can be used to deduce the time in the evolutionary past when two or more species diverged from one another. The hypothesis is based on the result from neutral theory that the substitution rate would equal the neutral mutation rate and be independent of population size. Populations and species would therefore diverge genetically at a constant rate according to theory.

Monophyletic group A grouping on a phylogenetic tree where all taxa are descendants from a common ancestor. For example, all Hominidae constitute a monophyletic group.

Mother's curse A mutation in maternally inherited mitochondrial DNA that increases in frequency in a population because of its detrimental effects on male fitness.

Muller's ratchet Irreversible accumulation of deleterious mutations in asexual organisms.

Multiple alignment The alignment of more than two sequences of DNA, RNA, or amino acids.

Multi-regional model A model proposed to explain the evolution of anatomically modern humans. It states that all modern human populations are descended from geographically isolated populations of *Homo erectus*, beginning about 2 million years ago. Since DNA evidence has provided support for an out-of-Africa model, it is no longer accepted.

Mutation A permanent change in the nucleotide sequence of a genome. Mutations are caused by errors during DNA replication, damage to DNA that fails to be repaired, and insertion or deletion of DNA segments through the action of transposable elements, replication slippage, and unequal crossing over.

Mutation theory of evolution A hypothesis once presented as an alternative to Darwinian selection, suggesting that evolution proceeds through discrete jumps caused by mutations with large phenotypic effects. The hypothesis was effectively rejected through the work of early population geneticists and the authors of the modern synthesis of evolution.

N50 statistic A statistic for assessing the condition of an assembly. The N50 can be used for contigs or scaffolds and is the smallest size of these structures on which at least 50 percent of all assembled sequences occurs.

Natural selection A deterministic difference in survival and/or reproductive output of biological entities that differ in one or more characteristics. Biological entities can, for instance, be individuals that differ in a genetically determined phenotypic trait.

Nearly neutral theory A modified version of neutral theory which considers that slightly deleterious mutations would predominantly evolve by genetic drift in small populations but would be removed by selection in large populations.

Negative epistasis An interaction between alleles at two or more loci in which the combined effect of the loci on phenotypic value or fitness is less than the sum of the loci taken separately.

Neighbor Joining A method of constructing a phylogenetic tree using a clustering approach on the genetic distances among taxa. Produces an unrooted, additive tree.

Nested models When two or more models are a special case of one another. For example, by altering a parameter in the more complicated model, it becomes equivalent to the simpler one.

Neutral theory of molecular evolution A theory that holds that evolution at the molecular level is largely driven by random genetic drift acting on neutral muta-tions (alleles with no bearing on fitness).

Newick format A way of using text to represent evolu-tionary relationships and phylogenetic trees. Commonly used in programming and phylogenetic packages.

Niche construction An organism's physical alteration of its environment.

Node The point on a tree where two branches meet, repre-senting a point in time where the two taxa, sequences, or tips shared a common ancestor.

Nonhomologous end joining A small insertion or deletion introduced as a result of DNA repair following a double strand break initiated by CRISPR/Cas9.

Nonsense mutation A point mutation that changes an amino acid specifying codon into a stop codon.

Non-synonymous mutation A mutation in a protein cod-ing gene that causes a change in the amino acid sequence of the resulting protein. They are also called replacement mutations, and contrast with synonymous or silent mutations that do not affect the amino acid sequence.

Nucleosome The structure formed when DNA wraps around a set of histone proteins.

Nucleotide The subunits that DNA and RNA are made of. In DNA one nucleotide consists of a deoxyribose sugar bound to a phosphate group and one of the four nitrogen-containing bases adenine, guanine, cytosine, and thy-mine. An RNA nucleotide is similar, except that the sugar is ribose and the base thymine is replaced by uracil.

Nucleotide diversity The average number of nucleotide differences per site between pairs of DNA sequences drawn randomly from a population (π).

Null hypothesis In statistics a null hypothesis is a general statement or a default position concerning the relation-ship between two phenomena. A typical null hypothesis would state that there is no relationship between phenomenon A and B, whereas an alternative hypothesis would state a causal relationship between the two.

Off-target effects Unintended effects of genome editing that extend beyond the target region. For example, CRISPR/Cas9 genome edits which affect genes that are not specifically targeted.

Oligonucleotides Short DNA sequences synthetically made in the laboratory. Oligonucleotides include PCR and sequencing primers that are constructed to have the complementary nucleotide sequence to the template DNA we wish to target and which will therefore bind to that template. Other examples include adapters, which are oligonucleotides that can be ligated to template DNA fragments, and anchors ligated to a glass surface, which can bind adapters through sequence complementarity.

Ontogenetic constraint Adaptive evolution hindered or slowed down as a consequence of how the organism develops during ontogenesis.

Open reading frame A continuous stretch of codons that does not contain a stop codon and thus a DNA sequence that can be translated into a protein. A mature mRNA contains an open reading frame between the start and stop codons.

Operational taxonomic units A means of clustering and grouping individuals based on their shared genetic variation.

Origin of replication A sequence motif that functions as an attachment site for a protein complex that unwinds double-stranded DNA and initiates DNA replication.

Orthologous Two genes that occur in two species and share a common ancestor.

Outbreeding A tendency for genetic relatives to inter-breed less often than expected from random mating.

Outgroup A more distantly related species than the focal ingroup that serves as a reference when determining the evolutionary relationships of the ingroup. Often used to root a tree.

Out-of-Africa model A model of the evolution of modern humans where humans have a recent origin in Africa and subsequently spread from that continent to popu-late much of the rest of the world.

Overdominance A form of dominance in which the phenotypic trait value and/or fitness of the heterozygote exceeds either homozygote.

Paired-end A sequencing mode common on high-throughput sequencing machines that allows the sequencing of both ends of short DNA fragments, typically < 1Kb.

PAM Protospacer adjacent motif, a short DNA sequence that must be present in the genome to allow the CRISPR/Cas9 complex to recognize the target region. Usually a simple GC or NGG nucleotide motif.

Pangenome A more recent approach to genome assembly that attempts to incorporate the variation seen in a spe-cies, typically by assembling a genome from multiple individuals.

Parallel evolution The simultaneous evolution of a genetic or phenotypic character state in two closely related lineages.

Paralogous When gene duplication causes two (or more) copies of a gene to segregate in a species, the duplicate is no longer orthologous but is now paralogous.

Parapatric speciation Speciation taking place between geographically structured populations that continue to exchange genes while diverging.

Parsimony The principle of Occam's razor such that when choosing between alternative hypotheses, we should choose the one that requires the fewest assumptions.

Parsimony score A count of the number of substitutions on a tree required for the sequence alignment to fit the proposed tree model. Also known as the tree length.

Pedigree A description or a chart of the family relationship between a group of individuals.

Perfect linkage disequilibrium Two loci are in perfect linkage disequilibrium when an allele at one of the loci is perfectly associated with an allele at the other locus, so that $D' = 1$.

Peripatric speciation The hypothesis that founder events are particularly important in speciation. The argument is that increased rate of genetic drift in small, isolated founder populations would tend to break down epistatic networks of co-adapted genes causing rapid evolution away from the larger parental population. Peripatric speciation is a special case of allopatric speciation.

Phenotypic plasticity The ability of one genotype to produce different phenotypes depending on environmental conditions.

Phylogenetics The study of evolutionary relationships among taxa and species.

Phylogenetic species concept A definition of a species as monophyletic groups of organisms of common ancestry distinguishable from other such groups of organisms.

Phylogenetic tree A visualization of the evolutionary relationships among a group of taxa.

Phylogenomics Identical to phylogenetics but exclusively using genomic data to reconstruct evolutionary relationships.

Phylogeny See phylogenetic tree.

Phylogeography The study of the spatial and temporal arrangement of genes at the intraspecific and interspecific level.

Plasmid A small circular DNA molecule in many bacteria and archaea that can replicate independently.

Pleiotropy One gene influences two or more different phenotypic traits.

Point mutation The substitution of one nucleotide with a different nucleotide.

Poly-A tail A long chain of adenine nucleotides that are attached to the 3′ end of a messenger RNA molecule (mRNA). The poly-A tail makes the molecule more stable and prevents degradation and aids in the transportation of the mRNA from the nucleus to a ribosome in the cytoplasm.

Polyploid hybrid speciation (PHS) Speciation resulting from hybridization between two divergent species facilitated by genome duplication in the hybrid.

Polyploidy The process by which cells or organisms obtain more than two chromosome sets. For instance, fertilization of non-reduced (diploid) gametes will result in four sets of each chromosome (tetraploid).

Population expansion A sharp increase in population size

Population mutation rate parameter θ (theta) A theoretical compound parameter in evolutionary genetics that incorporates the neutral mutation rate μ and effective population size N_e. For a diploid locus $\theta = 4\,N_e\,\mu$.

Positive epistasis An interaction between alleles at two or more loci in which the combined effect of the loci on phenotypic value or fitness is larger than the sum of the loci taken separately.

Posterior probability A key component of Bayesian probability, the posterior is the probability of a parameter estimate given our data. An alternative way to think of it is as our estimate of the probability of an event or parameter after we have considered the available evidence.

Postzygotic barrier A phenotypic trait that causes low fitness in hybrids between divergent populations or species.

Pre-mRNA An immature mRNA molecule that contains transcribed elements that are later excised (introns).

Prezygotic barrier A phenotypic trait that reduces the likelihood that mating or fertilization will take place between members of divergent populations or species.

Primary hybrid zone A hybrid zone that originated through divergent and/or disruptive selection between parapatric populations that have continued to interbreed. This is in contrast with secondary hybrid zones that originate when geographically isolated and divergent populations come into secondary contact.

Principal Components Analysis (PCA) A method for summarizing and visualizing the covariance among many measurements. In genetics, PCA is typically used to determine population structure.

Prior A probability distribution that specifies our prior knowledge of an event or parameter. Used in Bayesian probability.

Prisoner's dilemma A standard game in game theory that explores the limits of cooperative behavior.

Programming language A formal set of instructions for computers set out in a language format with grammatical form (syntax) and meaning (semantics). Programming languages are the basis of all computer programs and applications. In bioinformatics, a programming language can be used to create custom scripts and analysis pipelines.

Promoter A DNA sequence motif that functions as an attachment site for RNA polymerase, the protein that transcribes DNA to RNA.

Protein moonlighting The same protein performing different functions depending on the tissue it is produced in.

Proteomics The study of the proteome—i.e. the entire set of proteins encoded and transcribed from the genome.

Pseudogene A DNA segment that is related to a gene but is non-functional.

Punctuated equilibria An evolutionary hypothesis suggesting that evolutionary stasis characterizes most of the evolutionary history of organisms and that adaptive evolution is restricted to sudden bursts in connection with speciation events.

Quantitative genetics This is the branch of evolutionary genetics that deals with continuously varying phenotypic traits. A general assumption is that such phenotypic variation is affected by a large number of Mendelian genes, each with a small effect.

Recessive See dominant.

Reciprocal altruism A trait or behavior that increases fitness of a recipient at a temporal cost to the actor, but which is later repaid when the roles are reversed.

Reciprocal sign epistasis An interaction between alleles at two or more loci in which the combined effect of the loci on phenotypic value or fitness is opposite from the sum of the loci taken separately.

Recombination The collection of processes occurring during meiosis, which results in the production of offspring with a different combination of traits than those found in either parent. Recombination events such as unequal crossing over and gene conversion can result in mutations.

Red Queen hypothesis An evolutionary hypothesis suggesting that organisms must constantly adapt to keep track of coevolving natural enemies. A specific version of the hypothesis suggests that sexual reproduction is maintained to escape coevolving parasites.

Reference genome A representative example of a genome of an organism—i.e. its DNA sequence and genes. Often created from just a single individual, but some hybrid assemblies of multiple individuals exist and new methods incorporating population level variation are also emerging.

Regulatory sequence DNA sequence motifs that function as attachment sites for proteins (transcription factors) that regulate gene expression.

Reinforcement The hypothesis that natural selection against the production of unfit hybrids might favor traits that increase (reinforce) prezygotic barriers to gene exchange.

Relative fitness The fitness of a genotype relative to the genotype with the highest fitness in the population.

Replacement mutation A mutation that alters the amino acid sequence of the gene it specifies. The same as non-synonymous mutation.

Replication slippage An error during replication of short tandem repeats that results in the insertion or deletion of repeat units.

Replicative transposition Transposition by means of copying of the transposable element followed by insertion of the resulting copies elsewhere in the genome.

Reporter construct A construct that includes a gene which is easily visualized with fluorescence, such as the green-fluorescent protein derived from jellyfish. This is then inserted in the genome next to the target gene such that when the target is transcribed, both the target and the reporter construct are expressed. Since the reporter protein is fluorescent, it can be seen easily and the spatial distribution of gene expression quantified.

Reproductive success The *per capita* growth rate of a population or a biological entity, such as individuals with a specific genotype. This would normally correspond to the expected number of fertile, surviving offspring of that entity.

Resequencing Whole genome sequencing of individuals other than the reference. Used to quantify the variation at the population or species level.

RNA editing The process in which certain cells enzymatically alter the sequence of mRNAs to yield a different protein.

RNA gene A DNA sequence that specifies a noncoding RNA molecule.

RNAi RNA interference—a biological process in which RNA molecules such as microRNA and short interfering RNA inhibit gene expression or translation. Can be used as a research tool by introducing synthetic RNA to selectively suppress genes of interest.

RNA-seq Direct sequencing of RNA transcripts using high-throughput sequencing. Provides evidence of genes and also makes it possible to quantify gene expression.

Root The node on a tree from which all other taxa descend. Trees where the root is defined are described as **rooted**.

Scaffold A long stretch of sequence assembled from smaller contigs, typically with the help of mate-pair libraries.

Segregating sites Polymorphic sites in the genome.

Segregation distorter A gene complex that "cheats" during gametogenesis and becomes overrepresented among the mature gametes.

Selective sweep The reduction or elimination of genetic variation near a positively selected allele caused by linkage disequilibrium between the selected allele and polymorphisms in close physical proximity. Variants linked to the selected allele will therefore "hitchhike" to high frequency in the population as the selected allele becomes fixed by selection. Due to recombination, selective sweeps only affect a local region around the positively selected allele.

Sexual conflict Antagonistic selection on one or more traits arising from different fitness optima in the two sexes.

Sexual selection Differential mating success between members of the same sex in a population. Sexual selection affects traits that enhance mating success.

Shotgun sequencing The random shearing and sequencing of a genome with short read sequencing technology. Analogous to the wide, random firing of a shotgun. Genome assembly is then necessary to produce a reference genome.

Silencer Regulatory sequence that binds transcription factors that repress gene expression.

Silent mutation A mutation that does not alter the amino acid sequence of the gene it specifies. The same as synonymous mutation.

Sister species Two taxa on an evolutionary tree that share their most recent common ancestor with one another.

Sliding window The procedure in genome scan analysis in which various statistics (such as F_{ST} and Tajima's D) are calculated sequentially for windows of equal length along aligned chromosomes. Typically, the windows are a small fraction of the total chromosome length. Sliding windows can be overlapping or non-overlapping. When put together a genome scan results with information on how the statistic(s) varies along the genomes of the populations studied.

Speciation reversal Extinction through loss of phenotypic and genetic distinctiveness of two or more species, caused by fusion of their divergent genomes into one through extensive hybridization and backcrossing.

Species selection Different rate of extinction/speciation in evolving taxa.

Species tree A tree showing the evolutionary relationship among species; contrast with gene tree.

Sperm competition Competition between the sperm cells of two or more different males to fertilize the same egg.

Spliceosome An RNA-protein complex that catalyzes the excision of introns from pre-mRNA and ligation of exons.

Stabilizing selection A form of selection in which intermediate trait values are favored by selection.

Stem cell An undifferentiated cell that can transform into any kind of cell by epigenetic modification.

Stochastic error Random error. In phylogenetics, when an inadequate data source is used to estimate a phylogeny, there is a danger we will accept the wrong tree model. An example of this is using only a single gene tree that is discordant from the species tree. Can only be fixed by including more data.

Substitution The replacement of one nucleotide with a different one through mutation. Confusingly, it is not only used as a synonym for a point mutation (molecular genetics), but also to describe the process of replacement of an ancestral nucleotide through fixation of a derived mutation in a population (neutral theory/population genetics), and finally to describe a nucleotide polymorphism in an alignment of sequences (phylogenetics).

Substitution rate The rate at which new mutations become fixed in a population (neutral theory/population genetics).

Supergene A group of co-adapted neighboring genes on a chromosome which are inherited together and not broken up by recombination. Supergenes may arise in connection with chromosomal inversions that prevent the genes from recombining.

Supermatrix A phylogenomics approach where all genes and sequences are concatenated into one large sequence and treated as a single entity.

Supertree A phylogenomic method where a consensus tree is constructed from all the individual gene trees. Generally, the most common topology is assumed to be the species tree.

Sympatric speciation The process through which new species evolve from a single ancestral species within the same geographical region and in the face of gene flow.

Sympatry The literal meaning in Greek is "same fatherland." Two species live in sympatry where their breeding ranges overlap.

Synonymous mutation A mutation that does not alter the amino acid sequence of the gene it specifies. The same as silent mutation.

Systematic error Alternative to stochastic error. Systematic errors are consistent, repeatable errors that can be caused by flawed experimental design, wrongly calibrated instruments, and more. In phylogenetics, systematic error occurs when the model of evolution we have specified is wrong, causing us to estimate the wrong phylogeny. An issue irrespective of the amount of data.

Tajima's D A population genetic test statistic based on the allele frequency spectrum that can be used to infer deviations from neutral expectation in terms of selection or changes in population size.

Tandem repeats Segments of similar DNA sequences which are repeated in such a way that the repetitions are directly adjacent to each other.

Taxonomy The pursuit of a classification of living things on Earth.

Telomere Protecting caps at the ends of chromosomes consisting of repetitive sequence.

Three prime untranslated region (3′UTR) The section of messenger RNA (mRNA) that follows immediately downstream of the stop codon of the open reading frame. It contains various regulatory sequences for the translation of the open reading frame and for regulation of gene expression, and ends in the poly-A tail which is important for the stability of the mRNA and for its transport from the nucleus to the ribosome.

Time-lag Delayed response to a novel selective regime, for instance because of a lack of genetic variation selection could act upon.

Time to most recent common ancestor (TMRCA) The amount of time that has passed since two species shared a common ancestor.

Tit-for-tat A cooperative strategy in the repeated prisoner's dilemma game that won a famous computer tournament.

Trade-off Investment in one trait occurs at the expense of a different trait.

Transcription The process in which a particular DNA segment is copied into RNA by the enzyme RNA polymerase.

Transcription factor Proteins that regulate gene expression by binding to regulatory sequences. Different transcription factors can enhance or repress transcription and thus regulate when and how much gene product a cell produces.

Transcriptomics The high-throughput sequencing of the expressed proportion of the genome—i.e. gene transcript products. The transcriptome can therefore be thought of as the expressed genome.

Transduction The process in which foreign DNA is introduced to a bacterial cell by means of a virus (bacteriophage).

Transformation Uptake and incorporation of genetic material from the surroundings resulting in genetic change. Certain species of bacteria have this ability.

Transgenic organism An organism that has had its genome altered artificially using genetic engineering techniques. This includes anything from transplantation of genes from one organism into the genome of a targeted organism to the simple insertion of a reporter construct to measure gene expression. A transgenic organism is the same as a genetically modified organism (GMO).

Transgressive segregation The formation of extreme phenotypes among hybrids and backcrosses—i.e. trait values outside the range of variation found in either of the parent species.

Transition Substitution of one purine with another purine (A <–> G) or of one pyrimidine with another pyrimidine (C <–> T).

Translation The process in which the genetic code of DNA is translated into a protein via a messenger RNA intermediate.

Translocation Rearrangements of chromosomal parts between nonhomologous chromosomes.

Transposable element A genetic element with the ability to change genomic location (transpose) either by a cut-and-paste process (conservative transposition) or by replication and insertion of the resulting copies (replicative transposition).

Transposition The movement of a transposable element to a different genomic location.

Transversion Substitution of one purine with a pyrimidine or of one pyrimidine with a purine.

Tree topology The branching pattern and node order of a tree.

Truncation selection A standard method in selective breeding. Animals or plants are ranked according to their phenotypic value on a trait of interest, such as milk yield or fruit size, and then the top percentage in the favored direction of change is chosen for breeding.

Type specimen An individual of a species used to establish its formal binomial classification and recognize its phenotype.

Ultrametric An additive phylogenetic tree where all tips are of an equal distance from the root; typically used with the assumption of a molecular clock model—i.e. branch length represents the time since divergence from a common ancestor.

Underdominance A form of dominance in which the phenotypic trait value and/or fitness of the heterozygote is lower than that of either homozygote.

Unequal crossing over Crossing over at nonhomologous sites between chromosomes resulting in an insertion and a deletion in the resulting gametes.

Uninformative prior A prior distribution with no information on our expectation on an event or parameter. See also **prior**.

UPGMA A tree building method that uses genetic distance from a distance matrix and assumes a constant molecular clock. UPGMA produces ultrametric trees where all tips are equidistant from the root.

Viability The capability of living and surviving.

Virus Small infective agents that replicate inside cells of living organisms.

References

Abbot, R., Albach, D., Ansell, S., et al. (2013). Hybridization and speciation. *Journal of Evolutionary Biology*, **26**, 229–246.

Agrawal, A.F. (2001). Sexual selection and the maintenance of sexual reproduction. *Nature*, **411**, 692–695.

Ahrens, C.W., Rymer, P.D., Stow, A., et al. (2018). The search for loci under selection: trends, biases and progress. *Molecular Ecology*, **27**, 1342–1356.

Alatalo, R.V., Eriksson, D., Gustafsson, L., and Lundberg, A. (1990). Hybridization between pied and collared fly-catchers—sexual selection and speciation theory. *Journal of Evolutionary Biology*, **3**, 375–389.

Alatalo, R.V. and Gustafsson, L. (1988). Genetic component of morphological differentiation in coal tits under competitive release. *Evolution*, **42**, 200–203.

Albertin, W. and Marullo, P. (2012). Polyploidy in fungi: evolution after whole-genome duplication. *Proceedings of the Royal Society B*, **279**, 2497–2509.

Alcaide, M., Scordato, E.S.C., Price, T.D., and Irwin, D. (2014). Genomic divergence in a ring species complex. *Nature*, **511**, 83–85.

Alerstam, T., Nilsson, S.G., and Ulfstrand, S. (1974). Niche organization during winter in woodland birds in Southern Sweden and the island of Gotland. *Oikos*, **25**, 321–330.

Alexander, D.H., Novembre, J., and Lange, K. (2009). Fast model-based estimation of ancestry in unrelated individuals. *Genome Research*, **19**, 1655–1664.

Amambua-Ngwa, A., Tetteh, K.K.A., Manske, M., et al. (2012). Population genomic scan for candidate signatures of balancing selection to guide antigen characterization in malaria parasites. *PLoS Genetics*, **8**, e1002992-14.

Amoasii, L., Hildyard, J., Li, H., and Sanchez-Ortiz, E. (2018). Gene editing restores dystrophin expression in a canine model of Duchenne muscular dystrophy. *Science*, **362**, 86–91.

Amoasii, L., Long, C., Li, H., et al. (2017). Single-cut genome editing restores dystrophin expression in a new mouse model of muscular dystrophy. *Science Translational Medicine*, **9**, eaan8081.

Andersson, M. (1994). *Sexual Selection*. Princeton University Press, Princeton, NJ, USA.

Andersson, M.B. (1982). Female choice selects for extreme tail length in a widowbird. *Nature*, **299**, 818–820.

Andres, J.A., Maroja, L.S., and Harrison, R.G. (2008). Searching for candidate speciation genes using a proteomic approach: seminal proteins in field crickets. *Proceedings of the Royal Society B*, **275**, 1975–1983.

Andrews, K.R., Good, J.M., Miller, M.R., Luikart, G., and Hohenlohe, P.A. (2016). Harnessing the power of RADseq for ecological and evolutionary genomics. *Nature Reviews Genetics*, **17**, 81–92.

Arnegard, M.E., McGee, M.D., and Matthews, B., et al. (2014). Genetics of ecological divergence during speciation. *Nature*, **511**, 307–311.

Arnqvist, G. (2014). Cryptic female choice. In: D.M. Shuker and L.W. Simmons (eds), *The Evolution of Insect Mating Systems*, pp. 204–220. Oxford University Press, Oxford, UK.

Arnqvist, G. and Rowe, L. (2013). *Sexual Conflict*. Princeton University Press, Princeton, NJ, USA.

Avent, N.D. and Reid, M.E. (2000). The Rh blood group system: a review. *Blood*, **95**, 375–387.

Avery, O.T., MacLeod, C.M., and McCarty, M. (1944). Studies on the chemical nature of the substance inducing transformation of pneumococcal types: Induction of transformation by a desoxyribonucleic acid fraction isolated from pneumococcus type III. *Journal of Experimental Medicine*, **79**, 137–158.

Avise, J.C. (2000). *Phylogeography: The History and Formation of Species*. Harvard University Press, Cambridge, MA, USA.

Avise, J.C. (2009). Phylogeography: retrospect and prospect. *Journal of Biogeography*, **36**, 3–15.

Avise, J.C., Giblin-Davidson, C., Laerm, J., Patton, J.C., and Lansman, R.A. (1979). Mitochondrial DNA clones and matriarchal phylogeny within and among geographic populations of the pocket gopher, *Geomys pinetis*. *Proceedings of the National Academy of Sciences of the USA*, **76**, 6694–6698.

Axelrod, R. (1984). *The Evolution of Cooperation*. Basic Books, New York, USA.

Bachtrog, D., Hom, E., Wong, K.M., and de Jong, P. (2008). Genomic degradation of a young Y chromosome in *Drosophila miranda*. *Genome Biology*, **9**, R30.

Bailey, R.I., Tesaker, M.R., Trier, C.N., and Sætre, G.-P. (2015). Strong selection on male plumage in a hybrid zone between a hybrid bird species and one of its parents. *Journal of Evolutionary Biology*, **28**, 1257–1269.

Baird, N.A., Etter, P.D., Atwood, T.S., et al. (2008). Rapid SNP discovery and genetic mapping using sequenced RAD markers. *PLoS One*, **3**, e3376.

Baird, S.J.E. (2017). The impact of high-throughput sequencing technology on speciation research: maintaining perspective. *Journal of Evolutionary Biology*, **30**, 1482–1487.

Baird, S.J.E., Barton, N.H., and Etheridge, A.M. (2003). The distribution of surviving blocks of an ancestral genome. *Theoretical Population Biology*, **64**, 451–471.

Baldassarre, D.T., White, T.A., Karubian, J., and Webster, M.S. (2014). Genomic and morphological analysis of a semipermeable avian hybrid zone suggests asymmetrical introgression of a sexual signal. *Evolution*, **68**, 2644–2657.

Baldwin, J.M. (1896). A new factor in evolution. *The American Naturalist*, **30**, 441–451.

Baquero, F. and Blázquez, J. (1997). Evolution of antibiotic resistance. *Trends in Ecology & Evolution*, **12**, 482–487.

Barrett, R.D.H., Rogers, S.M., and Schluter, D. (2008). Natural selection on a major armor gene in threespine stickleback. *Science*, **322**, 255–257.

Bartlett, J.M.S. and Stirling, D. (2003). A short history of the polymerase chain reaction. In: J. M. S. Bartlett and D. Stirling (eds) *Methods in Molecular Biology, Vol 226: PCR Protocols, 2nd edition*. Humana Press Inc., Totowa, NJ, USA.

Barton, N.H. and Charlesworth, B. (1984). Genetic revolutions, founder effects, and speciation. *Annual Reviews in Ecology and Systematics*, **15**, 133–164.

Barton, N.H. and Gale, K.S. (1993). Genetic analysis of hybrid zones. In: R.G. Harrison *Hybrid Zones and the Evolutionary Process*, pp. 13–45. Oxford University Press, Oxford, UK.

Barton, N.H. and Hewitt, G.M. (1985). Analysis of hybrid zones. *Annual Reviews in Ecology and Systematics*, **16**, 113–148.

Bateson, W. (1909). Heredity and variation in modern lights. In: A.C. Seward (ed.) *Darwin and Modern Science*. Cambridge University Press, Cambridge, UK.

Baum, D.A. and Shaw, K.L. (1995). Genealogical perspectives on the species problem. In: P.C. Hoch and A.G. Stephenson (eds) *Experimental and Molecular Approaches to Plant Biosystematics*, pp. 289–303. Monographs in Systematic Botany, Missouri Botanical Garden, St. Louis, MO, USA.

Bazin, E., Glémin, S., and Galtier, N. (2006). Population size does not influence mitochondrial genetic diversity in animals. *Science*, **312**, 570–572.

Beaumont, M.A. and Balding, D.J. (2004). Identifying adaptive genetic divergence among populations from genome scans. *Molecular Ecology*, **13**, 969–980.

Beaumont, M.A. and Nichols, R.A. (1996). Evaluating loci for use in the genetic analysis of population structure. *Proceedings of the Royal Society B*, **263**, 1619–1626.

Belancio, V.P., Deininger, P.L., and Roy-Engel, A.M. (2009). LINE dancing in the human genome: transposable elements and disease. *Genome Medicine*, **1**, 97.

Belancio, V.P., Hedges, D.J., and Deininger, P.L. (2008). Mammalian non-LTR retrotransposons: for better or worse, in sickness and in health. *Genome Research*, **18**, 343–358.

Beldade, R., Holbrook, S.J., Schmitt, R.J., Planes, S., Malone, D., and Bernardi, G. (2012). Larger female fish contribute disproportionately more to self-replenishment. *Proceedings of the Royal Society B*, **279**, 2116–2121.

Bell, G. (1982). *The Masterpiece of Nature: The Evolution and Genetics of Sexuality*. Croom Helm, London, UK.

Bell, M. (1994). Paleobiology and evolution of threespine stickleback. In: M. Bell and S.A. Foster *The Evolutionary Biology of the Threespine Stickleback*. Oxford University Press, Oxford, UK.

Bell, M.A., Aguirre, W.E., and Buck, N.J. (2004). Twelve years of contemporary armor evolution in a threespine stickleback population. *Evolution*, **58**, 814–824.

Bell, S.D. and Jackson, S.P. (1998). Transcription and translation in Archaea: a mosaic of eukaryal and bacterial features. *Trends in Microbiology*, **6**, 222–228.

Belton, J.-M., McCord, R.P., Gibcus, J.H., et al. (2012). Hi-C: A comprehensive technique to capture the conformation of genomes. *Methods*, **58**, 268–276.

Benson, W.W. (1972). Natural selection for Müllerian mimicry in *Heliconius erato* in Costa Rica. *Science*, **176**, 936–939.

Berget, S.M., Moore, C., and Sharp, P.A. (1977). Spliced segments at the 5′ terminus of adenovirus 2 late mRNA. *Proceedings of the National Academy of Sciences of the USA*, **74**, 3173–3175.

Berner, D., Grandchamp, A.-C., and Hendry, A.P. (2009). Variable progress toward ecological speciation in parapatry: stickleback across eight lake-stream transitions. *Evolution*, **63**, 1740–1753.

Bernstein, H., Byerly, H.C., Hopf, F.A., and Michod, R.E. (1985). Genetic damage, mutation, and the evolution of sex. *Science*, **299**, 1277–1281.

Bertorelle, G., Benazzo, A., and Mona, S. (2010). ABC as a flexible framework to estimate demography over space and time: some cons, many pros. *Molecular Ecology*, **19**, 2609–2625.

Betto-Colliard, C., Hofmann, S., Sermier, R., Perrin, N., and Stöck, M. (2018). Profound genetic divergence and asymmetric parental genome contributions as hallmarks of hybrid speciation in polyploid toads. *Proceedings of the Royal Society B*, **285**, 20172667.

Birkhead, T.R. and Møller, A.P. (1998). *Sperm Competition and Sexual Selection*. Academic Press, New York.

Blake, C.C. (1979). Exons encode protein functional units. *Nature*, **277**, 598.

Boag, P.T. (1983). The heritability of external morphology in Darwin's ground finches (*Geospiza*) on Isla Daphne Major, Galápagos. *Evolution*, **37**, 877–894.

Bono, J.M., Olesnicky, E.C., and Matzkin, L.M. (2015). Connecting genotypes, phenotypes and fitness: harnessing the power of CRISPR/Cas9 genome editing. *Molecular Ecology*, **24**, 3810–3822.

Boveri, T.H. (1904). *Ergebnisse über die Konstitution der Chromatischen Substanz des Zellkerns*. Verlag von Gustav Fischer, Jena, Germany. University Press, Princeton, NJ, USA.

Boyle, E.A., Li, Y.I., and Pritchard, J.K. (2017). An expanded view of complex traits: from polygenic to omnigenic. *The Cell*, **169**, 1177–1186.

Brochmann, C., Brysting, A.K., and Alsos, I.G. (2004). Polyploidy in arctic plants. *Biological Society of the Linnean Society*, **82**, 521–536.

Broman, K. and Sen, S. (2009). *A Guide to QTL Mapping with R/qtl*. Springer, New York, USA.

Brown, W.L. and Wilson, E.O. (1956). Character displacement. *Systematic Zoology*, **5**, 49–64.

Browning, S.R. and Browning, B.L. (2007). Rapid and accurate haplotype phasing and missing data inference for whole genome association studies by use of localized haplotype clustering. *American Journal of Human Genetics*, **81**, 1084–1097.

Brunet, M., Guy, F., Pilbeam, D., et al. (2002) A new hominid from the Upper Miocene of Chad, Central Africa. *Nature*, **418**, 145–151.

Buenrostro, J.D., Giresi, P.G., Zaba, L.C., Chang, H.Y., and Greenleaf, W.J. (2013). Transposition of native chromatin for fast and sensitive epigenomic profiling of open chromatin, DNA-binding proteins and nucleosome position. *Nature Methods*, **10**, 1213–1218.

Burki, F. (2014). The Eukaryotic Tree of Life from a Global Phylogenomic Perspective. *Cold Spring Harbor Perspectives in Biology*, **6**, a016147.

Burri, R., Nater, A., Kawakai, T., et al. (2015). Linked selection and recombination rate variation drive the evolution of the genomic landscape of differentiation across the speciation continuum of *Ficedula* flycatchers. *Genome Research*, **25**, 1656–1665.

Bush, G.L. (1966). The taxonomy, cytology and evolution of the genus *Rhagoletis* in North America (Diptera, Tephritidae). *Bulletin of the Museum of Comparative Zoology, Harvard University*, **134**, 431–562.

Butlin, R.K. and Smadja, C.M. (2018). Coupling, reinforcement and speciation. *American Naturalist*, **191**, 155–172.

Cann, R.L., Stoneking, M., and Wilson, A.C. (1987). Mitochondrial-DNA and Human-Evolution. *Nature*, **325**, 31–36.

Carlborg, Ö., Jacobsson, L., Åhgren, P., Siegel, P., and Andersson, L. (2006). Epistasis and the release of genetic variation during long-term selection. *Nature Genetics*, **38**, 418–420.

Carlsson, M. and Carlsson, A. (1990). Interactions between glutamatergic and monoaminergic systems within the basal ganglia—implications for schizophrenia and Parkinson's disease. *Trends in Neuroscience*, **13**, 272–274.

Carneiro, M., Albert, F.W., Afonso, S., et al. (2014) The genomic architecture of population divergence between subspecies of the European rabbit. *PLoS Genetics*, **10**, e1003519.

Carter, A.J.R., Hermission, J., and Hansen, T.F. (2005). The role of epistatic gene interactions in the response to selection and the evolution of evolvability. *Theoretical Population Biology*, **68**, 179–196.

Chan, Y.F., Marks, M.E., Jones, F.C., et al. (2010). Adaptive evolution of pelvic reduction in sticklebacks by recurrent deletion of a pitx1 enhancer. *Science*, **327**, 302–305.

Chargaff, E. (1950). Chemical specificity of nucleic acids and mechanisms of their enzymatic degradation. *Experientia*, **6**, 201–209.

Charlesworth, B. (1998). Measures of divergence between populations and the effect of forces that reduce variability. *Molecular Biology and Evolution*, **15**, 538–543.

Charlesworth, B., Coyne, J.A., and Barton, N.H. (1987). The relative rates of evolution of sex chromosomes and autosomes. *The American Naturalist*, **130**, 113–146.

Charlesworth, B., Nordborg, M., and Charlesworth, D. (1997). The effects of local selection, balanced polymorphism and background selection on equilibrium patterns of genetic diversity in subdivided populations. *Genetics Research*, **70**, 155–174.

Chen, S.-H., Habib, G., Yang, C.-Y., et al. (1987). Apolipoprotein B-48 is the product of a messenger RNA with an organ-specific in-frame stop codon. *Science*, **238**, 363–366.

Chow, L.T., Gelinas, R.E., Broker, T.R., and Roberts, R.J. (1977). An amazing sequence arrangement at the 5′ ends of adenovirus 2 messenger RNA. *The Cell*, **12**, 1–8.

Ciomborowska, J., Rosikiewicz, W., Szklarczyk, D., Makalowski, W., and Makalowska, I. (2013). "Orphan" retrogenes in the human genome. *Molecular Biology and Evolution*, **30**, 384–396.

Clark, P.D. and Pazdernik, N.J. (2012). *Molecular Biology, 2nd edition*. Academic Press, Waltham MA, USA.

Clutton-Brock, T.H., Guinness, F.E., and Albon, S.E. (1982). *Red Deer: Behaviour and Ecology of Two Sexes*. Edinburgh University Press, Edinburgh, UK.

Colosimo, P.F., Hosemann, K.E., Balabhadra, S., et al. (2005). Widespread parallel evolution in sticklebacks by repeated fixation of Ectodysplasin alleles. *Science*, **307**, 1928–1933.

Colosimo, P.F., Peichel, C.L., Nereng, K., et al. (2004). The genetic architecture of parallel armor plate reduction in threespine sticklebacks. *PLoS Biology*, **2**, e109.

Comeron, J.M., Williford, A., Kliman, R.M. (2007). The Hill–Robertson effect: evolutionary consequences of weak selection and linkage in finite populations. *Heredity*, **100**, 19–31.

Compeau, P.E.C., Pevzner, P.A., and Tesler, G. (2011). How to apply de Bruijn graphs to genome assembly. *Nature Biotechnology*, **29**, 987–991.

Cook, L.M. (2003). The rise and fall of the *carbonaria* form of the peppered moth. *The Quarterly Review of Biology*, **78**, 399–417.

Cooney, C.R., Bright, J.A., Capp, E.J.R., et al. (2017). Mega-evolutionary dynamics of the adaptive radiation of birds. *Nature*, **542**, 1–16.

Corradi, N., Pombert, J.F., Farinelli, L., Didier, E.S., and Keeling, P.J. (2010). The complete sequence of the smallest known nuclear genome from the microsporidian *Encephalitozoon intestinalis*. *Nature Communications*, **1**, 77.

Cortijo, S., Wardenaar, R., Colomé-Tatché, M., et al. (2014). Mapping the epigenetic basis of complex traits. *Science*, **343**, 1145–1148.

Coyne, J.A. and Orr, H.A. (1989). Two rules of speciation. In: D. Otte and J.A. Endler (eds) *Speciation and its Consequences*, pp. 180–207. Sinauer Associates, Sunderland, MA, USA.

Coyne, J.A. and Orr, H.A. (2004). *Speciation*. Oxford University Press, Oxford, UK.

Cracraft, J. (1989). Speciation and its ontology: The empirical consequences of alternative species concepts for understanding patterns and processes of differentiation. In: D. Otte and J.A. Endler (eds) *Speciation and its Consequences*, pp. 29–59. Sinauer Associates, Sunderland, MA, USA.

Crick, F.H.C. (1958). On protein synthesis. *Symposia of the Society for Experimental Biology*, **12**, 138–163.

Crick, F.H.C. (1988). *What Mad Pursuit: A Personal View on Scientific Discovery*. Basic Books, New York, USA.

Cruickshank, T.E. and Hahn, M.W. (2014). Reanalysis suggests that genomic islands of speciation are due to reduced diversity, not reduced gene flow. *Molecular Ecology*, **23**, 3133–3157.

Currie, C.R., Scott, J.A., Summerbell, R.C., and Malloch, D. (1999). Fungus-growing ants use antibiotic-producing bacteria to control garden parasites. *Nature*, **398**, 701–704.

Cutter, A.D. and Payseur, B.A. (2013) Genomic signatures of selection at linked sites: unifying the disparity among species. *Nature Reviews Genetics*, **14**, 262–264.

Darvasi, A. (1998). Experimental strategies for the genetic dissection of complex traits in animal models. *Nature Genetics*, **18**, 19–24.

Darwin, C. (1859). *On the Origin of Species by Means of Natural Selection, or the Preservation of Favored Races in the Struggle for Life*. Modern Library, New York, USA.

Darwin, C. (1871). *The Descent of Man, and Selection in Relation to Sex*. Jon Murray, London, UK.

Darwin, C. (1876). *Cross and Self-fertilization of Plants*. John Murray, London, UK.

Darwin, C. and Wallace, A.R. (1858). On the tendency of species to form varieties; and on the perpetuation of varieties and species by natural means of selection. *Journal of the Proceedings of the Linnean Society of London, Zoology*, **3**, 45–62.

Davey, J.W., Hohenlohe, P.A., Etter, P.D., Boone, J.Q., Catchen, J.M., and Blaxter, M.L. (2011). Genome-wide genetic marker discovery and genotyping using next-generation sequencing. *Nature Reviews Genetics*, **12**, 499–510.

Dawkins, R. (1976). *The Selfish Gene*. Oxford University Press, Oxford, UK.

Dawkins, R. (1986). *The Blind Watchmaker*. Norton & Company Ltd, London, UK.

Day, R.L., Laland, K.N., and Odling-Smee, J. (2003). Rethinking adaptation: The niche- construction perspective. *Perspectives in Biology and Medicine*, **46**, 80–95.

de Queiroz, K. and Donoghue, M.J. (1990). Phylogenetic systematics or Nelson's version of cladistics? *Cladistics*, **6**, 61–75.

de Sousa Abreu, R., Penalva, L.O., Marcotte, E.M., and Vogel, C. (2009). Global signatures of protein and mRNA expression levels. *Molecular BioSystems*, **5**, 1512–1526.

de Vries, H. (1900). *The Mutation Theory: Experiments and Observation on the Origin of Species in the Vegetable, Vol. 1*. Classical Reprint 2012, Forgotten Books, London.

de Vries, H. (1903). *The Mutation Theory: Experiments and Observation on the Origin of Species in the Vegetable, Vol. 2*. Classical Reprint 2013, Forgotten Books, London.

Dean, A.M. and Thornton, J.W. (2007). Mechanistic approaches to the study of evolution: the functional synthesis. *Nature Reviews Genetics*, **8**, 675–688.

Dekker, J., Belmont, A.S., Guttman, M., et al. (2017). The 4D nucleome project. *Nature*, **549**, 219–226.

Delaneau, O., Howie, B., Cox, A., Zagury, J.-F., and Marchini, J. (2013). Haplotype estimation using sequence reads. *American Journal of Human Genetics*, **93**, 687–696.

Delmore, K.E. and Irwin, D.E. (2014). Hybrid songbirds employ intermediate routes in a migratory divide. *Ecology Letters*, **17**, 1211–1218.

Delsuc, F., Brinkmann, H., and Philippe, H. (2005). Phylogenomics and the reconstruction of the tree of life. *Nature Reviews Genetics*, **6**, 361–375.

Der Sarkissian, C., Ermini, L., Schubert, M., et al. (2015). Evolutionary genomics and conservation of the endangered Przewalski's horse. *Current Biology*, **25**, 2577–2583.

Diz, A.P., Martínez-Fernández, M., and Rolán-Alvarez, E. (2012). Proteomics in evolutionary ecology: linking the genotype with the phenotype. *Molecular Ecology*, **21**, 1060–1080.

Dobzhansky, T. (1934). Studies on hybrid sterility. I. spermatogenesis in pure and hybrid *Drosophila pseudoobscura*. *Zeitschrift für Zellforschung und Mikroskopische Anatomie*, **21**, 169–221.

Dobzhansky, T. (1937). *Genetics and the Origin of Species*. Reprint, 1982 by Columbia University Press, New York, USA.

Dobzhansky, T. (1940). Speciation as a stage in evolutionary divergence. *American Naturalist*, **74**, 312–321.

Donohue, K. (2013). Development in the wild: Phenotypic plasticity. *Annual Plant Reviews*, **45**, 321–356.

Doolittle, W.F. (1978). Genes in pieces: were they ever together? *Nature*, **272**, 581–582.

Doolittle, W.F. (2013). Is junk DNA bunk? A critique of ENCODE. *Proceedings of the National Academy of Sciences of the USA*, **110**, 5294–5300.

Doolittle, W.F. and Stoltzfus, A. (1993) Molecular evolution. Genes-in-pieces revisited. *Nature*, **361**, 403.

Dulai, K.S., von Dornum, M., Mollon, J.D., and Hunt, D.M. (1999). The evolution of trichromatic color vision by opsin gene duplication in New World and Old World primates. *Genome Research*, **9**, 629–638.

Dunn, C.W., Hejnol, A., Matus, D.Q., et al. (2008). Broad phylogenomic sampling improves resolution of the animal tree of life. *Nature*, **452**, 745–749.

Dunn, C.W. and Munro, C. (2016). Comparative genomics and the diversity of life. *Zoologica Scripta*, **45**, 5–13.

Dunning Hotopp, J.C. (2011). Horizontal gene transfer between bacteria and animals. *Trends in Genetics*, **27**, 157–163.

Eberhard, W.G. (1996). *Female Control: Sexual Selection by Cryptic Female Choice*. Princeton University Press, Princeton, NJ, USA.

Edelson, E. (1999). *Gregor Mendel: And the Roots of Genetics*. Oxford University Press, New York, USA.

Edwards, S.V., Potter, S., Schmitt, C.J., Bragg, J.G., and Moritz, C. (2016). Reticulation, divergence, and the phylogeography–phylogenetics continuum. *Proceedings of the National Academy of Sciences of the USA*, **113**, 8025–8032.

Ehrlich, P.R. and Raven, P.H. (1964). Butterflies and plants: a study in coevolution. *Evolution*, **18**, 586–608.

Eichler, E.E., Flint, J., Gibson, G., Kong, A., Leal, S.M., Moore, J.H., and Nadeau, J.H. (2010). Missing heritability and strategies for finding the underlying causes of complex disease. *Nature Reviews Genetics*, **11**, 446–450.

Eisen, J.A. and Fraser, C.M. (2003). Phylogenomics: Intersection of evolution and genomics. *Science*, **300**, 1706–1707.

Ekblom, R. and Wolf, J.B.W. (2014). A field guide to whole-genome sequencing, assembly and annotation. *Evolutionary Applications*, **7**, 1026–1042.

Eldredge, N. and Gould, S.J. (1972). Punctuated equilibria: an alternative to phyletic gradualism. In: T.J.M. Schopf (ed.) *Models in Paleobiology*, pp 82–115. Freeman Cooper, San Francisco, USA.

Elgvin, T.O., Trier, C.N., Tørresen, O.K., et al. (2017). The genomic mosaicism of hybrid speciation. *Science Advances*, **3**, e1602996.

Ellegren, H. (2011). Sex chromosome evolution: recent progress and the influence of male and female heterogamety. *Nature Reviews Genetics*, **12**, 157–166.

Ellegren, H. (2014). Genome sequencing and population genomics in non-model organisms. *Trends in Ecology & Evolution*, **29**, 51–63.

Ellegren, H., Smeds, L., Burri, R., et al. (2012). The genomic landscape of species divergence in *Ficedula* flycatchers. *Nature*, **491**, 756–760.

Enattah, N.S., Sahi, T., Savilahti, E., Terwilliger, J.D., Peltonen, L., and Järvelä, I. (2002). Identification of a variant associated with adult-type hypolactasia. *Nature Genetics*, **30**, 233–237.

Engel, M.S., Panian, J., Wiltschko, D., and Grimaldi, D.A. (2004). New light shed on the oldest insect. *Nature*, **427**, 627–630.

English, A.C., Richards, S., Han, Y., et al. (2012). Mind the gap: Upgrading genomes with Pacific Biosciences RS long-read sequencing technology. *PLoS One*, **7**, e47768.

Erickson, P.A., Ellis, N.A., and Miller, C.T. (2016). Microinjection for Transgenesis and Genome Editing in Threespine Sticklebacks. *Journal of Visualized Experiments*, **111**, e54055.

Erwin, A.A., Galdos, M.A., Wickersheim, M.L., et al. (2015). piRNAs are associated with diverse transgenerational effects of gene and transposon expression in a hybrid dysgenic syndrome of *D. virilis*. *PLoS Genetics*, **11**, e1005332.

Esvelt, K.M., Smidler, A.L., Catteruccia, F., and Church, G.M. (2014). Emerging technology: Concerning RNA-guided gene drives for the alteration of wild populations. *eLife*, **3**, e03401.

Evanno, G., Regnaut, S., and Goudet, J. (2005). Detecting the number of clusters of individuals using the software STRUCTURE: a simulation study. *Molecular Ecology*, **14**, 2611–2620.

Excoffier, L., Dupanloup, I., Huerta-Sánchez, E., Sousa, V.C., and Foll, M. (2013). Robust demographic inference from genomic and SNP data. *PLoS Genetics*, **9**, e1003905–17.

Excoffier, L., Foll, M., Petit, R.J. (2009). Genetic consequences of range expansions. *Annual Review of Ecology, Evolution, and Systematics*, **40**, 481–501.

Falconer, D.S. and Mackay, T.F.C. (1996). *Introduction to Quantitative Genetics, 4th edition*. Longmans Green, Harlow, Essex, UK.

Falster, D.S. and Westoby, M. (2003). Plant height and evolutionary games. *Trends in Ecology & Evolution*, **18**, 337–343.

Faria, R., Chaube, P., Morales, H.E., et al. (2018). Multiple chromosomal rearrangements in a hybrid zone between *Littorina saxatilis* ecotypes. *Molecular Ecology*. Accepted Author Manuscript. doi:10.1111/mec.14972

Feder, J.L., Egan, S.P., and Nosil, P. (2012). The genomics of speciation-with-gene-flow. *Trends in Genetics*, **28**, 342–350.

Feldman, M.W. and Otto, S.P. (1991). A comparative approach to the population-genetic theory of segregation distortion. *The American Naturalist*, **137**, 443–456.

Felsenstein, J. (1974). The evolutionary advantage of recombination. *Genetics*, **78**, 737–756.

Felsenstein, J. (1981). Skepticism towards Santa Rosalia, or why are there so few kinds of animals? *Evolution*, **35**, 124–138.

Field, Y., Boyle, E.A., Telis, N., et al. (2016). Detection of human adaptation during the past 2000 years. *Science*, **354**, 760–764.

Findlay, G.D., Yi, X., MacCoss, M.J., and Swanson, W.J. (2008). Proteomics reveals novel *Drosophila* seminal fluid proteins transferred at mating. *PLoS Biology*, **6**, e178.

Finnegan, D.J. (2012). Retrotransposons. *Current Biology*, **22**, R432–R437.

Firman, R.C. (2018). Postmating sexual conflict and female control over fertilization during gamete interaction. *Annals of the New York Academy of Sciences*, **1422**, 48–64.

Fisher, R.A. (1918). The correlation between relatives on the supposition of Mendelian inheritance. *Transactions of the Royal Society of Edinburgh*, **52**, 399–433.

Fisher, R.A. (1930). *The Genetical Theory of Natural Selection—A Complete Variorum Edition, 1999*. Oxford University Press, Oxford, UK.

Fishman, L., Aagaard, J., and Tuthill, J.C. (2008). Toward the evolutionary genomics of gametophytic divergence: patterns of transmission ratio distortion in monkeyflowers (*Mimulus*) hybrids reveal a complex genetic basis for conspecific pollen precedence. *Evolution*, **62**, 2958–2970.

Flicek, P. and Birney, E. (2009). Sense from sequence reads: methods for alignment and assembly. *Nature Methods*, **6**, S6–S12.

Forterre, P. and Prangishvili, M. (2009). The origin of viruses. *Research in Microbiology*, **160**, 466–472.

Fountain, T., Ravinet, M., Naylor, R., Reinhardt, K., and Butlin, R.K. (2016). A linkage map and QTL analysis for pyrethroid resistance in the bed bug *Cimex lectularis*. *Genes, Genomes, Genetics*, **6**, 4059–4066.

Francioli, L.C., Polak, P.P., Koren, A., et al. (2015). Genome-wide patterns and properties of *de novo* mutations in humans. *Nature Genetics*, **47**, 822–826.

Fu, Q., Hajdinjak, M., Moldovan, O.T., et al. (2015). An early modern human from Romania with a recent Neanderthal ancestor. *Nature*, **524**, 216–219.

Fuiman, L.A. (1997). What can flatfish ontogenies tell us about pelagic and benthic lifestyles? *Journal of Sea Research*, **37**, 257–267.

Galagan, J.E., Nusbaum, C., Roy, A., et al. (2002). The genome of *M. acetivorans* reveals extensive metabolic and physiological diversity. *Genome Research*, **12**, 532–542.

Gardner, D. (2010). *Future Babble: Why Expert Predictions Fail and Why We Believe Them Anyway*. McClelland & Stewart, Toronto, Canada.

Gaunitz, C., Fages, A., Hanghoj, K., and Albrechtsen, A. (2018). Ancient genomes revisit the ancestry of domestic and Przewalski's horses. *Science*, **360**, 111–114.

Gautier, M., Klassmann, A., and Vitalis, R. (2016). REHH 2.0: a reimplementation of the R package REHH to detect positive selection from haplotype structure. *Molecular Ecology Resources*, **17**, 78–90.

Gavrilets, S. (2004). *Fitness Landscapes and the Origin of Species*. Princeton University Press, Princeton, NJ, USA.

Gavrilets, S. (2014). Is sexual conflict an "engine of speciation"? *Cold Spring Harbor Perspectives in Biology*, **6**, a017723.

Gemmell, N.J., Metcalf, V.J., and Allendorf, F.W. (2004). Mother's curse: the effect of mtDNA on individual fitness and population viability. *Trends in Ecology & Evolution*, **19**, 238–244.

Gigord, L.D.B., Macnair, M.R., and Smithson, A. (2001). Negative frequency-dependent selection maintains a dramatic flower color polymorphism in the rewardless orchid *Dactylorhiza sambucina* (L.) Soò. *Proceedings of the National Academy of Sciences of the USA*, **98**, 6253–6255.

Gilad, Y., Wiebe, V., Przeworski, M., Lancet, D., and Pääbo, S. (2004). Loss of olfactory receptor genes coincides with the acquisition of full trichromatic vision in primates. *PLoS Biology*, **2**, e5.

Gilbert, W. (1978). Why genes in pieces? *Nature*, **271**, 501.

Girirajan, S. (2017). Missing heritability and where to find it. *Genome Biology*, **18**, 89.

Gladyshev, E.A., Meselson, M., and Arkhipova, I.R. (2008). Massive horizontal gene transfer in bdelloid rotifers. *Science*, **320**, 1210–1213.

Glazer, A.M., Killingbeck, E.E., Mitros, T., Rokshar, D.S., and Miller, C.T. (2015). Genome assembly improvement and mapping convergently evolved skeletal traits in stickleback with genotyping-by-sequencing. *Genes, Genomes, Genetics*, **5**, 1463–1472.

Goldberg, E.E., Kohn, J.R., Lande, R., et al. (2010). Species selection maintains self- incompatibility. *Science*, **330**, 493–495.

Golding, G.B. and Strobeck, C. (1980). Linkage disequilibrium in a finite population that is partially selfing. *Genetics*, **94**, 777–789.

Gompert, Z. and Buerkle, C.A. (2016). What, if anything, are hybrids: enduring truths and challenges associated with population structure and gene flow. *Evolutionary Applications*, **9**, 909–923.

Gompert, Z., Mandeville, E.G., and Buerkle, C.A. (2017). Analysis of population genomic data from hybrid zones. *Annual Review of Ecology, Evolution, and Systematics*, **48**, 207–229.

Goodwin, S., McPherson, J.D., and McCombie, W.R. (2016). Coming of age: ten years of next-generation sequencing technologies. *Nature Reviews Genetics*, **17**, 333–351.

Gould, S.J. (2002). *The Structure of Evolutionary Theory*. The Belknap Press of Harvard University Press, Cambridge, MA, USA.

Grabherr, M.G., Haas, B.J., Yassour, M., et al. (2011). Full-length transcriptome assembly from RNA-Seq data without a reference genome. *Nature Biotechnology*, **29**, 644–652.

Grafen, A. (1990). Biological signals as handicaps. *Journal of Theoretical Biology*, **144**, 517–546.

Grahame, J.W., Wilding, C.S., and Butlin, R.K. (2006). Adaptation to a steep environmental gradient and an associated barrier to gene exchange in *Littorina saxatilis*. *Evolution*, **60**, 268–278.

Grant, P.R. (1972). Convergent and divergent character displacement. *Biological Journal of the Linnean Society*, **4**, 36–98.

Grant, P.R. and Grant, B.R. (2006). Evolution of character displacement in Darwin's finches. *Science*, **313**, 224–226.

Grant, P.R. and Grant, B.R. (2011). *How and Why Species Multiply—The Radiation of Darwin's Finches*. Princeton University Press, Princeton, NJ, USA.

Grasby, K.L., Verweij, K.J.H., Mosing, M.A., Zietch, B.P., and Medland, S.E. (2017). Estimating heritability from twin studies. In: R. Elston (ed.) *Statistical Human Genetics. Methods in Molecular Biology, vol 1666*. Humana Press, New York, USA.

Green, R.E., Krause, J., Briggs, A.W., et al. (2010). A draft sequence of the Neandertal genome. *Science*, **328**, 710–722.

Greenman, C., Stephens, P., Smith, R., et al. (2007). Patterns of somatic mutation in human cancer genomes. *Nature*, **446**, 153–158.

Griffin, A.S. (2004). Social learning about predators: a review and prospectus. *Learning & Behavior*, **32**, 131–140.

Guerrero, R.F. and Hahn, M.W. (2017). Speciation as a sieve for ancestral polymorphism. *Molecular Ecology*, **26**, 5262–5368.

Guimaraes, P.R., Jr, Galetti, M., and Jordano, P. (2008). Seed dispersal anachronisms: Rethinking the fruits extinct megafauna ate. *PLoS One*, **3**, e1745.

Gutenkunst, R.N., Hernandez, R.D., Williamson, S.H., and Bustamante, C.D. (2009). Inferring the joint demographic history of multiple populations from multidimensional SNP frequency data. *PLoS Genetics*, **5**, e1000695–11.

Hackett, S.J., Kimball, R.T., Reddy, S., et al. (2008). A phylogenomic study of birds reveals their evolutionary history. *Science*, **320**, 1763–1768.

Haldane, J.B.S. (1922). Sex ratio and unisexual sterility in hybrid animals. *Journal of Genetics*, **12**, 101–109.

Haldane, J.B.S. (1932). *The Causes of Evolution*. Reprint by Princeton University Press 1993, Princeton, NJ, USA.

Hamilton, W.D. (1964a). The genetical evolution of social behaviour. I. *Journal of Theoretical Biology*, **7**, 1–16.

Hamilton, W.D. (1964b). The genetical evolution of social behaviour. II. *Journal of Theoretical Biology*, **7**, 17–52.

Hamilton, W.D. (1980). Sex versus non-sex versus parasite. *Oikos*, **35**, 282–290.

Hamilton, W.D., Axelrod, R., and Tanese, R. (1990). Sexual reproduction as an adaptation to resist parasites (a review). *Proceedings of the National Academy of Sciences of the USA*, **87**, 3566–3573.

Hansen, T.F. and Houle, D. (2008). Measuring and comparing evolvability and constraint in multivariate characters. *Journal of Evolutionary Biology*, **21**, 1201–1219.

Hansen, T.F., Pélabon, C., and Houle, D. (2011). Heritability is not evolvability. *Evolutionary Biology*, **38**, 258.

Hansen, T.F. and Wagner, G.P. (2001). Modelling genetic architecture: a multilinear model of gene interaction. *Theoretical Population Biology*, **59**, 61–86.

Hardy, G.H. (1908). Mendelian proportions in a mixed population. *Science*, **18**, 49–50.

Harold, D., Abraham, R., Hollingworth, P., et al. (2009). Genome-wide association study identifies variants at CLU and PICALM associated with Alzheimer's disease. *Nature Genetics*, **41**, 1088–1093.

Harris, K. and Nielsen, R. (2013). Inferring demographic history from a spectrum of shared haplotype lengths. *PLoS Genetics*, **9**, e1003521.

Hart, J.C. and Miller, C.T. (2017). Sequence-based mapping and genome editing reveal mutations in stickleback hps5 cause oculocutaneous albinism and the casper phenotype. *Genes, Genomes, Genetics*, **7**, 3123–3131.

Hasegawa, M., Kishino, H., and Yano, T.-A. (1985). Dating of the human–ape splitting by a molecular clock of mitochondrial DNA. *Journal of Molecular Evolution*, **22**, 160–174.

Haynes, K.F., Gemeno, C., Yeargan, K.V., and Johnson, K.M. (2002). Aggressive chemical mimicry of moth pheromones by a bolas spider: how does this specialist predator attract more than one species of prey? *Chemoecology*, **12**, 99–105.

Hedges, S.B., Marin, J., Suleski, M., Paymer, M., and Kumar, S. (2015). Tree of life reveals clock-like speciation and diversification. *Molecular Biology and Evolution*, **32**, 835–845.

Hedrick, P.W. and Garcia-Dorado, A. (2016). Understanding inbreeding depression, purging and genetic rescue. *Trends in Ecology & Evolution*, **31**, 940–952.

Helbig, A.J. (1991a). Inheritance of migratory direction in a bird species: a cross-breeding experiment with SE-and SW-migrating blackcaps (*Sylvia atricapilla*). *Behavioural Ecology and Sociobiology*, **28**, 9–12.

Helbig, A.J. (1991b). SE-and SW-migrating blackcap (*Sylvia atricapilla*) populations in Central Europe: Orientation of birds in the contact zone. *Journal of Evolutionary Biology*, **4**, 657–670.

Hellenthal, G., Busby, G., Band, G.B.J., et al. (2014). A genetic atlas of human admixture history. *Science*, **343**, 747–751.

Hendry, A.P., Bolnick, D.I., Berner, D., and Peichel, C.L. (2009). Along the speciation continuum in sticklebacks. *Journal of Fish Biology*, **75**, 2000–2036.

Hendry, A.P., Taylor, E.B., and McPhail, J.D. (2002). Adaptive divergence and the balance between selection and gene flow: Lake and stream stickleback in the Misty system. *Evolution*, **56**, 1199–1216.

Henig, R.M. (2009). *The Monk in the Garden: The Lost and Found Genius of Gregor Mendel, the Father of Genetics*. Houghton Mifflin Company, Boston, New York, USA.

Hermansen, J.S., Haas, F., Trier, C.N., et al. (2014). Hybrid speciation through sorting of parental incompatibilities in Italian sparrows. *Molecular Ecology*, **23**, 5831–5842.

Hewitt, G.M. (2000). The genetic legacy of the Quaternary ice ages. *Nature*, **405**, 907–913.

Hewitt, G.M. (2001). Speciation, hybrid zones and phylogeography—or seeing genes in space and time. *Molecular Ecology*, **10**, 537–549.

Hickerson, M.J. (2014). All models are wrong. *Molecular Ecology*, **23**, 2887–2889.

Hickerson, M.J., Carstens, B.C., Cavender-Bares, J., et al. (2010). Phylogeography's past, present, and future: 10 years after Avise, 2000. *Molecular Phylogenetics and Evolution*, **54**, 291–301.

Higuchi, R., Bowman, B., Freiberger, M., Ryder, O.A., and Wilson, A.C. (1984). DNA-Sequences from the quagga, an extinct member of the horse family. *Nature*, **312**, 282–284.

Hill, W.G. and Robertson, A. (1966). The effect on linkage on limits to artificial selection. *Genetic Research*, **8**, 269–294.

Hirose, M., Hasegawa, A., Mochida, K., et al. (2017). CRISPR/Cas9-mediated genome editing in wild-derived mice: generation of tamed wild-derived strains by mutation of the a (nonagouti) gene. *Scientific Reports*, **7**, 42476.

Ho, S.Y.W., Duchene, S., Molak, M., and Shapiro, B. (2015). Time-dependent estimates of molecular evolutionary rates: evidence and causes. *Molecular Ecology*, **24**, 6007–6012.

Ho, S.Y.W., Shapiro, B., Phillips, M.J., Cooper, A., and Drummond, A.J. (2007). Evidence for time dependency of molecular rate estimates. *Systematic Biology*, **56**, 512–522.

Hoban, S., Kelley, J.L., and Lotterhos, K.E. (2016). Finding the genomic basis of local adaptation: Pitfalls, practical solutions, and future directions. *American Naturalist*, **188**, 379–397.

Hobolth, A., Dutheil, J.Y., Hawks, J., Schierup, M.H., and Mailund, T. (2011). Incomplete lineage sorting patterns among human, chimpanzee, and orangutan suggest recent orangutan speciation and widespread selection. *Genome Research*, **21**, 349–356.

Hodgkinson, A. and Eyre-Walker, A. (2011). Variation in the mutation rate across mammalian genomes. *Nature Reviews Genetics*, **12**, 756–766.

Hoekstra, H.E. and Coyne, J.A. (2007). The locus of evolution: evo devo and the genetics of adaptation. *Evolution*, **61**, 995–1016.

Hoekstra, H.E., Drumm, K.E., and Nachman, M.W. (2004). Ecological genetics of adaptive color polymorphism in pocket mice: geographic variation in selected and neutral genes. *Evolution*, **58**, 1329–1341.

Hohenlohe, P.A., Bassham, S., Etter, P.D., et al. (2010). Population genomics of parallel adaptation in threespine stickleback using sequenced RAD tags. *PLoS Genetics*, **6**, e1000862–23.

Hollingworth, P., Harold, D., Sims, R., et al. (2011). Common variants at ABVA7, MS4A6A/MS4A4E, EPHA1, CD33 and CD2AP are associated with Alzheimer's disease. *Nature Genetics*, **43**, 429–435.

Holmquist, R., Jukes, T.H., and Moise, H. (1976). The evolution of the globin family genes: Concordance of stochastic and augmented maximum parsimony genetic distances for α hemoglobin, β hemoglobin and myoglobin phylogenies. *Journal of Molecular Biology*, **105**, 39–74.

Hopkins, R., Guerrero, R.F., Rausher, M.D., and Kirkpatrick, M. (2014). Strong reinforcing selection in a Texas wildflower. *Current Biology*, **24**, 1–5.

Hopkins, R. and Rausher, M,D, (2011), Identification of two genes causing reinforcement in the Texas wildflower *Phlox drummondii*. *Nature*, **469**, 411–414.

Hopkins, R. and Rausher, M.D. (2012). Pollinator-mediated selection on flower color alleles drives reinforcement. *Science*, **335**, 1090–1092.

Hori, M. (1993). Frequency-dependent natural selection in the handedness of scale-eating cichlid fish. *Science*, **260**, 216–219.

Hori, M., Ochi, H., and Kohda, M. (2007). Inheritance pattern of lateral dimorphism in two cichlids (a scale eater, *Perissodus microlepis*, and a herbivore, *Neolamprologus moorii*) in Lake Tanganyika. *Zoological Science*, **24**, 486–492.

Hornett, E.A. and Wheat, C.W. (2012). Quantitative RNA-Seq analysis in non-model species: assessing transcriptome assemblies as a scaffold and the utility of evolutionary divergent genomic reference species. *BMC Genomics*, **13**, 361.

Hosken, D.J. and Ward, P.I. (2001). Experimental evidence for testis size evolution via sperm competition. *Ecology Letters*, **4**, 10–13.

Houle, D. (1992). Comparing evolvability and variability of quantitative traits. *Genetics*, **130**, 195–204.

Howard, D.J. (1999). Conspecific sperm and pollen precedence and speciation. *Annual Review of Ecology and Systematics*, **30**, 109–132.

Huber, H., Hohn, M.J., Rachel, R., Fuchs, T., Wimmer, V.C., and Stetter, K.O. (2002). A new phylum of Archaea represented by a nanosized hyperthermophilic symbiont. *Nature*, **417**, 63–67.

Hudson, R.E., Aukema, J.E., Rispe, C., and Roze, D. (2002). Altruism, cheating and anticheater adaptations in cellular slime molds. *The American Naturalist*, **160**, 31–43.

Hudson, R.R., Kreitman, M., and Aguadé, M. (1987). A test of neutral molecular evolution based on nucleotide data. *Genetics*, **116**, 153–159.

Hug, L.A., Baker, B.J., Anantharaman, K., et al. (2016). A new view of the tree of life. *Nature Microbiology*, **1**, 1–6.

Husband, B.C. (2000). Constraints on polyploid evolution: a test of the minority exclusion principle. *Proceedings of the Royal Society B*, **267**, 217–223.

Huxley, J. (1942). *Evolution—The Modern Synthesis*. Reprint, 2010 by MIT Press, Cambridge, MA, USA.

Ibbotson, R.E., Hunt, D.M., Bowmaker, J.K., and Mollon, J.D. (1992). Sequence divergence and copy number of the middle- and longwave photopigment genes in Old World monkeys. *Proceedings of the Royal Society B*, **247**, 145–154.

International Human Genome Sequencing Consortium (2001). Initial sequencing and analysis of the human genome. *Nature*, **409**, 860–921.

Irwin, D.E., Bensch, S., Irwin, J.H., and Price, T.D. (2005). Speciation by distance in a ring species. *Science*, **307**, 414–416.

Jacobs, G.H., Neitz, M., Deegan, J.F., and Neitz, J. (1996). Trichromatic colour vision in new world monkeys. *Nature*, **382**, 156–158.

Jarvis, E.D., Mirarab, S., Aberer, A.J., et al. (2014). Whole-genome analyses resolve early branches in the tree of life of modern birds. *Science*, **346**, 1320–1331.

Jetz, W., Thomas, G.H., Joy, J.B., Hartmann, K., and Mooers, A.O. (2012). The global diversity of birds in space and time. *Nature*, **491**, 444–448.

Jia, J., Zhao, S., Kong, X., et al. (2013). *Aegilops tauschii* draft genome sequence reveals a gene repertoire for wheat adaptation. *Nature*, **496**, 91–95.

Jiang, W., Zhou, H., Bi, H., et al. (2013). Demonstration of CRISPR/Cas9/sgRNA-mediated targeted gene modification in *Arabidopsis*, tobacco, sorghum and rice. *Nucleic Acids Research*, **41**, e188.

Jiggins, C.D., Salazar, C., Linares, M., and Mavárez, J. (2008). Hybrid trait speciation and *Heliconius* butterflies. *Philosophical Transactions of the Royal Society B*, **363**, 3047–3054.

Johannesson, K., Butlin, R.K., Panova, M., and Westram, A.M. (2017). Mechanisms of divergence and speciation in *Littorina saxatilis*: Integrating knowledge from ecology and genetics with new data emerging from genomic studies. In: *Population Genomics*. Springer, Cham.

Johnson, J.M., Castle, J., Garret-Engele, P., et al. (2003). Genome-wide survey of human alternative pre-mRNA splicing with exon junction microarrays. *Science*, **302**, 2141–2144.

Jones, F.C., Chan, Y.F., Schmutz, J., et al. (2012). A genome-wide SNP genotyping array reveals patterns of global and repeated species-pair divergence in sticklebacks. *Current Biology*, **22**, 83–90.

Jones, M.R., Mills, L.S., Alves, P.C., et al. (2018). Adaptive introgression underlies polymorphic seasonal camouflage in snowshoe hares. *Science*, **360**, 1355–1358.

Joyce, D.A., Lunt, D.H., Genner, M.J., Turner, G.F., Bills, R., and Seehausen, O. (2011). Repeated colonization and hybridization in Lake Malawi cichlids. *Current Biology*, **21**, R108–R109.

Jukema, J. and Piersma, T. (2006). Permanent female mimics in a lekking shorebird. *Biology Letters*, **2**, 161–164.

Kalisz, S., Vogler, D.W., and Hanley, K.M. (2004). Context-dependent autonomous self-fertilization yields reproductive assurance and mixed mating. *Nature*, **430**, 884–887.

Kamm, J.A., Terhorst, J., and Song, Y.S. (2017). Efficient computation of the joint sample frequency spectra for multiple populations. *Journal of Computational and Graphical Statistics*, **26**, 182–194.

Kashi, Y. and King, D.G. (2006). Simple sequence repeats as advantageous mutators in evolution. *Trends in Genetics*, **22**, 253–259.

Kayser, M., Brauer, S., and Stoneking, M. (2003). A genome scan to detect candidate regions influenced by local natural selection in human populations. *Molecular Biology and Evolution*, **20**, 893–900.

Kelleher, J., Etheridge, A.M., and McVean, G. (2016). Efficient coalescent simulation and genealogical analysis for large sample sizes. *PLoS Computational Biology*, **12**, e1004842–22.

Kern, A.D. and Hahn, M.W. (2018). The neutral theory in light of natural selection. *Molecular Biology and Evolution*, **35**, 1366–1371.

Ketting, R.F. (2011). The many faces of RNAi. *Developmental Cell*, **20**, 148–161.

Kim, K.-W., Jackson, B.C., Zhang, H., et al. (2018). Black or red: A sex-linked colour polymorphism in a song bird is maintained by balancing selection. *BioRxiv*. doi: http://dx.doi.org/10.1101/437111

Kimura, M. (1968). Evolutionary rate at the molecular level. *Nature*, **217**, 624–626.

Kimura, M. (1983). *The Neutral Theory of Molecular Evolution*. Cambridge University Press, Cambridge, UK.

King, J.L. and Jukes, T.H. (1969). Non-Darwinian evolution. *Science*, **164**, 788–798.

Kingman, J.F.C. (1982a). On the genealogy of large populations. *Journal of Applied Probability*, **19**, 27–43.

Kingman, J.F.C. (1982b). The coalescent. *Stochastic Processes and their Applications*, **13**, 235–248.

Kingman, J.F.C. (2000). Origins of the coalescent 1974–1982. *Genetics*, **156**, 1461–1463.

Kirk, G.S., Raven, J.E., and Schofield, M. (1983). *The Presocratic Philosophers—A Critical History with a Selection of Texts, 2nd edition*. Cambridge University Press, Cambridge, UK.

Kirkpatrick, M. (1982). Sexual selection and the evolution of female choice. *Evolution*, **36**, 1–12.

Kirkpatrick, M. (2010). How and why chromosomal inversions evolve. *PLoS One*, **8**, e1000501.

Kitano, J., Bolnick, D.I., Beauchamp, D.A., et al. (2008). Reverse evolution of armor plates in the threespine stickleback. *Current Biology*, **18**, 769–774.

Knowles, D.G. and McLysagth, A. (2009). Recent *de novo* origin of human protein-coding genes. *Genome Research*, **19**, 1752–1759.

Knowles, L.L. (2009). Statistical phylogeography. *Annual Reviews of Ecology, Evolution, and Systematics*, **40**, 593–612.

Kohn, M.H., Murphy, W.J., Ostrander, E.A., and Wayne, R.K. (2006). Genomics and conservation genetics. *Trends in Ecology & Evolution*, **21**, 629–637.

Kohne, D.E. (1970). Evolution of higher-organism DNA. *Quarterly Review of Biophysics*, **3**, 327–375.

Kondrashov, A. (1988). Deleterious mutations and the evolution of sexual reproduction. *Nature*, **336**, 435–440.

Koonin, E.V. (2006). The origin of introns and their role in eukaryogenesis: a compromise solution to the introns-early versus introns-late debate? *Biology Direct*, **1**, 22.

Koonin, E.V., Krupovic, M., and Yutin, N. (2015). Evolution of double-stranded DNA viruses of eukaryotes: from bacteriophages to transposons to giant viruses. *Annals of the New York Academy of Sciences*, **1341**, 10–24.

Kubatko, L.S. and Degnan, J.H. (2007). Inconsistency of phylogenetic estimates from concatenated data under coalescence. *Systematic Biology*, **56**, 17–24.

Kucharski, R., Maleszka, J., Foret, S., and Maleszka, R. (2008). Nutritional control of reproductive status in honeybees via DNA methylation. *Science*, **319**, 1827–1830.

Küpper, C., Stocks, M., Risse, J.E., et al. (2016). A super-gene determines highly divergent male reproductive morphs in the ruff. *Nature Genetics*, **48**, 79–83.

Kusakabe, M., Ishikawa, A., Ravinet, M., et al. (2017). Genetic basis for variation in salinity tolerance between stickleback ecotypes. *Molecular Ecology*, **26**, 304–319.

Kylafis, G. and Loreau, M. (2008). Ecological and evolutionary consequences of niche construction for its agent. *Ecology Letters*, **11**, 1072–1081.

Kyriacou, C.P., Peixoto, A.A., Sandrelli, F., Costa, R., and Tauber, E. (2008). Clines in clock genes: fine tuning circadian rhythms to the environment. *Trends in Genetics*, **24**, 124–132.

Kyrou, K., Hammond, A.M., Galizi, R., et al. (2018). A CRISPR–Cas9 gene drive targeting doublesex causes complete population suppression in caged *Anopheles gambiae* mosquitoes. *Nature Biotechnology*, **36**, 1062–1066.

Laird, C.D., McConaughy, B.L., and McCarthy, B.J. (1969). Rate of fixation of nucleotide substitutions in evolution. *Nature*, **224**, 149–154.

Lamarck, J.B.P.A.M. (1809). *Zoological Philosophy: An Exposition with Regard to the Natural History of Animals*. Reprint 2012, Cornell University Library Digital Collections.

Lambert, J.C., Heath, S., Even, G., et al. (2009). Genome-wide association study identifies variants at CLU and CR1 associated with Alzheimer's disease. *Nature Genetics*, **41**, 1094–1099.

Lamichhaney, S., Berglund, J., Almén, M.S., et al. (2015). Evolution of Darwin's finches and their beaks revealed by genome sequencing. *Nature*, **518**, 371–375.

Lamichhaney, S., Fan, G., Widemo, F., et al. (2016). Structural genomic changes underlie alternative reproductive strategies in the ruff (*Philomachus pugnax*). *Nature Genetics*, **48**, 84–88.

Lande, R. (1981). Models of speciation by sexual selection on polygenic traits. *Proceedings of the National Academy of Sciences of the USA*, **78**, 3721–3725.

Lander, E.S., Linton, L.M., Birren, B., et al. (2001). Initial sequencing and analysis of the human genome. *Nature*, **409**, 860–921.

Langer, D., Hain, J., Thuriaux, P., and Zillig, W. (1995). Transcription in archaea: similarity to that in eukarya. *Proceedings of the National Academy of Sciences of the USA*, **92**, 5768–5772.

Langerhans, B.R. and DeWitt, T.J. (2002). Plasticity constrained: over-generalized induction cues cause maladaptive phenotypes. *Evolutionary Ecology Research*, **4**, 857–870.

Larracuente, A.M. and Presgraves, D.C. (2012). The selfish segregation distorter gene complex of *Drosophila melanogaster*. *Genetics*, **192**, 33–53.

Lawson, D.J., Hellenthal, G., Myers, S., and Falush, D. (2012). Inference of population structure using dense haplotype data. *PLoS Genetics*, **8**, e1002453–16.

Lemmon, E.M. and Lemmon, A.R. (2013). High-throughput genomic data in systematics and phylogenetics. *Annual Review of Ecology, Evolution, and Systematics*, **44**, 99–121.

Leonard, W.R., Snodgrass, J.J., and Robertson, M.L. (2007). Effects of brain evolution on human nutrition and metabolism. *Annual Review of Nutrition*, **27**, 311–327.

Leslie, S., Winney, B., Hellenthal, G., et al. (2015). The fine-scale genetic structure of the British population. *Nature*, **519**, 309–314.

Levene, P.A. (1917). The structure of yeast nucleic acid. *Journal of Biological Chemistry*, **31**, 591–598.

Lewontin, R.C. (1964). The interaction of selection and linkage I. General considerations; heterotic models. *Genetics*, **49**, 49–67.

Lewontin, R.C. and Kojima, K.-I. (1960). The evolutionary dynamics of complex polymorphisms. *Evolution*, **14**, 458–472.

Lewontin, R.C. and Krakauer, J. (1973). Distribution of gene frequency as a test of the theory of the selective neutrality of polymorphisms. *Genetics*, **74**, 175–195.

Lexer, C., Welch, M.E., Durphy, J.L., and Rieseberg, L.H. (2003). Natural selection for salt tolerance quantitative trait loci (QTLs) in wild sunflower hybrids: implications for the origin of *Helianthus paradoxus*, a diploid hybrid species. *Molecular Ecology*, **12**, 1225–1235.

Librado, P., Fages, A., Gaunitz, C., et al. (2016). The evolutionary origin and genetic makeup of domestic horses. *Genetics*, **204**, 423–434.

Librado, P., Gamba, C., Gaunitz, C., et al. (2017). Ancient genomic changes associated with domestication of the horse. *Science*, **356**, 442–445.

Lifschytz, E. and Lindsley, D.L. (1972). The role of X-chromosome inactivation during spermatogenesis. *Proceedings of the National Academy of Sciences of the USA*, **69**, 182–186.

Lindblad-Toh, K., Wade, C.M., Mikkelsen, T.S., et al. (2005). Genome sequence, comparative analysis and haplotype structure of the domestic dog. *Nature*, **438**, 803–819.

Linn, C., Jr, Feder, J.L., Nojima, S., Dambroski, H.R., Berlocher, S.H., and Roelofs, W. (2003). Fruit odor discrimination and sympatric host race formation in *Rhagoletis*. *Proceedings of the National Academy of Sciences of the USA*, **100**, 11490–11493.

Long, T.A.F., Pischedda, A., Stewart, A.D., and Rice, W.R. (2009). A cost of sexual attractiveness to high-fitness females. *PLoS Biology*, **7**, e1000254.

Lotterhos, K.E. and Whitlock, M.C. (2014). Evaluation of demographic history and neutral parameterization on the performance of F_{ST} outlier test. *Molecular Ecology*, **23**, 2178–2192.

Lowry, D.B. and Willis, J.H. (2010). A widespread chromosomal inversion polymorphism contributes to a major life history transition, local adaptation, and reproductive isolation. *PLoS Biology*, **8**, e1000500.

Luce, R.D. and Raiffa, H. (1957). *Games and Decisions; Introduction and Critical Survey*. Wiley, New York, USA.

Lumley, A.J., Michalczyk, L., Kitson, J.J.N., et al. (2015). Sexual selection protects against extinction. *Nature*, **522**, 470–473.

Lush, J.L. (1937). *Animal Breeding Plans*. Iowa State College Press, Ames, Iowa, USA.

Lynch, M. (2010). Evolution of the mutation rate. *Trends in Genetics*, **28**, 345–352.

Lynch, M., Ackerman, M.S., Gout, J.-F., et al. (2016). Genetic drift, selection and the evolution of the mutation rate. *Nature Reviews Genetics*, **17**, 704–714.

MacHugh, D.E., Larson, G., and Orlando, L. (2017). Taming the past: ancient DNA and the study of animal domestication. *Annual Review of Animal Biosciences*, **5**, 329–351.

Maddison, W.P. (1997). Gene trees in species trees. *Systematic Biology*, **46**, 523–536.

Maher, B. (2008). Personal genomes: The case of missing heritability. *Nature*, **456**, 18–21.

Mallet, J. (1989). The evolution of insecticide resistance: Have the insects won? *Trends in Ecology & Evolution*, **11**, 336–340.

Mallet, J. (1995). A species definition for the Modern Synthesis. *Trends in Ecology & Evolution*, **10**, 294–299.

Mallet, J. (2007). Hybrid speciation. *Nature*, **446**, 279–283.

Man, P.Y.W., Turnbull, D.M., and Chinnery, P.F. (2002). Leber hereditary optic neuropathy. *Journal of Medical Genetics*, **39**, 162–169.

Mank, J.E., Vicoso, B., Berlin, S., and Charlesworth, B. (2010). Effective population size and the faster-X effect. Empirical results and their interpretation. *Evolution*, **64**, 663–674.

Manolio, T.A., Collins, F.S., Cox, N.J., et al. (2010). Finding the missing heritability of complex diseases. *Nature*, **461**, 747–753.

Marchini, J., Cutler, D., Patterson, N., et al. (2006). A comparison of phasing algorithms for trios and unrelated individuals. *American Journal of Human Genetics*, **78**, 437–450.

Marchini, J., Howie, B., Myers, S., McVean, G., and Donelly, P. (2007). A new multipoint method for genome-wide association studies by imputation of genotypes. *Nature Genetics*, **39**, 906–913.

Marchinko, K.B. and Schluter, D. (2007). Parallel evolution by correlated response: lateral plate reduction in three-spine stickleback. *Evolution*, **61**, 1084–1090.

Marcussen, T., Heier, L., Brysting, A.K., Oxelman, B., and Jakobsen, K.S. (2014). From gene trees to dated allopolyploid network: insights from the Angiosperm genus *Viola* (Violaceae). *Systematic Biology*, **64**, 84–101.

Mardis, E.R. (2008). Next-generation DNA sequencing methods. *Annual Review of Genomics and Human Genetics*, **9**, 387–402.

Maretty, L., Jensen, J.M., Petersen, B., et al. (2017). Sequencing and de novo assembly of 150 genomes from Denmark as a population reference. *Nature*, **548**, 87–91.

Margulis, L. (1970). *Origin of Eukaryotic cells: Evidence and Research Implications for a Theory of the Origin and Evolution of Microbial, Plant, and Animal Cells on the Precambrian Earth*. New Haven, London.

Marler, P. and Peters, S. (1977). Selective vocal learning in a sparrow. *Science*, **198**, 519–521.

Marraffini, L.A. and Sontheimer, E.J. (2010). CRISPR interference: RNA-directed adaptive immunity in bacteria and archaea. *Nature Reviews Genetics*, **11**, 181–190.

Marshall, J.A.R. (2011). Group selection and kin selection: formally equivalent approaches. *Trends in Ecology & Evolution*, **26**, 325–332.

Martin, S.H., Dasmahapatra, K.K., Nadeau, N.J., et al. (2013). Genome-wide evidence for speciation with gene flow in *Heliconius* butterflies. *Genome Research*, **23**, 1817–1828.

Martin, S.H., Davey, J.W., and Jiggins, C.D. (2014). Evaluating the use of ABBA–BABA statistics to locate introgressed loci. *Molecular Biology and Evolution*, **32**, 244–257.

Mason, A.S. and Pires, J.C. (2015). Unreduced gametes: meiotic mishap or evolutionary mechanism? *Trends in Genetics*, **31**, 5–10.

Matute, D.R. (2010). Reinforcement of gametic isolation in *Drosophila*. *PLoS Biology*, **8**, e1000341.

Mavárez, J., Salazar, C.A., Bermingham, E., Salcedo, C., Jiggins, C.D., and Linares, M. (2006). Speciation by hybridization in *Heliconius* butterflies. *Nature*, **441**, 868–871.

Maynard Smith, J (1964). Group selection and kin selection. *Nature*, **201**, 1145–1147.

Maynard Smith, J. (1978). *The Evolution of Sex*. Cambridge University Press, Cambridge, UK.

Maynard Smith, J. (1982). *Evolution and the Theory of Games*. Cambridge University Press, Cambridge, UK.

Maynard Smith, J. (1989). *Evolutionary Genetics*. Oxford University Press, Oxford, UK.

Maynard Smith, J. and Price, G.R. (1973). The logic of animal conflict. *Nature*, **246**, 15–18.

Mayr, E. (1942). *Systematics and the Origin of Species*. Columbia University Press, New York, USA.

Mayr, E. (1954). Change of genetic environment and evolution. In: J. Huxley, A.C. Hardy, E.B. Ford (eds) *Evolution as a Process*, pp 157–180. Allen & Unwin, London, UK.

Mayr, E. (1982). *The Growth of Biological Thought: Diversity, Evolution and Inheritance*. The Belknap Press of Harvard University Press, Cambridge, MA, USA.

McCauley, D.E. (2013). Paternal leakage, heteroplasmy, and the evolution of plant mitochondrial genomes. *New Phytologist*, **200**, 966–977.

McClintock, B. (1950). The origin and behavior of mutable loci in maize. *Proceedings of the National Academy of Sciences of the USA*, **36**, 344–355.

McCormack, J.E., Harvey, M.G., Faircloth, B.C., et al. (2013). A phylogeny of birds based on over 1,500 loci collected by target enrichment and high-throughput sequencing. *PLoS One*, **8**, e54848–11.

McCutcheon, J.P. and Moran, N.A. (2012). Extreme genome reduction in symbiotic bacteria. *Nature Reviews Microbiology*, **10**, 13–26.

McDonald, J.H. and Kreitman, M. (1991). Adaptive protein evolution at the Adh locus in *Drosophila*. *Nature*, **351**, 652–654.

McKinnon, J.S. and Rundle, H.D. (2002). Speciation in nature: the threespine stickleback model systems. *Trends in Ecology & Evolution*, **17**, 480–488.

McLysaght, A., Hokamp, K., and Wolfe, K.H. (2002). Extensive genomic duplication during early chordate evolution. *Nature Genetics*, **31**, 200–204.

Mehdiabadi, N.J. and Schultz, T.R. (2010). Natural history and phylogeny of the fungus-farming ants (Hymenoptera: Formicidae: Myrmicinae: Attini). *Myrmecological News*, **13**, 37–55.

Meier, J.I., Marques, D.A., Mwaiko, S., Wagner, C.E., Excoffier, L., and Seehausen, O. (2017b). Ancient hybridization fuels rapid cichlid fish adaptive radiations. *Nature Communications*, **8**, 14363.

Meier, J.I., Marques, D.A., Wagner, C.E., Excoffier, L., and Seehausen, O. (2018). Genomics of parallel ecological speciation in Lake Victoria cichlids. *Molecular Biology and Evolution*, **35**, 1489–1506.

Meier, J.I., Sousa, V.C., Marques, D.A., et al. (2017a). Demographic modelling with whole-genome data reveals parallel origin of similar *Pundamilia* cichlid species after hybridization. *Molecular Ecology*, **26**, 123–141.

Meier, J.I., Vranken, N., Marques, D.A., et al. (2017c). Ancient hybridization fuels rapid cichlid fish adaptive radiations. *Nature Communications*, **8**, 14363.

Meltz Steinberg, K., Schneider, V.A., and Church, D.M. (2017). Building and improving reference genome assemblies. *Proceedings of the IEEE*, **105**, 422–435.

Mendel, G. (1865). *Versuche über Pflanzenhybriden—Zwei Abhandlungen, 1865*. Classic Reprint Series, Forgotten Books, London.

Menozzi, P., Piazza, A., and Cavalli-Sforza, L. (1978). Synthetic maps of human gene frequencies in Europeans. *Science*, **201**, 786–792.

Metzker, M.L. (2010). Sequencing technologies—the next generation. *Nature Reviews Genetics*, **11**, 31–46.

Meyer, C.A. and Liu, X.S. (2014). Identifying and mitigating bias in next-generation sequencing methods for chromatin biology. *Nature Reviews Genetics*, **15**, 709–721.

Michel, A.P., Sim, S., Powell, T.H.Q., Taylor, M.S., Nosil, P., and Feder, J.L. (2010). Widespread genomic divergence during sympatric speciation. *Proceedings of the National Academy of Sciences of the USA*, **107**, 9724–9729.

Michod, R.E. (2007). Evolution of individuality during the transition from unicellular to multicellular life. *Proceedings of the National Academy of Sciences of the USA*, **104**, 8613–8618.

Mills, R.E., Luttig, C.T., Larkins, C.E., Beauchamp, A., Tsui, C., Pittard, W.S., Devine, S.W. (2006). An initial map of insertion and deletion (INDEL) variation in the human genome. *Genome Research*, **16**, 1182–1190.

Misof, B., Liu, S., Meusemann, K., et al. (2014). Phylogenomics resolves the timing and pattern of insect evolution. *Science*, **346**, 763–767.

Mock, D.W. and Parker, G.A. (1997). *The Evolution of Sibling Rivalry*. Oxford Series in Ecology and Evolution, Oxford University Press, Oxford, UK.

Moran, J.V., DeBerardinis, R.J., and Kazazian, H.H., Jr (1999). Exon shuffling by L1 retrotransposition. *Science*, **283**, 1530–1534.

Morgan, L.T., Schmidt, O.G., Gelarden, I.A., Parrish, II, R.C., and Lively, C.M. (2011). Running with the Red Queen: Host-parasite coevolution selects for biparental sex. *Science*, **333**, 216–218.

Morgan, T.H. (1911). The origin of five mutations in eye color in *Drosphila* and their modes of inheritance. *Science*, **33**, 534–537.

Morgan, T.H., Sturtevant, A.H., Muller, H.J., and Bridges, C.B. (1915). *The Mechanism of Mendelian Heredity*. Henry Holt and Company, New York, USA.

Morgante, M., Brunner, S., Pea, G., et al. (2005). Gene duplication and exon shuffling by helitron-like transposons generate intraspecies diversity in maize. *Nature Genetics*, **37**, 997–1002.

Morozova, O., Hirst, M., and Marra, M.A. (2009). Applications of new sequencing technologies for transcriptome analysis. *Annual Review of Genomics and Human Genetics*, **10**, 135–151.

Mueller, U.G., Gerardo, N.M., Aanen, D.K., Six, D.L., and Schultz, T.R. (2005). The evolution of agriculture in insects. *Annual Review of Ecology, Evolution, and Systematics*, **36**, 563–595.

Müller, F. (1878). Über die Vortheile der Mimicry bei Schmetterlingen. *Zoologischer Anzeiger*, **1**, 54–55.

Muller, H.J. (1932). Some genetic aspects of sex. *American Naturalist*, **66**, 118–138.

Muller, H.J. (1940). Bearing of the *Drosophila* work on systematics. In: J. Huxley (ed.) *The New Systematics*, pp. 185–268. Clarendon Press, Oxford, UK.

Muller, H.J. (1964). The relation of recombination to mutational advance. *Mutation Research*, **1**, 2–9.

Myhre, F. and Klepaker, T. (2009). Body armour and lateral-plate reduction in freshwater three-spined stickleback *Gasterosteus aculeatus*: adaptations to a different buoyancy regime? *Journal of Fish Biology*, **75**, 2062–2074.

Nachman, M.W. (2001). Single nucleotide polymorphisms and recombination rate in humans. *Trends in Genetics*, **17**, 481–485.

Nachman, M.W., Hoekstra, H.E., D'Agostino, S.L. (2003). The genetic basis of adaptive melanism in pocket mice. *Proceedings of the National Academy of Sciences of the USA*, **100**, 5268–5273.

Nadachowska-Brzyska, K., Burri, R., Olason, P.I., et al. (2013). Demographic divergence history of pied flycatcher and collared flycatcher inferred from whole-genome re-sequencing data. *PLoS Genetics*, **9**, e1003942–14.

Nagarajan, N. and Pop, M. (2013). Sequence assembly demystified. *Nature Reviews Genetics*, **14**, 157–167.

Naisbit, R.E., Jiggins, C.D., and Mallet, J. (2001). Disruptive sexual selection against hybrids contributes to speciation between *Heliconius cydno* and *Heliconius melpomene*. *Proceedings of the Royal Society B*, **268**, 1849–1854.

Nakayama, H., Nakayama, N., Seiki, S., et al. (2014). Regulation of the KNOX-GA gene module induces heterophyllic alteration in North American lake cress. *The Plant Cell*, **26**, 4733–4748.

Narasimhan, V.M., Rahbari, R., Scally, A., et al. (2017). Estimating the human mutation rate from autozygous segments reveals population differences in human mutational processes. *Nature Communications*, **8**, 303.

Nei, M. (1987). *Molecular Evolutionary Genetics*. Columbia University Press, New York, USA.

Nei, M. and Li, W.-H. (1979). Mathematical model for studying genetic variation in terms of restriction endonucleases. *Proceedings of the National Academy of Sciences of the USA*, **76**, 5269–5273.

Nelson, K.E., Clayton, R.A., Gill, S.R., et al. (1999). Evidence for lateral gene transfer between Archaea and Bacteria from genome sequence of *Thermotoga maritima*. *Nature*, **399**, 323–329.

Nicotra, A.B., Atkin, O.K., Bonser, S.P., et al. (2010). Plant phenotypic plasticity in a changing climate. *Trends in Plant Science*, **15**, 684–692.

Nielsen, R., Akey, J.M., Jakobsson, M., et al. (2017). Tracing the peopling of the world through genomics. *Nature*, **541**, 302–310.

Noonan, J.P., Coop, G., Kudaravalli, S., et al. (2006). Sequencing and analysis of Neanderthal genomic DNA. *Science*, **314**, 1113–1118.

Noor, M.A.F. and Bennett, S.M. (2009). Islands of speciation or mirages in the desert? Examining the role of restricted recombination in maintaining species. *Heredity*, **103**, 439–444.

Nosil, P. (2012). *Ecological Speciation*. Oxford Series in Ecology and Evolution. Oxford University Press, Oxford, UK.

Nosil, P., Funk, D.J., and Ortiz-Barrientos, D. (2009). Divergent selection and heterogeneous genomic divergence. *Molecular Ecology*, **18**, 375–402.

Novembre, J. and Di Rienzo, A. (2009). Spatial patterns of variation due to natural selection in humans. *Nature Reviews Genetics*, **10**, 745–755.

Novembre, J., Johnson, T., Bryc, K., et al. (2008). Genes mirror geography within Europe. *Nature*, **456**, 98–101.

Novembre, J. and Stephens, M. (2008). Interpreting principal component analyses of spatial population genetic variation. *Nature Genetics*, **40**, 646–649.

Nowak, M.A., Tarnita, C.E., and Wilson, E.O. (2010). The evolution of eusociality. *Nature*, **466**, 1057–1062.

O'Brien, S.J., Haussler, D., and Ryder, O. (2014). The birds of Genome10K. *GigaScience*, **3**, 1214–1215.

O'Brown, N.M., Summers, B.R., Jones, F.C., Brady, S.D., and Kingsley, D.M. (2015). A recurrent regulatory change underlying altered expression and Wnt response of the stickleback armor plates gene EDA. *eLife*, **4**, e05290.

O'Conell, J., Gurdasani, D., Delaneau, O., et al. (2014). A general approach for haplotype phasing across the full spectrum of relatedness. *PLoS Genetics*, **10**, e1004234-21.

O'Dea, A., Lessios, H.A., Coates, A.G., et al. (2016). Formation of the Isthmus of Panama. *Science Advances*, **2**, e1600883.

O'Donovan, M.C. (1993). A novel gene containing a trinucleotide repeat that is expanded and unstable on Huntington's disease chromosomes. *Cell*, **72**, 971–983.

Obbard, D.J., Gordon, K.H.J., Buck, A.H., and Jiggins, F.M. (2009). The evolution of RNAi as a defense against viruses and transposable elements. *Philosophical Transactions of the Royal Society B*, **364**, 99–115.

Ohno, S. (1970). *Evolution by Gene Duplication*. Springer-Verlag, New York, USA.

Ohta, T. (1973). Slightly deleterious mutant substitutions in evolution. *Nature*, **246**, 96–98.

Orlando, L., Gilbert, M.T.P., and Willerslev, E. (2015). Reconstructing ancient genomes and epigenomes. *Nature Reviews Genetics*, **16**, 395–408.

Orr, H.A. (1996). Dobzhansky, Bateson, and the genetics of speciation. *Genetics*, **144**, 1331–1335.

Otto, S.P. and Whitton, J. (2000). Polyploid incidence and evolution. *Annual Review of Genetics*, **34**, 401–437.

Owens, I.P.F., Bennet, P.M., and Harvey, P.H. (1999). Species richness among birds: body size, life history, sexual selection or ecology? *Proceedings of the Royal Society B*, **266**, 933–939.

Ozsolak, F. and Milos, P.M. (2010). RNA sequencing: advances, challenges and opportunities. *Nature Reviews Genetics*, **12**, 87–98.

Pääbo, S. (1985). Molecular-cloning of ancient Egyptian mummy DNA. *Nature*, **314**, 644–645.

Pace, N.R. (2006). Time for a change. *Nature*, **441**, 289.

Page, R. and Holmes, E.C. (1998). *Molecular Evolution: A Phylogenetic Approach*. Blackwell Publishing, Oxford, UK.

Palacio-López, K., Beckage, B., Scheiner, S., and Molofsky, J. (2015). The ubiquity of phenotypic plasticity in plants: a synthesis. *Ecology & Evolution*, **5**, 3389–3400.

Palopoli, M.F. (2000). Genetic partners in crime: Evolution of an ultraselfish supergene that specializes in sperm sabotage. In: J.B. Wolf, E.D. Brodie III, and M.J. Wade (eds) *Epistasis and the Evolutionary Process*. Oxford University Press, New York, USA.

Palumbi, S.R. (1998). Species formation and the evolution of gamete recognition loci. In: D.J. Howard and S.F. Berlocher (eds), *Endless Forms: Species and Speciation*, pp 271–278. Oxford University Press, Oxford, UK.

Panhuis, T.M., Butlin, R., Zuk, M., and Tregenza, T. (2001). Sexual selection and speciation. *Trends in Ecology & Evolution*, **16**, 364–371.

Pardo-Diaz, C., Salazar, C., Baxter, S.W., et al. (2012). Adaptive introgression across species boundaries in *Heliconius* butterflies. *PLoS Genetics*, **8**, e1002752.

Park, P.J. (2009). ChIP–seq: advantages and challenges of a maturing technology. *Nature Reviews Genetics*, **10**, 669–680.

Parker, G.A. (1970). Sperm competition and its evolutionary consequence in the insects. *Biological Reviews*, **45**, 525–567.

Paten, B., Novak, A.M., Eizenga, J.M., and Garrison, E. (2017). Genome graphs and the evolution of genome inference. *Genome Research*, **27**, 665–676.

Patterson, N., Price, A.L., and Reich, D. (2006). Population structure and eigenanalysis. *PLoS Genetics*, **2**, e190.

Patthy, L. (1999). Genome evolution and the evolution of exon-shuffling—a review. *Gene*, **238**, 103–114.

Payseur, B.A., Peicheng, J., and Haasl, R.J. (2011). A genomic portrait of human microsatellite variation. *Molecular Biology and Evolution*, **28**, 303–312.

Payseur, B.A., Presgraves, D.C., and Filatov, D.A. (2018). Sex chromosomes and speciation. *Molecular Ecology*, **27**, 3745–3748.

Peichel, C.L., Nereng, K.S., Ohgi, K.A., et al. (2001). The genetic architecture of divergence between threespine stickleback species. *Nature*, **414**, 901–905.

Pellicer, J., Fay, M.F., and Leitch, I.J. (2010). The largest eukaryotic genome of them all? *Botanical Journal of the Linnean Society*, **164**, 10–15.

Pérez-Palma, E., Bustos, B.I., Villamán, C.F., et al. (2014). Overrepresentation of glutamate signaling in Alzheimer's disease: network-based pathway enrichment using meta-analysis of genome-wide association studies. *PLoS One*, **9**, e95413.

Pfennig, D.W. and Pfennig, K.S. (2012). Development and evolution of character displacement. *Annals of the New York Academy of Sciences*, **1256**, 89–107.

Pfenninger, M. and Schwenk, K. (2007). Cryptic animal species are homogeneously distributed among taxa and biogeographical regions. *BMC Evolutionary Biology*, **7**, 121.

Phadnis, N. and Orr, H.A. (2009). A single gene causes both male sterility and segregation distortion in *Drosophila* hybrids. *Science*, **323**, 376–379.

Phillips, P.C. (2008). Epistasis—the essential role of gene interactions in the structure and evolution of genetic systems. *Nature Reviews Genetics*, **9**, 855–867.

Poelstra, J.W., Vijay, N., Bossu, C.M., et al. (2014). The genomic landscape underlying phenotypic integrity in the face of gene flow in crows. *Science*, **344**, 1410–1414.

Poelstra, J.W., Vijay, N., Hoeppner, M.P., and Wolf, J.B.W. (2015). Transcriptomics of colour patterning and coloration shifts in crows. *Molecular Ecology*, **24**, 4617–4628.

Pop, M. (2009). Genome assembly reborn: recent computational challenges. *Briefings in Bioinformatics*, **10**, 354–366.

Posada, D. and Crandall, K. (2001). Intraspecific gene genealogies: trees grafting into networks. *Trends in Ecology & Evolution*, **16**, 37–45.

Poulter, M., Hollox, E., Harvey, C.B., et al. (2003). The causal element for the lactase persistence/non-persistence polymorphism is located in a 1 Mb region of linkage disequilibrium in Europeans. *Annals of Human Genetics*, **67**, 298–311.

Prangishvili, D., Albers, S.-V., Holz, S., et al. (1998). Conjugation in Archaea: frequent occurrence of conjugative plasmids in *Sulfolobus*. *Plasmid*, **40**, 190–202.

Presgraves, D.C. (2008). Sex chromosomes and speciation in *Drosophila*. *Trends in Genetics*, **24**, 336–343.

Presgraves, D.C. (2010). The molecular evolutionary basis of species formation. *Nature Reviews Genetics*, **11**, 175–180.

Price, T.D., Qvarnström, A., and Irwin, D.E. (2003). The role of phenotypic plasticity in driving genetic evolution. *Proceedings of the Royal Society B*, **270**, 1433–1440.

Pritchard, J.K., Stephens, M., and Donnelly, P. (2000). Inference of population structure using multilocus genotype data. *Genetics*, **155**, 945–959.

Provan, J. and Bennett, K. (2008). Phylogeographic insights into cryptic glacial refugia. *Trends in Ecology & Evolution*, **23**, 564–571.

Prum, R.O., Berv, J.S., Dornburg, A., et al. (2015). A comprehensive phylogeny of birds (Aves) using targeted next-generation DNA sequencing. *Nature*, **526**, 569–573.

Pryke, S. (2010). Sex chromosome linkage of mate preference and color signal maintains assortative mating between interbreeding finch morphs. *Evolution*, **64**, 1301–1310.

Pusey, A.E. (1987). Sex-biased dispersal and inbreeding avoidance in birds and mammals. *Trends in Ecology & Evolution*, **2**, 295–299.

Quinn, G.P. and Keough, M.J. (2002). *Experimental Design and Data Analysis for Biologists*. Cambridge University Press, Cambridge, UK.

Qvarnström, A. and Bailey, R.I. (2009). Speciation through linkage of sex-linked genes. *Heredity*, **102**, 4–15.

Racimo, F., Sankararaman, S., Nielsen, R., and Huerta-Sánchez, E. (2015). Evidence for archaic adaptive introgression in humans. *Nature Reviews Genetics*, **16**, 359–371.

Raeymaekers, J.A.M., van Houdt, J.K.J., Larmuseau, M.H.D., Geldof, S., Volckaert, F.A.M. (2006). Divergent selection as revealed by P_{ST} and QTL-based F_{ST} in three-spined stickleback (*Gasterosteus aculeatus*) populations along a coastal-inland gradient. *Molecular Ecology*, **16**, 891–905.

Raeymaekers, J.A.M., Konijnendijk, N., Larmuseau, M.H.D., Hellemans, B., De Meester, L., and Volckaert, F.A.M. (2014). A gene with major phenotypic effects as a target for selection vs. homogenizing gene flow. *Molecular Ecology*, **23**, 162–181.

Rand, D.M. and Kann, L.M. (1996). Excess amino acid polymorphism in mitochondrial DNA: Contrasts among genes from *Drosophila*, mice, and humans. *Molecular Biology and Evolution*, **13**, 735–748.

Randler, C. (2007). Assortative mating of carrion *Corvus corone* and hooded crows *C. cornix* in the hybrid zone in Eastern Germany. *Ardea*, **95**, 143–149.

Rannala, B. and Yang, Z. (2008). Phylogenetic inference using whole genomes. *Annual Review of Genomics and Human Genetics*, **9**, 217–231.

Ravindran, P. (2012). Barbara McClintock and the discovery of jumping genes. *Proceedings of the National Academy of Sciences of the USA*, **109**, 20198–20199.

Ravinet, M., Elgvin, T.O., Trier, C.N., Aliabadian, M., Gavilov, A., and Sætre, G.-P. (2018b). Signatures of human-commensalism in the house sparrow genome. *Proceedings of the Royal Society B*, **285**, 20181246.

Ravinet, M., Faria, R., Butlin, R.K., et al. (2017). Interpreting the genomic landscape of speciation: a road map for finding barriers to gene flow. *Journal of Evolutionary Biology*, **30**, 1450–1477.

Ravinet, M., Hynes, R., Poole, R., et al. (2015). Where the lake meets the sea: strong reproductive isolation is associated with adaptive divergence between lake resident and anadromous three-spined sticklebacks. *PLoS One*, **10**, e0122825.

Ravinet, M., Westram, A., Johannesson, K., Butlin, R., André, C., Panova, M. (2016). Shared and nonshared genomic divergence in parallel ecotypes of *Littorina saxatilis* at a local scale. *Molecular Ecology*, **25**, 287–305.

Ravinet, M., Yoshida, K., Shigenobu, S., Toyoda, A., Fujiyama, A., and Kitano, J. (2018a). The genomic landscape at a late stage of stickleback speciation: High genomic divergence interspersed by small localized regions of introgression. *PLoS Genetics*, **14**, e1007358.

Reich, D., Green, R.E., Kircher, M., et al. (2010). Genetic history of an archaic hominin group from Denisova Cave in Siberia. *Nature*, **468**, 1053–1060.

Reiley, M.T., Faulkner, G.J., Dubnau, J., Ponomarev, I., and Gage, F.H. (2013). The role of transposable elements in health and diseases of the central nervous system. *Journal of Neuroscience*, **33**, 17577–17586.

Repping, S., Skaletsky, H., Brown, L., et al. (2003). Polymorphism for a 1.6-Mb deletion of the human Y chromosome persists through balance between recurrent mutation and haploid selection. *Nature Genetics*, **35**, 247–251.

Rice, S.H. (2004). *Evolutionary Theory—Mathematical and Conceptual Foundations*. Sinauer Associates, Sunderland, MA, USA.

Rice, W.A. (1996). Sexually antagonistic male adaptation triggered by experimental arrest of female evolution. *Nature*, **361**, 232–234.

Rice, W.A. and Holland, B. (1997). The enemies within: intergenomic conflict, interlocus contest evolution (ICE) and the intraspecific red queen. *Behavioral Ecology and Sociobiology*, **41**, 1–10.

Riesch, R., Muschick, M., and Lindke, D. (2017). Transitions between phases of genomic differentiation during stick-insect speciation. *Nature Ecology & Evolution*, **1**, 0082.

Rieseberg, L.H. (1991). Homoploid reticulate evolution in *Helianthus* (Asteraceae): evidence from ribosomal genes. *American Journal of Botany*, **78**, 1218–1237.

Rieseberg, L.H., Raymond, O., Rosenthal, D.M., et al. (2003). Major ecological transitions in wild sunflowers facilitated by hybridization. *Science*, **301**, 1211–1216.

Rieseberg, L.H. and Willis, J.H. (2007). Plant speciation. *Science*, **317**, 910–914.

Riyahi, S., Hammer, Ø., Arbabi, T., et al. (2013). Beak and skull shapes of human commensal and non-commensal house sparrows *Passer domesticus*. *BMC Evolutionary Biology*, **13**, 200.

Robertson, H.M. and Paterson, H.E.H. (1982). Mate recognition and mechanical isolation in *Enallagma* damselflies (Odonata: Coenagrionidae). *Evolution*, **36**, 243–250.

Rockman, M.V. (2012). The QTN program and the alleles that matter for evolution: all that's gold does not glitter. *Evolution*, **66**, 1–17.

Roesti, M., Hendry, A.P., Salzburger, W., and Berner, D. (2012). Genome divergence during evolutionary diversification as revealed in replicate lake-stream stickleback population pairs. *Molecular Ecology*, **21**, 2852–2862.

Rokas, A. and Abbot, P. (2009). Harnessing genomics for evolutionary insights. *Trends in Ecology & Evolution*, **24**, 192–200.

Rolán-Alvarez, E., Johannesson, K., and Erlandsson, J. (1997). The maintenance of a cline in the marine snail *Littorina saxatilis*: The role of home site advantage and hybrid fitness. *Evolution*, **51**, 1838–1847.

Ronquist, F., van der Mark, P., and Huelsenbeck, J. (2009). Bayesian phylogenetic analysis using MrBayes. In: P. Lemey, M. Salemi, and A.-M. Vandamme (eds) *The Phylogenetic Handbook: A Practical Approach to Phylogenetic Analysis and Hypothesis Testing*, pp. 210–265. Cambridge University Press, Cambridge, UK.

Rowe, L., Chenoweth, S.F., and Agrawal, A.F. (2018). The genomics of sexual conflict. *American Naturalist*, **192**, 274–286.

Rudkin, D.M., Young, G.A., and Nowlan, G.S. (2008). The oldest horseshoe crab: a new xiphosurid from late Ordovician Konservat-Lagerstätten deposits, Manitoba, Canada. *Palaeontology*, **51**, 1–9.

Runemark, A., Hey, J., Hansson, B., and Svensson, E.I. (2011). Vicariance divergence and gene flow among islet populations of an endemic lizard. *Molecular Ecology*, **21**, 117–129.

Runemark, A., Trier, C.N., Eroukhmanoff, F., et al. (2018). Variation and constraints in hybrid genome formation. *Nature Ecology & Evolution*, **2**, 549–556.

Runnegar, B. (1982). A molecular-clock date for the origin of animal phyla. *Lethaia*, **15**, 199–205.

Sabeti, P.C., Reich, D.E., Higgins, J.M., et al. (2002). Detecting recent positive selection in the human genome from haplotype structure. *Nature*, **419**, 832–837.

Sabeti, P.C., Varilly, P., Fry, B., et al. (2007). Genome-wide detection and characterization of positive selection in human populations. *Nature*, **449**, 913–918.

Sackton, T.B., Corbett-Detig, R.B., Nagaraju, J., Vaishna, L., Arunkuma, K.P., and Hartl, D.L. (2014). Positive selection drives faster-Z evolution in silkmoths. *Evolution*, **68**, 2331–2342.

Sæther, S.A., Sætre, G.-P., Borge, T., et al. (2007). Sex chromosome-linked species recognition and evolution of reproductive isolation in flycatchers. *Science*, **318**, 95–97.

Sætre, G.-P. (2013). Hybridization is important in evolution, but is speciation? *Journal of Evolutionary Biology*, **26**, 256–258.

Sætre, G.-P., Borge, T., Lindroos, K., et al. (2003). Sex chromosome evolution and speciation in *Ficedula* flycatchers. *Proceedings of the Royal Society B*, **270**, 53–59.

Sætre, G.-P., Cuevas, A., Hermansen, J.S., et al. (2017). Rapid polygenic response to secondary contact in a hybrid species. *Proceedings of the Royal Society B*, **284**, 20170365.

Sætre, G.-P., Riyahi, S., Aliabadian, M., et al. (2012). Single origin of human commensalism in the house sparrow. *Journal of Evolutionary Biology*, **25**, 788–796.

Sætre, G.-P., Moum, T., Bureš, S., Král, M., Adamjan, M., and Moreno, J. (1997). A sexually selected character displacement in flycatchers reinforces premating isolation. *Nature*, **387**, 589–592.

Salzberg, S.L., Phillippy, A.M., Zimin, A., et al. (2012). GAGE: A critical evaluation of genome assemblies and assembly algorithms. *Genome Research*, **22**, 557–567.

Sander, J.D. and Joung, J.K. (2014). CRISPR-Cas systems for editing, regulating and targeting genomes. *Nature Biotechnology*, **32**, 347–355.

Sanderson, M.J. (2008). Phylogenetic signal in the eukaryotic tree of life. *Science*, **321**, 121–123.

Sanger, F., Nicklen, S., and Coulson, A.R. (1977). DNA sequencing with chain-terminating inhibitors. *Proceedings of the National Academy of Sciences of the USA*, **74**, 5463–5467.

Santos, F.P., Santos, F.C., and Pacheco, J.M. (2018). Social norm complexity and past reputations in the evolution of cooperation. *Nature*, **555**, 242–245.

Scally, A. and Durbin, R. (2012). Revising the human mutation rate: implications for understanding human evolution. *Nature Reviews Genetics*, **13**, 745–753.

Scally, A., Dutheil, J.Y., Hillier, L.W., et al. (2012). Insights into hominid evolution from the gorilla genome sequence. *Nature*, **483**, 169–175.

Schaefer, H.M. and Ruxton, G.D. (2009). Deception in plants: mimicry or perceptual exploitation? *Trends in Ecology & Evolution*, **24**, 676–685.

Schemske, D.W. and Bradshaw, H.D. (1999). Pollinator preference and the evolution of floral traits in monkey flowers (*Mimulus*). *Proceedings of the National Academy of Sciences of the USA*, **96**, 11910–11915.

Schielzeth, H., Kempenaers, B., Ellegren, H., and Forstmeier, W. (2012). QTL linkage mapping of zebra finch beak color shows an oligogenic control of a sexually selected trait. *Evolution*, **66**, 18–30.

Schlichting, C.D. (2002). Phenotypic plasticity in plants. *Plant Species Biology*, **17**, 85–88.

Schlichting, C.D. and Wund, M.A. (2014). Phenotypic plasticity and epigenetic marking: an assessment of evidence for genetic accommodation. *Evolution*, **68**, 656–672.

Schluter, D. (2000a). Ecological character displacement in adaptive radiation. *American Naturalist*, **156**, S4–S16.

Schluter, D. (2000b). *The Ecology of Adaptive Radiation*. Oxford University Press, Oxford, UK.

Schluter, D. and Conte, G.L. (2009). Genetics and ecological speciation. *Proceedings of the National Academy of Sciences of the USA*, **106**, 9955–9962.

Schmitz, R.J. (2014). The secret garden—epigenetic alleles underlie complex traits. *Science*, **343**, 1082–1083.

Schmitz, R.J., Schultz, M.D., Urich, M.A., et al. (2013). Patterns of population epigenomic diversity. *Nature*, **495**, 193–198.

Schrider, D.R. and Kern, A.D. (2016). S/HIC: Robust identification of soft and hard sweeps using machine learning. *PLoS Genetics*, **12**, e1005928–31.

Schrider, D.R. and Kern, A.D. (2018). Supervised machine learning for population genetics: a new paradigm. *Trends in Genetics*, **34**, 301–312.

Schubert, M., Jónsson, H., Chang, D., et al. (2014). Prehistoric genomes reveal the genetic foundation and cost of horse domestication. *Proceedings of the National Academy of Sciences of the USA*, **111**, E5661–E5669.

Searcy, W.A., Marler, P., Peters, S.S. (1981). Species song discrimination in adult female song and swamp sparrows. *Animal Behaviour*, **29**, 997–1003.

Seehausen, O. (2000). Explosive speciation rates and unusual species richness in haplochromine cichlid fishes: Effects of sexual selection. *Advances in Ecological Research*, **31**, 237–274.

Seehausen, O. (2004). Hybridization and adaptive radiation. *Trends in Ecology & Evolution*, **19**, 198–207.

Seehausen, O., Butlin, R.K., Keller, I., et al. (2014). Genomics and the origin of species. *Nature Reviews Genetics*, **15**, 176–192.

Seehausen, O., Terai, Y., Magalhaes, I.S., et al. (2008). Speciation through sensory drive in cichlid fish. *Nature*, **455**, 620–626.

Seehausen, O. and Wagner, C.E. (2014). Speciation in freshwater fishes. *Annual Reviews of Ecology, Evolution, and Systematics*, **45**, 621–651.

Servedio, M.R. and Boughman, J.W. (2017). The role of sexual selection in local adaptation and speciation. *Annual Review of Ecology, Evolution, and Systematics*, **48**, 85–109.

Servedio, M.R. and Noor, M.A.F. (2003). The role of reinforcement in speciation: theory and data. *Annual Review of Ecology, Evolution and Systematics*, **34**, 339–364.

Servedio, M.R., Van Doorn, G.S., Kopp, M., Frame, A.M., and Nosil, P. (2011). Magic traits in speciation: "magic" but not rare? *Trends in Ecology & Evolution*, **26**, 389–397.

Shalem, O., Sanjana, N.E., and Zhang, F. (2015). High-throughput functional genomics using CRISPR–Cas9. *Nature Reviews Genetics*, **16**, 299–311.

Sharma, M.D., Wilson, A.J., and Hosken, D.J. (2016). Fisher's sons' effect in sexual selection: absent, intermittent or just low experimental power? *Journal of Evolutionary Biology*, **29**, 2464–2470.

Shen, H., Li, J., Zhang, J., et al. (2013). Comprehensive characterization of human genome variation by high coverage whole-genome sequencing of forty four Caucasians. *PLoS One*, **8**, e59494.

Shendure, J. and Ji, H. (2008). Next-generation DNA sequencing. *Nature Biotechnology*, **26**, 1135–1145.

Siller, S. (2001). Sexual selection and the maintenance of sex. *Nature*, **411**, 689–692.

Silventoinen, K., Kaprio, J., Lahelma, E., Viken, R.J., and Rose, R.J. (2003). Assortative mating by body height and BMI: Finnish twins and their spouses. *American Journal of Human Biology*, **15**, 620–627.

Simpson, G.G. (1944). *Tempo and Mode in Evolution*. Reprint 1984 by Columbia University Press, New York, USA.

Slagsvold, T. and Wiebe, K.L. (2007). Learning the ecological niche. *Proceedings of the Royal Society B*, **274**, 19–23.

Slatkin, M. (1987). Gene flow and the geographic structure of natural populations. *Science*, **236**, 787–792.

Slatkin, M. (1994). Linkage disequilibrium in growing and stable populations. *Genetics*, **137**, 331–336.

Slatkin, M. (2008). Linkage disequilibrium—understanding the evolutionary past and mapping the medical future. *Nature Reviews Genetics*, **9**, 477–485.

Slon, V., Mafessoni, F., Vernot, B., et al. (2018). The genome of the offspring of a Neanderthal mother and a Denisovan father. *Nature*, **561**, 1–12.

Smadja, C.M. and Butlin, R.K. (2011). A framework for comparing processes of speciation in the presence of gene flow. *Molecular Ecology*, **20**, 5123–5140.

Sober, E. and Wilson, D.S. (1998). *Unto Others – The Evolution and Psychology of Unselfish Behavior*. Harvard University Press, Cambridge, MA, USA.

Soh, Y.Q.S., Alföldi, J., Pyntikova, T., et al. (2014). Sequencing the mouse y chromosome reveals convergent gene acquisition and amplification on both sex chromosomes. *Cell*, **159**, 800–813.

Sousa, V. and Hey, J. (2013). Understanding the origin of species with genome-scale data: modelling gene flow. *Nature Reviews Genetics*, **14**, 404–414.

Spielmann, M., Lupiáñez, D.G., and Mundlos, S. (2018). Structural variation in the 3D genome. *Nature Reviews Genetics*, **19**, 453–467.

Stebbins, G.L. (1950). *Variation and Evolution in Plants*. Columbia University Press, New York, USA.

Stebbins, G.L. (1971). *Processes of Organic Evolution*. Prentice-Hall, Englewood Cliffs, NJ, USA.

Stelkens, R.B., Schmid, C., and Seehausen, O. (2015). Hybrid breakdown in cichlid fish. *PLoS One*, **10**, e0127207.

Stoletzki, N. and Eyre-Walker, A. (2010). Estimation of the neutrality index. *Molecular Biology and Evolution*, **28**, 63–70.

Stork, N.E. (2018). How many species of insects and other terrestrial arthropods are there on earth? *Annual Review of Entomology*, **63**, 31–45.

Storz, J.F. (2005). Invited Review: Using genome scans of DNA polymorphism to infer adaptive population divergence. *Molecular Ecology*, **14**, 671–688.

Stronen, A.V., Tessier, N., Jolicoeur, H., et al. (2012). Canid hybridization: contemporary evolution in human-modified landscapes. *Ecology & Evolution*, **2**, 2128–2140.

Summers-Smith, D. (1988). *The Sparrows*. T & AD Poyser, Calton, UK.

Sun, L., Wang, J., Zhu, X., et al. (2018). HpQTL: a geometric morphometric platform to compute the genetic architecture of heterophylly. *Briefings in Bioinformatics*, **19**, 603–612.

Sutton, W.S. (1902). On the morphology of the chromosome group in *Brachystola magna*. *Biological Bulletin*, **4**, 24–39.

Suzuki, M.M. and Bird, A. (2008). DNA methylation landscapes: provocative insights from epigenomics. *Nature Reviews Genetics*, **9**, 465–476.

Swanson, W.J. and Vacquier, V.D. (2002). The rapid evolution of reproductive proteins. *Nature Reviews Genetics*, **3**, 137–144.

Számadó, S. and Penn, D.J. (2015). Why does costly signalling evolve? Challenges with testing the handicap hypothesis. *Animal Behaviour*, **110**, e9–e12.

Taberlet, P., Fumagalli, L., Wust-Saucy, A.G., and Cosson, J.F. (1998). Comparative phylogeography and postglacial colonization routes in Europe. *Molecular Ecology*, **7**, 453–464.

Tajima, F. (1989). Statistical method for testing the neutral mutation hypothesis by DNA polymorphism. *Genetics*, **123**, 585–595.

Takahashi, T. and Hori, M. (2008). Evidence of disassortative mating in a Tanganyikan cichlid fish and its role in the maintenance of intrapopulation dimorphism. *Biology Letters*, **4**, 497–499.

Takayama, S. and Isogai, A. (2005). Self-incompatibility in plants. *Annual Reviews in Plant Biology*, **56**, 467–489.

Tang, S. and Presgraves, D.C. (2015). Lineage-specific evolution of the complex Nup160 hybrid incompatibility between *Drosophila melanogaster* and its sister species. *Genetics*, **200**, 1245–1254.

Taylor, C.C.W., Hare, R.M., and Barnes, J. (1999). *Greek Philosophers—Socrates, Plato, and Aristotle*. Oxford University Press, Oxford, UK.

Taylor, P.D., Wild, G., and Gardner, A. (2007). Direct fitness or inclusive fitness: how shall we model kin selection? *Journal of Evolutionary Biology*, **20**, 301–309.

Templeton, A.R. (2008). The reality and importance of founder speciation in evolution. *BioEssays*, **30**, 470–479.

Terns, R.M. and Terns, M.P. (2014). CRISPR-based technologies: prokaryotic defense weapons repurposed. *Trends in Genetics*, **30**, 111–118.

Tishkoff, S.A., Reed, F.A., and Ranciaro, A., et al. (2007). Convergent adaptation of human lactase persistence in Africa and Europe. *Nature Genetics*, **39**, 31–40.

Toomey, M.B., Marques, C.I., Andrade, P., et al. (2018). A non-coding region near *Follistatin* controls head colour polymorphism in the Gouldian finch. *Proceedings of the Royal Society B*, **285**, 20181788.

Tørresen, O.K., Star, B., Jentoft, S., et al. (2017). An improved genome assembly uncovers prolific tandem repeats in Atlantic cod. *BMC Genomics*, **18**, 311.

Trapnell, C., Roberts, A., Goff, L., et al. (2012). Differential gene and transcript expression analysis of RNA-seq experiments with TopHat and Cufflinks. *Nature Protocols*, **7**, 562–578.

Trautwein, M.D., Wiegmann, B.M., Beutel, R., Kjer, K.M., and Yeates, D.K. (2012). Advances in insect phylogeny at the dawn of the postgenomic era. *Annual Review of Entomology*, **57**, 449–468.

Trier, C.N., Hermansen, J.S., Sætre, G.-P., and Bailey, R.I. (2014). Evidence for mito-nuclear and sex-linked reproductive barriers between the hybrid Italian sparrow and its parent species. *PLoS Genetics*, **10**, e1004075.

Trivers, R.L. (1971). The evolution of reciprocal altruism. *Quarterly Review of Biology*, **46**, 35–37.

Trivers, R.L. (1972). Parental investment and sexual selection. In: B. Campbell (ed.) *Sexual Selection and the Descent of Man*, pp. 136–179. Heinemann, London, UK.

Trucchi, E., Mazzarella, A.B., Gilfillan, G.D., et al. (2016). BsRADseq: screening DNA methylation in natural populations of non-model species. *Molecular Ecology*, **25**, 1697–1713.

Turcotte, M.M. and Levine, J.M. (2016). Phenotypic plasticity and species coexistence. *Trends in Ecology & Evolution*, **31**, 803–813.

Turelli, M. and Orr, H.A. (1995). The dominance theory of Haldane's rule. *Genetics*, **140**, 389–402.

Turner, L.M. and Harr, B. (2014). Genome-wide mapping in a house mouse hybrid zone reveals hybrid sterility loci and Dobzhansky–Muller interactions. *eLife*, **3**, e02504.

Turner, T.L., Hahn, M.W., and Nuzhdin, S.V. (2005). Genomic islands of speciation in *Anopheles gambiae*. *PLoS Biology*, **3**, e285–7.

Turner, W.C., Kausrud, K.L., Krishnappa, Y.S., et al. (2014). Fatal attraction: vegetation responses to nutrient inputs attract herbivores to infectious anthrax carcass sites. *Proceedings of the Royal Society B*, **281**, 20141785.

Twyford, A.D. and Friedman, J. (2015). Adaptive divergence in the monkey flower *Mimulus guttatus* is maintained by a chromosomal inversion. *Evolution*, **69**, 1476–1486.

Udovic, D. (1980). Frequency-dependent selection, disruptive selection and the evolution of reproductive isolation. *The American Naturalist*, **116**, 621–641.

Vacquier, V.D. and Swanson, W.J. (2011). Selection in the rapid evolution of gamete recognition proteins in marine invertebrates. *Cold Spring Harbor Perspectives in Biology*, **3**, a002931.

Valcu, C.-M. and Kempenaers, B. (2015). Proteomics in behavioral ecology. *Behavioral Ecology*, **26**, 1–15.

van Valen, L. (1973). A new evolutionary law. *Evolutionary Theory*, **1**, 1–30.

van't Hof, A.E., Campagne, P., Rigden, D.J., et al. (2016). The industrial melanism mutation in British peppered moths is a transposable element. *Nature*, **534**, 102–105.

Veeramah, K.R. and Hammer, M.F. (2014). The impact of whole-genome sequencing on the reconstruction of human population history. *Nature Reviews Genetics*, **15**, 149–162.

Venter, J.C., Adams, M.D., Myers, E.W., et al. (2001). The sequence of the human genome. *Science*, **291**, 1304–1351.

Verhoeven, K.J.F., vonHoldt, B.M., and Sork, V.L. (2016). Epigenetics in ecology and evolution: what we know and what we need to know. *Molecular Ecology*, **25**, 1631–1638.

Verzijden, M.N., ten Cate, C., Servedio, M.R., et al. (2012). The impact of learning on sexual selection and speciation. *Trends in Ecology & Evolution*, **27**, 511–519.

Via, S. and West, J. (2008). The genetic mosaic suggests a new role for hitchhiking in ecological speciation. *Molecular Ecology*, **17**, 4334–4345.

Vicoso, B. and Charlesworth, B. (2006). Evolution on the X chromosome: unusual patterns and processes. *Nature Reviews Genetics*, **7**, 645–653.

Vijay, N., Poelstra, J.W., Künstner, A., and Wolf, J.B.W. (2013). Challenges and strategies in transcriptome assembly and differential gene expression quantification. A comprehensive *in silico* assessment of RNA-seq experiments. *Molecular Ecology*, **22**, 620–634.

Vitti, J.J., Grossman, S.R., and Sabeti, P.C. (2013). Detecting natural selection in genomic data. *Annual Review of Genetics*, **47**, 97–120.

Vogel, C. and Marcotte, E.M. (2012). Insights into the regulation of protein abundance from proteomic and transcriptomic analyses. *Nature Reviews Genetics*, **13**, 227–232.

Voight, B.F., Kurdaravalli, S., Wen, X., and Pritchard, J.K. (2006). A map of recent positive selection in the human genome. *PLoS Biology*, **4**, e72.

Voje, K.L. (2016). Tempo does not correlate with mode in the fossil record. *Evolution*, **70**, 2678–2689.

Vonlanthen, P., Bittner, D., Hudson, A.G., et al. (2012). Eutrophication causes speciation reversal in whitefish adaptive radiations. *Nature*, **482**, 357–362.

Wagh, K., Bhatia, A., Alexe, G., Reddy, A., Ravikumar, V., Seiler, M., Boemo, M., Yao, M., Cronk, L., Naqvi, A., Ganesan, S., Levine, A.J., and Bhanot, G. (2012). Lactase persistence and lipid pathway selection in the Maasai. *PLoS One*, **7**, e44751.

Wagner, C.E., Keller, I., Wittwer, S., et al. (2013). Genome-wide RAD sequence data provide unprecedented resolution of species boundaries and relationships in the Lake Victoria cichlid adaptive radiation. *Molecular Ecology*, **22**, 787–798.

Wall, J.D., Lohmueller, K.E., and Plagnol, V. (2009). Detecting ancient admixture and estimating demographic parameters in multiple human populations. *Molecular Biology and Evolution*, **26**, 1823–1827.

Wang, Z., Gerstein, M., and Snyder, M. (2009). RNA-Seq: a revolutionary tool for transcriptomics. *Nature Reviews Genetics*, **10**, 57–63.

Warzluff, W.F., Gongidi, P., Woods, K.R., and Maltais, L.J. (2002). The human and mouse replication-dependent histone genes. *Genomes*, **80**, 487–498.

Watakabe, I., Hashimoto, H., Kimura, Y., et al. (2018). Highly efficient generation of knock-in transgenic medaka by CRISPR/Cas9-mediated genome engineering. *Zoological Letters*, **4**, 3.

Watanabe, Y., Yokobori, S., Inabe, T., et al. (2002). Introns in protein coding genes in Archaea. *FEBS Letters*, **510**, 27–30.

Waterhouse, P.M., Wang, M.-B., and Lough, T. (2001). Gene silencing as an adaptive defence against viruses. *Nature*, **411**, 834–842.

Watson, J.D. and Crick, F.H.C. (1953). A structure for deoxyribose nucleic acid. *Nature*, **171**, 737–738.

Watterson, G.A. (1975). On the number of segregating sites in genetical models without recombination. *Theoretical Population Biology*, **7(2)**, 256–276.

Webber, B.L., Raghu, S., and Edwards, O.R. (2015). Opinion: Is CRISPR-based gene drive a biocontrol silver bullet or global conservation threat? *Proceedings of the National Academy of Sciences of the USA*, **112**, 10565–10567.

Weigand, H. and Leese, F. (2018). Detecting signatures of positive selection in non-model species using genomic data. *Zoological Journal of the Linnean Society*, **184(2)**, 528–583.

Weinberg, W. (1908). Über den nachweis der vererbung beim menschen. *Jahreshefte des Vereins für Vaterländische Naturkunde in Württemberg*, **64**, 368–382.

Weir, B.S. and Cockerham, C.C. (1969). Group inbreeding with 2 linked loci. *Genetics*, **63**, 711–742.

West, S.A., Griffin, A.S., and Gardner, A. (2007). Evolutionary explanations for cooperation. *Current Biology*, **17**, R661–R672.

West-Eberhard, J. (1989). Phenotypic plasticity and the origins of diversity. *Annual Reviews in Ecology and Systematics*, **20**, 249–178.

West-Eberhard, J. (2014). Darwin's forgotten idea: the social essence of sexual selection. *Neuroscience & Biobehavioral Reviews*, **46**, 501–508.

Westram, A.M., Galindo, J., Rosenblad, M.A., Grahame, J.W., and Butlin, R.K. (2014). Do the same genes underlie parallel phenotypic divergence in different *Littorina saxatilis* populations? *Molecular Ecology*, **23**, 4603–4616.

Westram, A.M., Panova, M., Galindo, J., Butlin, R.K. (2016). Targeted resequencing reveals geographical patterns of differentiation for loci implicated in parallel evolution. *Molecular Ecology*, **25**, 3169–3186.

Westram, A.M., Rafajlović, M., Chaube, P. et al. (2018). Clines on the seashore: The genomic architecture underlying rapid divergence in the face of gene flow. *Evolution Letters*, **2**, 297–309.

Westram, A.M. and Ravinet, M. (2017). Land ahoy? Navigating the genomic landscape of speciation while avoiding shipwreck. *Journal of Evolutionary Biology*, **30**, 1522–1525.

White, M.A., Ané, C., Dewey, C.N., Larget, B.R., and Payseur, B.A. (2009). Fine-scale phylogenetic discordance across the house mouse genome. *PLoS Genetics*, **5**, e1000729.

Whitlock, M.C., Phillips, P.C., Moore, F.B.G., and Tonsor, S.J. (1995). Multiple fitness peaks and epistasis. *Annual Review of Ecology and Systematics*, **26**, 601–621.

Whitt, S.R., Wilson, L.M., Tenaillon, M.I., Gaut, B.S., and Buckler, E.S. (2002). Genetic diversity and selection in the maize starch pathway. *Proceedings of the National Academy of Sciences of the USA*, **99**, 12959–12962.

Widemo, F. (1998). Alternative mating strategies in the ruff: a mixed ESS? *Animal Behaviour*, **56**, 329–336.

Wilding, C.S., Butlin, R.K., and Grahame, J. (2001). Differential gene exchange between parapatric morphs of *Littorina saxatilis* detected using AFLP markers. *Journal of Evolutionary Biology*, **14**, 611–619.

Wiley, C., Qvarnström, A., Andersson, G., Borge, T., and Sætre, G.-P. (2009). Postzygotic isolation over multiple generations of hybrid descendants in a natural hybrid zone: how well do single-generation estimates reflect reproductive isolation? *Evolution*, **63**, 1731–1739.

Wilkinson, G.S. (1984). Reciprocal food sharing in the vampire bat. *Nature*, **308**, 181–184.

Wilkinson, G.S. (1988). Reciprocal altruism in bats and other mammals. *Ethology and Sociobiology*, **9**, 85–100.

Willerslev, E. and Cooper, A, (2005), Ancient DNA. *Proceedings of the Royal Society B*, **272**, 3–16.

Williams, G.C. (1966). *Adaptation and Natural Selection*. Princeton University Press, Princeton, NJ, USA.

Williams, G.C. (1975). *Sex and Evolution*. Princeton University Press, Princeton, NJ, USA.

Williams, T.N., Mwangi, T.W., Wambua, S., Alexander, N.D., Kortok, M., Snow, R.W., and Marsh, K. (2005). Sickle cell trait and the risk of *Plasmodium falciparum* malaria and other childhood diseases. *Journal of Infectious Diseases*, **192**, 178–186.

Wilson, E.O. (2005). Kin selection as the key to altruism. *Social Research*, **72**, 159–166.

Wistow, G.J., Mulders, J.W., and de Jong, W.W. (1987). The enzyme lactate dehydrogenase as a structural protein in avian and crocodilian lenses. *Nature*, **326**, 622–624.

Woese, C.R., Kandler, O., Wheelis, M.L. (1990). Towards a natural system of organisms: Proposal for the domains Archaea, Bacteria, and Eucarya. *Proceedings of the National Academy of Sciences of the USA*, **87**, 4576–4579.

Wolf, J.B.W., Bayer, T., Haubold, B., Scilhabel, M., Rosenstiel, P., and Tautz, D. (2010). Nucleotide divergence vs. gene expression differentiation: comparative transcriptome sequencing in natural isolates from the carrion crow and its hybrid zone with the hooded crow. *Molecular Ecology*, **19**, 162–175.

Wolf, J.B.W. and Ellegren, H. (2017). Making sense of genomic islands of differentiation in light of speciation. *Nature Reviews Genetics*, **18**, 87–100.

Wolfe, K.H. (2001). Yesterday's polyploids and the mystery of diploidization. *Nature Reviews Genetics*, **2**, 331–341.

Wood, T.E., Takebayashi, N., Barker, M.S., Mayrose, I., Greenspoon, P.B., and Rieseberg, L.H. (2009). The frequency of polyploid speciation in vascular plants. *Proceedings of the National Academy of Sciences of the USA*, **106**, 13875–13879.

Woodward, S.R., Weyand, N.J., and Bunnell, M. (1994). DNA-sequence from Cretaceous period bone fragments. *Science*, **266**, 1229–1232.

Wrangham, R. and Conklin-Brittain, N. (2003). Cooking as a biological trait. *Comparative Biochemistry and Physiology Part A: Molecular & Integrative Physiology*, **136**, 35–46.

Wrangham, R., Jones, J.H., Laden, G., Pilbeam, D., and Conklin-Brittain, N. (1999). The raw and the stolen: Cooking and ecology of human origins. *Current Anthropology*, **40**, 567–594.

Wright, A.E., Darolti, I., Bloch, N.I., et al. (2017). Convergent recombination suppression suggests role of sexual selection in guppy sex chromosome formation. *Nature Communications*, **8**, 14251.

Wright, A.E., Fumagalli, M., Cooney, C.R., et al. (2018). Male-biased gene expression resolves sexual conflict through the evolution of sex-specific genetic architecture. *Evolution Letters*, **2**, 52–61.

Wright, G.A., Baker, D.D., Palmer, M.J., et al. (2013). Caffeine in floral nectar enhances a pollinator's memory of reward. *Science*, **339**, 1202–1204.

Wright, S. (1922). Coefficient of inbreeding and relationship. *American Naturalist*, **51**, 330–338.

Wright, S. (1931). Evolution in Mendelian populations. *Genetics*, **16**, 97–159.

Wright, S. (1932). The roles of mutation, inbreeding, crossbreeding and selection in evolution. *Proceedings of the 6th International Congress of Genetics*, **1**, 356–366.

Wright, S. (1937). The distribution of gene frequencies in populations. *Proceedings of the National Academy of Sciences of USA*, **23**, 307–320.

Wright, S. (1938). Size of population and breeding structure in relation to evolution. *Science*, **87**, 430–431.

Wu, C.I. (2001). The genic view of the process of speciation. *Journal of Evolutionary Biology*, **14**, 851–865.

Wynne-Edwards, V.C. (1962). *Animal Dispersion in Relation to Social Behavior*. Oliver & Boyd, London, UK.

Xu, J., Bauer, D.E., Kerenyi, M.A., et al. (2013). Corepressor-dependent silencing of fetal hemoglobin expression by BCL11A. *Proceedings of the National Academy of Sciences of the USA*, **110**, 6518–6523.

Yandell, M. and Ence, D. (2012). A beginner's guide to eukaryotic genome annotation. *Nature Reviews Genetics*, **13**, 329–342.

Yang, C., Liang, W., Cai, Y., et al. (2010). Coevolution in action: Disruptive selection on egg colour in an avian brood parasite and its host. *PLoS One*, **5**, e10816.

Yang, F., Fu, B., O'Brian, P.C.M., Nie, W., Ryder, O.A., and Ferguson-Smith, M.A. (2004). Refined genome-wide comparative map of the domestic horse, donkey and human based on cross-species chromosome painting: insight into the occasional fertility of mules. *Chromosome Research*, **12**, 65–76.

Yang, Z. (2006). *Computational Molecular Evolution*. Oxford University Press, Oxford, UK.

Yang, Z. (2014). *Molecular Evolution: A Statistical Approach*. Oxford University Press, Oxford, UK.

Yang, Z., Goldman, N., and Friday, A. (1994). Comparison of models for nucleotide substitution used in maximum likelihood phylogenetic estimation. *Molecular Biology and Evolution*, **11**, 316–324.

Yeates, S.E., Diamond, S.E., Einum, S., et al. (2013). Cryptic choice of conspecific sperm controlled by the impact of ovarian fluid on sperm swimming behavior. *Evolution*, **67**, 3523–2536.

Yoshida, K., Ishikawa, A., Toyoda, A., et al. (2019). Functional divergence of a heterochromatin-binding protein during stickleback speciation. *Molecular Ecology*. https://doi.org/10.1111/mec.14841

Young, D.L., Huyen, Y., and Allard, M.W. (1995). Testing the validity of the cytochrome B sequence from Cretaceous period bone fragments as dinosaur DNA. *Cladistics*, **11**, 199–209.

Zahavi, A. (1975). Mate selection—a selection for a handicap. *Journal of Theoretical Biology*, **53**, 204–214.

Zahavi, A. and Zahavi, A. (1997). *The Handicap Principle: A Missing Piece of Darwin's Puzzle*. Oxford University Press, New York, USA.

Zhang, L., Mazo-Vargas, A., and Reed, R.D. (2017). Single master regulatory gene coordinates the evolution and development of butterfly color and iridescence. *Proceedings of the National Academy of Sciences of the USA*, **114**, 10707–10712.

Zhang, X., Wang, H., Li, M., et al. (2014). Isolation of doublesex- and mab-3-related transcription factor 6 and its involvement in spermatogenesis in tilapia1. *Biology of Reproduction*, **91**, 1–10.

Zimorski, V., Ku, C., Martin, W.F., and Gould, S.B. (2014). Endosymbiotic theory for organelle origins. *Current Opinion in Microbiology*, **22**, 38–48.

Zuckerkandl, E. and Pauling, L.B. (1962). Molecular disease, evolution, and genic heterogeneity. In: M. Kasha and B. Pullman (eds) *Horizons in Biochemistry*, pp. 189–225. Academic Press, New York, USA.

Index

Species index